Antenna Theory and Microstrip Antennas

D. G. Fang

Science Press

CRC Press
Taylor & Francis Group
Boca Raton London New York

CRC Press is an imprint of the
Taylor & Francis Group, an **informa** business

CRC Press
Taylor & Francis Group
6000 Broken Sound Parkway NW, Suite 300
Boca Raton, FL 33487-2742

First issued in paperback 2019

ISBN-13: 978-1-4398-0727-9 (hbk)
ISBN-13: 978-0-367-38467-8 (pbk)

Library of Congress Cataloging-in-Publication Data

Fang, D. G.
   Antenna theory and microstrip antennas / author, D.G. Fang.
     p. cm.
   "A CRC title."
   Includes bibliographical references and index.
   ISBN 978-1-4398-0727-9 (hardcover : alk. paper)
   1. Microstrip antennas. I. Title.

TK7871.67.M5F36 2010
621.382'4--dc22
                                                         2009031426

To the memory of my father and mother, and to my family.

Contents

Preface

There are many excellent textbooks and handbooks on antenna theory and design. Recently, due to the increasing importance of microstrip antennas, several excellent textbooks and handbooks on them have been published and this topic has become a separate course. The purpose of this book is to serve either as a textbook to cover both antenna fundamentals and microstrip antennas or a self-study book for those attempting to plunge into this area. This book tries to make a good balance between the antenna fundamentals and microstrip antennas.

For the full wave analysis and optimization of antenna designs, there are many excellent books on computational electromagnetics and especially several powerful commercial software packages are available. This book is devoted to the introduction of techniques useful for the effective application of the software. The spectral domain approach is an important tool in analyzing the microstrip structures including the microstrip antennas, near-field measurement and high-frequency method. This book gives a systematic introduction to this approach. The main purpose of introducing this approach is not for numerical computation but for the understanding of some significant concepts.

The literature on antennas is vast and there are a variety of mathematical formalisms and numerical schemes, which often intimidate those who attempt to enter this field. In universities, it is a great challenge for professors to cover sufficient fundamentals with reasonably in-depth practical knowledge in a one-term course. In industry, one may find it hard to get a general understanding of the field before engaging in any specialized techniques. I believe that a concise and readable book, with both scope and depth, theoretical background, application materials and recent progress will be a welcome addition to the arsenal of books on this subject.

In this book, the presentation principle is to explain profound and abstract concepts in simple and concise terms. I tried to organize the contents logically and uniformly to lead the readers to draw inferences about other cases from one instance. Problems and answers provide a necessary supplement to the text and are used as an instrument to help the readers to gain insights and to facilitate understanding of the subtle points, usefulness of the principles and the techniques discussed in the text. This book includes my teaching and research experiences in this area over many years. Moreover, this book contains some recent developments including our own research results published in international journals. The contents of this book have been used as teaching materials in Laval University (Canada) in 1987 and in Chinese University of Hong Kong (Hong Kong) in 2002 and in Nanjing University of Science and Technology for more than a decade. For senior undergraduate-level course (two credits), the materials in Chapter 1 through Chapter 3 should be covered. The materials in Chapter 4 through Chapter 6 are suitable for a graduate-level course with two credits and those in Chapter 4 through Chapter 8 are suitable for a graduate course with three credits.

I have benefited from many experts through their excellent books and papers. I would like to express my sincere thanks to them, especially to Professors C. A. Balanis, W. L. Stutzman and G. A. Thiele, R. S. Elliot, R. E. Collin and F. J. Zucker, and S. M. Lin for their antenna books; J. A. Kong, N. H. Fang for their electromagnetic theory books; K. F. Lee and W. Chen, S. S. Zhong for their microstrip antenna books; R. H. Clarke, J.

Brown and E. V. Jull for their diffraction theory book; Y. R. Samii and E. Michielssen for their genetic algorithm book; Q. J. Zhang and K. C. Gupta for their neural network book; J. Litva and T. K. Y. Lo for their digital beamforming book; G. Y. Zhang for his phased array antenna book; Y. L. Chow, my former mentor when I was a visiting scholar at the University of Waterloo in 1981, for his deep insight to the complicated electromagnetic phenomena and his fuzzy electromagnetics which have been giving me a lot of enlightenment; G. Y. Delisle, my former mentor when I was a visiting scholar in Laval University in 1980, for his continuous help and support during the past years; R. Mittra, K. A. Michalski and J. R. Mosig, J. W. Bandler, E. K. Miller, C. H. Chan, K. M. Luk, E. K. N. Yung, J. Huang, K. L. Wu, W.P. Huang, D. M. Fu and R. S. Chen for their papers. Professor N. H. Fang, as the reviewer of this book, gave many invaluable comments on both the scientific content and the writing and Paul Bulger from the United States and project editor Karen Simon from Taylor & Francis Group did the final polishing; I gratefully acknowledge their contributions. The guidance and support from my former supervisor, Academician of Chinese Academy of Science Professor P. D. Ye and Academician of Chinese Academy of Science Professor S. G. Liu, and Academicians of Chinese Academy of Engineering Professor G. Y. Zhang and Professor Z. L. Sun are also very much appreciated.

I also wish to thank my graduate students who attended my course. Their active feedback, suggestions and corrections to the manuscripts, especially careful proofreading by Y. Lu, were very helpful. Some of them made important contributions to the research work involved in this book. I especially would name J. J. Yang, W. X. Sheng, Y. P. Xi, L. P. Shen, Y. X. Sun, Y. Ding, L. L. Wang, R. Zhang, C. Z. Luan, B. Chen, H. Wang, X. G. Chen, Y. Xiong, L. Zhao, W. M. Zhang, G. R. Zhou, N. Shahid, N. N. Feng, H. Q. Tao, G. B. Han, F. Ling, Y. M. Tao, K. Sha, Z. Li, Y. X. Guo, Y. Xu, Y. J. Zhou, J. Chen, C. Zhang, X. J. Zhang, W. M. Yu, J. S. Xu, and Y. Guo. Mr. W. M. Yu undertook the heavy duty of drawing all the figures and doing most of the typing. Without his effective help, this book could not have been completed so easily.

The support from the Ministry of Education through its listing of this book in the nationally scheduled textbooks, from the Bureau of Education in Jiangsu Province through its awarding of the antenna course as a Provincial Distinguished Graduate Course, and the financial support from the graduate school, the division of education and the division of international exchanges and cooperation of our University, the support from Professor Z. Liu, the Dean of School of Electronic Engineering and Optoelectronic Technology, and from the Defense Key Antenna and Microwave Laboratory through the grants: 00JS07.1.IBQ0201 and 51437080104BQ0206 are also very much acknowledged.

My wife, a physician, took care of my health and almost all family chores, in addition to her own busy practice to support my writing. I wish the publication of this book would partly pay back my debt of gratitude to her.

Finally, the comments and the criticisms from the readers will be very much appreciated. (E-mail: fangdg@mail.njust.edu.cn)

D.G.Fang
Nanjing University of Science and Technology
Nanjing

About the Author

Professor Da-Gang Fang was born in Shanghai, China, in 1937. He graduated from the graduate school of Beijing Institute of Posts and Telecommunications, Beijing, China, in 1966.

From 1980 to 1982, he was a visiting scholar at Laval University (Quebec, Canada), and the University of Waterloo (Ontario, Canada). Since 1986, he has been a Professor at the Nanjing University of Science and Technology (NUST), Nanjing, China. Since 1987, he has been a Visiting Professor with six universities in Canada and in Hong Kong. He has authored and co-authored two books, two book chapters and more than 360 papers. He is also the owner of three patents. His research interests include computational electromagnetics, microwave integrated circuits and antennas, and EM scattering.

Prof. Fang is a Fellow of IEEE and CIE (Chinese Institute of Electronics), an associate editor of two Chinese journals and is on the Editorial or Review Board of several international and Chinese journals. He was TPC chair of ICMC 1992, vice general chair of PIERS 2004, and a member of the International Advisory Committee of six international conferences, TPC co-chair of APMC 2005 and general co-chair of ICMMT 2008. He was also the recipient of the National Outstanding Teacher Award and People's Teacher Medal, and the Provincial Outstanding Teacher Award. His name was listed in Marquis *Who's Who in the World* (1995) and in the *International Biographical Association Directory* (1995).

CHAPTER 1

Basic Concepts of Antennas

1.1 Introduction

For wireless systems, the antenna is one of the critical components. A good design of the antenna can relax system requirements and improve overall system performance. The wireless systems include a large variety of different kinds, such as radar, navigation, landing systems, direct broadcast TV, satellite communications, mobile communications and so on. An antenna could be as large as 100m by 100m for radio telescope or as small as the order of centimeters in built-in handsets. All of them play an important role in science and daily life. Today we enjoy much benefit from wireless, and the significant contributions of antennas should not be underestimated.

An antenna is an electromagnetic transducer, used to convert, in the transmitting mode, guided waves within transmission lines to radiated free-space waves, or to convert, in the receiving mode, free-space waves to guided waves.

In 1886, Hertz demonstrated the first wireless electromagnetic system. In 1901, Marconi succeeded in sending signals over large distance from England to Newfoundland. Since Marconi's invention, through the 1940s antenna technology was primarily focused on wire related radiation elements and their operation frequencies up to about UHF. It was not until World War II that modern antenna technology was born and new elements, such as waveguide aperture, horns, reflectors, lenses, etc. were first introduced. The first use of phased array was reported in 1937. Most of the major advances in the theory of phased array antennas and their implementation occurred in 1960s. This kind of antenna can accomplish functions which the conventional one cannot do. Because the antenna beam in phased arrays can be steered to a new direction in microseconds and it may be widened or narrowed in microseconds as well, it provides much agility. Prior to 1950s, antennas with broadband patterns and impedance characteristics had bandwidths not much greater than about 2:1. In the 1950s, a breakthrough in antenna development occurred extending the maximum bandwidth to as great as 40:1 or more by using equiangular spiral or logarithmically periodic structures. Because the geometries of these antennas are specified by angle instead of linear dimensions, they have theoretically an infinite bandwidth. Therefore, they are referred to as frequency independent. The idea of the microstrip antenna was introduced in the 1950s by G. A. Deschamps, but it was not until 1970s that serious attention was paid to this element. To a large extent, the development of microstrip antennas has been driven by system requirements for antennas with low-profile, low-weight, low-cost, easy integrability into arrays and with microwave integrated circuits, or polarization diversity. Disadvantages of the original microstrip antenna configurations include narrow bandwidth, spurious feed radiation, poor polarization purity, limited power handling capacity and tolerance problems. Much of the development work in microstrip antennas has thus gone into efforts to overcome these problems so as to satisfy increasingly stringent system requirements. This effort has resulted in the development of novel microstrip antenna configurations and the development of accurate and versatile analytical models for the understanding of the inherent limitation of microstrip antennas, as well as for their design and optimization[1–4].

The good marriage between signal processing and electromagnetics results in a signal processing antenna that makes use of the all information on the aperture completely and adaptively. This kind of antenna is capable of generating independently controllable multi-beam and may be thought of as a smart antenna. The digital beamforming(DBF) antenna is a good solution to this purpose and the microstrip antenna is a good candidate to serve as the antenna element[5, 6]. Although the smart antenna is recognized as the ultimate antenna in the sense of making full use of the information on the antenna aperture, it will never close the way to further antenna development. The history from the Hertz dipole in 1886 to the smart antenna in recent years shows that the application requirements have always been the motivation for the development of antennas. Both in the present time and in the future, there are many challenging problems facing the antenna scientists and engineers.

Advances made in computer technology during the 1960s–1980s have had a major impact on the advance of modern antenna technology, and they are expected to have an even greater influence on antenna engineering in the 21st century and beyond. Beginning primarily in the early 1960s, numerical methods were introduced that allowed previously intractable complex antenna system configurations to be analyzed and designed very accurately. While in the past antenna design may have been considered a secondary issue in overall system design, today it plays a critical role. In fact, many system successes rely on the design and performance of the antenna. Also, while in the first half of the 20th century antenna technology may have been considered almost a trial-and-error operation, today it is truly an engineering art. Analysis and design methods are such that antenna system performance can be predicted with remarkable accuracy. In fact, many antenna designs proceed directly from the initial design stage to the prototype without intermediate testing. The level of confidence has increased tremendously.

1.2   Radiation Mechanism

Now let us explain the mechanism by which the electric lines of force are detached from the antenna to form the free-space waves. Figure 1.1(a) shows the lines of force created between the arms of a small center-fed dipole in the first quarter of the period during which the charge has reached its maximum value (assuming a sinusoidal time variation) and the lines have traveled outwardly a radial distance $\lambda/4$. During the next quarter of the period, the original lines travel an additional $\lambda/4$ (a total of $\lambda/2$ from the initial point) and the charge density on the conductors begins to diminish. This can be thought of as being accomplished by introducing opposite charges, which at the end of the first half of the period have neutralized the charges on the conductors. The lines of force created by the opposite charges travel a distance $\lambda/4$ during the second quarter of the first half and they are shown as dashed lines in Figure 1.1(b). The end result is that there are lines of force pointed upward in the first $\lambda/4$ distance and the same number of lines directed downward in the second $\lambda/4$. Since there is no net charge on the dipole, the lines of force must have been forced to detach themselves from the conductors and to unite together to form closed loops as shown in Figure 1.1(c). In the remaining second half of the period, the same process is followed but in the opposite direction. After that, the process continues and forms the propagation of electromagnetic wave.

1.3   Two Kinds of Linear Elementary Sources and Huygens' Planar Element[8]

In antenna problems, one is interested in determining the fields at points remote from the source. One type of the elementary source is the infinitesimal dipole, which may form the wire radiator. The wire radiator could be an electric current source, a fictitious electric

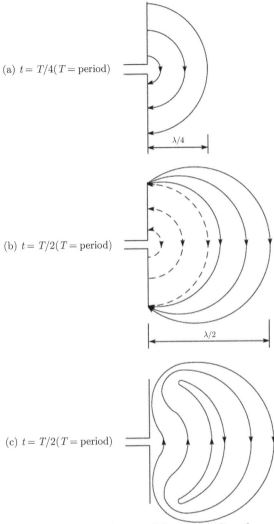

(a) $t = T/4 (T = \text{period})$

$\lambda/4$

(b) $t = T/2 (T = \text{period})$

$\lambda/2$

(c) $t = T/2 (T = \text{period})$

Figure 1.1   Formation and detachment of electric field lines for a short dipole.
(After Balanis [7], © 1997 Wiley)

current source or a fictitious magnetic source. The far field generated by them may be obtained by integrating the contributions of the infinitesimal dipoles along the wire. It is suggested that we should have the solution of an electric infinitesimal dipole first. The solution of the magnetic one then can be obtained by the duality. For an aperture antenna such as a horn, to calculate the far field by electric current, it is necessary to know the current distribution on all the walls of the horn, including the currents on the feeder. It is quite complicated. Alternatively one may use the equivalence principle. Usually the current on the outer wall is negligible. Based on this principle, the radiated fields may be calculated by using the fictitious electric and magnetic currents, that is the tangential electric and magnetic fields on the aperture. The elementary planar source on the aperture is the Huygens' element formed by a pair of orthogonal fictitious infinitesimal electric and magnetic dipoles.

1.3.1   Radiation Fields Generated by an Infinitesimal Electric Dipole

Consider the Maxwell's equations

$$\nabla \times \mathbf{E} = -j\omega\mu\mathbf{H}$$

$$\nabla \times \mathbf{H} = j\omega\epsilon\mathbf{E} + \mathbf{J} \tag{1.3.1}$$

Introduce

$$\mathbf{H}_A = \frac{1}{\mu}\nabla \times \mathbf{A} \tag{1.3.2}$$

where $\mathbf{A}$ is called the magnetic vector potential and subscript A indicates the field due to $\mathbf{A}$. Substituting (1.3.2) into the first equation of (1.3.1) results in

$$\nabla \times (\mathbf{E}_A + j\omega\mathbf{A}) = 0 \tag{1.3.3}$$

From the vector identify

$$\nabla \times (-\nabla\Phi_e) = 0 \tag{1.3.4}$$

and (1.3.3), it follows that

$$\mathbf{E}_A = -j\omega\mathbf{A} - \nabla\Phi_e \tag{1.3.5}$$

The scalar function Φ_e represents an arbitrary electric scalar potential, which is a function of position. Introducing the Lorentz condition $\nabla \cdot \mathbf{A} = -j\omega\epsilon\mu\Phi_e$, (1.3.5) becomes

$$\mathbf{E}_A = -j\omega\mathbf{A} - j\frac{\nabla\nabla \cdot \mathbf{A}}{\omega\epsilon\mu} \tag{1.3.6}$$

Substituting (1.3.2), (1.3.6) into the second equation of (1.3.1) reduces it into

$$\nabla \times \nabla \times \mathbf{A} - k^2\mathbf{A} = \nabla\nabla \cdot \mathbf{A} + \mu\mathbf{J} \tag{1.3.7}$$

Using the vector identity $\nabla \times \nabla \times \mathbf{A} = \nabla(\nabla \cdot \mathbf{A}) - \nabla^2\mathbf{A}$, (1.3.7) reduces to

$$\nabla^2\mathbf{A} + k^2\mathbf{A} = -\mu\mathbf{J} \tag{1.3.8}$$

where $k^2 = \omega^2\epsilon\mu$. The above equation is called the Helmholtz equation. To solve (1.3.8), let us assume that a source with current density J_z, which in the limit is an infinitesimal source, is located at the origin of a Cartesian coordinate system as shown in Figure 1.2. Since the current density is directed along the z-axis, only an A_z component will exist. Thus we can write (1.3.8) as

$$\nabla^2 A_z + k^2 A_z = -\mu J_z \tag{1.3.9}$$

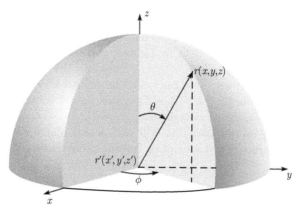

Figure 1.2   Coordinate system for point source at origin.

At points outside the source ($J_z=0$), the wave equation reduces to

$$\nabla^2 A_z + k^2 A_z = 0 \tag{1.3.10}$$

Since in the limit the source is a point, it requires that A_z is not a function of direction(θ and ϕ), that is, $A_z = A_z(r)$. Thus (1.3.10) may be written as

$$\nabla^2 A_z(r) + k^2 A_z(r) = \frac{d^2 A_z(r)}{dr^2} + \frac{2}{r}\frac{dA_z(r)}{dr} + k^2 A_z(r) = 0$$

which has two independent solutions

$$C_1 \frac{e^{-jkr}}{r}, \qquad C_2 \frac{e^{jkr}}{r}$$

We shall choose the outward wave as the solution

$$A_z = C_1 \frac{e^{-jkr}}{r} \tag{1.3.11}$$

In the static case ($k=0$) and with the presence of the source, we have

$$\nabla^2 A_z = -\mu J_z \tag{1.3.12}$$

which has the solution

$$A_z = \frac{C_1}{r} \tag{1.3.13}$$

This equation is the well known Poisson's equation. For electric potential Φ_e and charge density ρ, it has the following form and solution

$$\nabla^2 \Phi_e = -\frac{\rho}{\epsilon} \tag{1.3.14}$$

$$\Phi_e = \frac{1}{4\pi\epsilon} \iiint_V \frac{\rho}{r} dv' \tag{1.3.15}$$

Since equations of A_z and Φ_e have the same form, we have the solution for A_z as

$$A_z = \frac{\mu}{4\pi} \iiint_V \frac{J_z}{r} dv' \tag{1.3.16}$$

The same solutions can be obtained for A_x, A_y with the source J_x and J_y. The time-varying solution can be obtained by multiplying the static solution by $\exp(-jkr)$ as shown in (1.3.11) and (1.3.13), thus allowing us to write the time-varying solution in vector form

$$\mathbf{A} = \frac{\mu}{4\pi} \iiint_V \mathbf{J} \frac{e^{-jkr}}{r} dv' \tag{1.3.17}$$

If $\mathbf{J}$ represents linear density (m^{-1}), then (1.3.17) reduces to surface integral

$$\mathbf{A} = \frac{\mu}{4\pi} \iint_S \mathbf{J}_s \frac{e^{-jkr}}{r} ds' \tag{1.3.18}$$

If I_e represents current, then (1.3.18) reduces to line integrals of the form

$$\mathbf{A} = \frac{\mu}{4\pi} \int_C \mathbf{I}_e \frac{e^{-jkr}}{r} dl' \tag{1.3.19}$$

When the source point $\mathbf{r}'(x', y', z')$ is not at the origin, the distance between source point $\mathbf{r}'(x', y', z')$ and field point $\mathbf{r}(x, y, z)$ will be

$$\mathbf{R}(x', y', z'|x, y, z) = \mathbf{r}(x, y, z) - \mathbf{r}'(x', y', z')$$

In the above equations, r should be replaced by R.

Next, let us consider an infinitesimal electric dipole located at the origin along z axis, with a constant linear current I and length l. From (1.3.19)

$$\mathbf{A} = \hat{\mathbf{a}}_z \frac{\mu I}{4\pi r} e^{-jkr} \int_{-l/2}^{l/2} dz' = \hat{\mathbf{a}}_z \frac{\mu I l}{4\pi r} e^{-jkr} \tag{1.3.20}$$

In this case, $A_x = A_y = 0$, so $A_r = A_z \cos\theta$, $A_\theta = -A_z \sin\theta$. From (1.3.2) and (1.3.6), we have

$$H_r = H_\theta = E_\phi = 0 \tag{1.3.21}$$

$$H_\phi = j\frac{kIl\sin\theta}{4\pi r}\left[1 + \frac{1}{jkr}\right] e^{-jkr} \tag{1.3.22}$$

$$E_r = \eta\frac{Il\cos\theta}{2\pi r^2}\left[1 + \frac{1}{jkr}\right] e^{-jkr} \tag{1.3.23}$$

$$E_\theta = j\eta\frac{kIl\sin\theta}{4\pi r}\left[1 + \frac{1}{jkr} - \frac{1}{(kr)^2}\right] e^{-jkr} \tag{1.3.24}$$

when $kr \ll 1$, in the above solutions, the last terms dominate. As a consequence, E_r and E_θ, are in time-phase but they are in time phase quadrature with H_ϕ; therefore there is no time-average power flow associated with them. This region is called the reactive near field region. On the contrary, where $kr \gg 1$ ($r \gg \lambda$) the first terms dominate. In addition, because E_r is proportional to $1/r^2$, E_θ to $1/r$, E_r will be smaller than E_θ and can be neglected. Then (1.3.21)–(1.3.24) reduce to

$$E_\theta \approx -j\omega A_\theta = j\eta\frac{kIle^{-jkr}}{4\pi r}\sin\theta \tag{1.3.25}$$

$$E_r \approx E_\phi = H_r = H_\theta = 0 \tag{1.3.26}$$

$$H_\phi \approx \frac{E_\theta}{\eta} \tag{1.3.27}$$

where $\eta = \sqrt{\mu_0/\epsilon_0} = 120\pi = 377$ ohms is the wave impedance in free space. The **E**- and **H**- field components are perpendicular to each other, transverse to the radial direction of propagation, and the r variations are separable from those of θ and ϕ. The shape of the pattern is not a function of r, and the fields form a transverse electromagnetic (TEM) wave, E_θ and H_ϕ are in phase. In contrast to the reactive fields, these fields are radiation ones.

From the above results, we can see that the condition $kr \gg 1$ also allows the neglecting of E_r components which are proportional to $1/r^2$ and $1/r^3$. If we use "keeping only $1/r$ terms" as the criterion for the far field, it can be proved from (1.3.2) and (1.3.6) that the following relationship holds

$$\mathbf{E} = -j\omega A_\theta \hat{\mathbf{a}}_\theta, \quad \mathbf{H} = \hat{\mathbf{a}}_r \times \frac{\mathbf{E}}{\eta} \tag{1.3.28}$$

Example 1.1:

The radiating field by finite length electric dipole is shown in Figure 1.3. Assume that the electric current on the dipole is $\mathbf{I}_e(0,0,z') = \hat{\mathbf{a}}_z I_e$. The finite dipole antenna is subdivided into a number of infinitesimal dipoles of length $\Delta z'$. As the number of subdivisions is increased, each infinitesimal dipole approaches the length of dz'. Because the fields are symmetrical about z, only the zoy plane is considered. To calculate the field at the observation point P, all contributions from the infinitesimal dipoles are summed. Actually the superposition is the interference of waves with different electrical phase, assumed to be in phase, and space phase kd. The path-length difference $d = |\mathbf{r} - \mathbf{R}|$, where $|\mathbf{R}|$ is the distance between field point P and the source point, and $|\mathbf{r}|$ is the distance between the field point P and the origin.

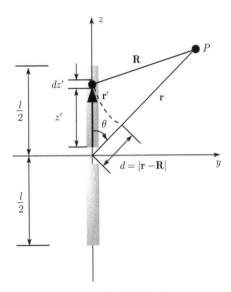

Figure 1.3   Finite electric dipole geometry.

Solution:

From Figure 1.3
$$R^2 - r^2 = (R - r)(R + r) = (z')^2 - 2rz'\cos\theta$$
let $R + r \approx 2r$, we have[9]
$$d = -\frac{(z')^2}{2r} + z'\cos\theta \tag{1.3.29}$$

To reduce the mathematical complexities, it will be assumed that the dipole has a negligible diameter (ideally zero). This is a good approximation provided the diameter is considerably smaller than the operating wavelength.

Based on the above statement and (1.3.25), (1.3.29), we have

$$E_\theta = \int_{-l/2}^{l/2} dE_\theta = \frac{j\eta k e^{-jkr}}{4\pi r}\sin\theta \left[\int_{-l/2}^{l/2} I_e(x', y', z')e^{-jk(z')^2/(2r)}e^{jkz'\cos\theta}dz'\right] \tag{1.3.30}$$
$$= AF_1 F_2 = AF$$

where $A = j\eta k\exp(-jkr)/(4\pi r)$ is a constant, $F_1 = \sin\theta$, $F = F_1 F_2$

$$F_2 = \int_{-l/2}^{l/2} I_e(x', y', z')e^{-jk(z')^2/(2r)}e^{jkz'\cos\theta}dz' \tag{1.3.31}$$

F_2 is called the Fresnel transform. The region corresponding to the approximation in (1.3.29) is called the Fresnel radiation field region. The usual requirement for the far field pattern is that the maximum error in path length due to finite range is less than $\lambda/16$ for radiation from all parts of the wire. For the length of l, the criterion becomes $l^2/8r < \lambda/16$ or

$$r > \frac{2l^2}{\lambda} \tag{1.3.32}$$

For this approximation, we only keep the second term in (1.3.29). Therefore d is the projection of the source vector $\mathbf{r}'$ onto $\mathbf{r}$. Consequently, (1.3.31) reduces to

$$F_2 = \int_{-l/2}^{l/2} I_e(x', y', z')e^{jkz'\cos\theta}dz' \tag{1.3.33}$$

The region corresponding to (1.3.32) is called the Fraunhofer radiation field region or far field region. It can be seen from (1.3.33) that the current distribution and the far field are a Fourier transform pair. The path length difference being $kz'\cos\theta$ shows that it is the parallel ray approximation. The phase compensation in (1.3.33) may be considered as the physical meaning of Fourier transform being focused at infinity.

The boundaries for separating the far-field (Fraunhofer), the radiating near-field (Fresnel) and the reactive near-field regions are not very rigid. There are some different criteria. Also the fields, as the boundaries from one region to the other are crossed, do not change abruptly but undergo a very gradual transition.

Usually the radiation property characterized by (1.3.30) is described as a radiation pattern or antenna pattern, which is defined as "a mathematical function or a graphical representation of the radiation properties of the antenna as a transform of the directional coordinates." In fact, the radiation pattern F may be written as

$$F = F_1 F_2 \tag{1.3.34}$$

where F_1 is the element factor and F_2 in (1.3.33) is the space factor. Both of them are independent of r, but F_2 as defined in (1.3.31) is not. (1.3.34) is referred to as the pattern multiplication. For this antenna, the element factor is the radiation pattern of a unit length infinitesimal dipole. The radiation pattern $\sin\theta$ is not due to interference, but due to the coordinate projection. However, the space factor is due to the interference. The element factor itself could be the radiation of an arbitrary antenna, where the same antennas are used to form a uniform linear array and the space factor is the array factor. This problem will be discussed later.

For a half wavelength dipole ($l = \lambda/2$) along z, the current distribution is a cosine function

$$I_e(0, 0, z') = \cos\left(\frac{\pi}{l}z'\right) = \cos(kz'), \quad -\frac{l}{2} \leqslant z' \leqslant \frac{l}{2}$$

The space factor F_2 is given by

$$F_2 = \int_{-l/2}^{l/2} \cos(kz')\exp(jkz'\cos\theta)dz' = \frac{2}{k}\frac{\cos(\pi\cos\theta/2)}{\sin^2\theta}$$

The radiation pattern for this popular half wavelength dipole is

$$F = F_1 F_2 = \frac{2}{k}\frac{\cos(\pi\cos\theta/2)}{\sin\theta}$$

1.3.2 Radiation Fields Generated by an Infinitesimal Magnetic Dipole

Although magnetic currents appear to be physically unrealizable, equivalent magnetic currents appear when we use the equivalence theorem. We may introduce the vector potential $\mathbf{F}$ and magnetic scalar potential Φ_m, to find the solutions $\mathbf{H}_F$ and $\mathbf{E}_F$. Following the same procedure for the electric source ($\mathbf{J} \neq 0, \mathbf{M} = 0$), we may have a set of equations for the magnetic source ($\mathbf{J} = 0, \mathbf{M} \neq 0$). They are listed in Table 1.1.

Table 1.1 Equations for electric (J) and magnetic (M) current sources.

Electric Sources ($\mathbf{J} \neq 0, \mathbf{M} = 0$)	Magnetic Sources ($\mathbf{J} = 0, \mathbf{M} \neq 0$)
$\nabla \times \mathbf{E}_A = -j\omega\mu\mathbf{H}_A$	$\nabla \times \mathbf{H}_F = j\omega\epsilon\mathbf{E}_F$
$\nabla \times \mathbf{H}_A = \mathbf{J} + j\omega\epsilon\mathbf{E}_A$	$-\nabla \times \mathbf{E}_F = \mathbf{M} + j\omega\mu\mathbf{H}_F$
$\nabla^2\mathbf{A} + k^2\mathbf{A} = -\mu\mathbf{J}$	$\nabla^2\mathbf{F} + k^2\mathbf{F} = -\epsilon\mathbf{M}$
$\mathbf{A} = \dfrac{\mu}{4\pi}\iiint\limits_V \mathbf{J}\dfrac{e^{-jkR}}{R}dv'$	$\mathbf{F} = \dfrac{\epsilon}{4\pi}\iiint\limits_V \mathbf{M}\dfrac{e^{-jkR}}{R}dv'$
$\mathbf{H}_A = \dfrac{1}{\mu}\nabla \times \mathbf{A}$	$\mathbf{E}_F = -\dfrac{1}{\epsilon}\nabla \times \mathbf{F}$
$\mathbf{E}_A = -j\omega\mathbf{A} - j\dfrac{1}{\omega\mu\epsilon}\nabla(\nabla \cdot \mathbf{A})$	$\mathbf{H}_F = -j\omega\mathbf{F} - j\dfrac{1}{\omega\mu\epsilon}\nabla(\nabla \cdot \mathbf{F})$
$\mathbf{E} = \mathbf{E}_A + \mathbf{E}_F$	
$\mathbf{H} = \mathbf{H}_A + \mathbf{H}_F$	

When two equations that describe the behavior of two different variables are of the same mathematic form, their solutions will also be identical. The variables in the two equations that occupy identical positions are known as dual quantities and a solution of one can be obtained by a systematic interchange of symbols to the other. This concept is known as the duality theorem. The dual quantities in Table 1.1 are listed in Table 1.2.

Table 1.2 Dual quantities for electric (J) and magnetic (M) current sources.

Electric Sources ($\mathbf{J} \neq 0, \mathbf{M} = 0$)	Magnetic Sources ($\mathbf{J} = 0, \mathbf{M} \neq 0$)
$\mathbf{E}_A$	$\mathbf{H}_F$
$\mathbf{H}_A$	$-\mathbf{E}_F$
$\mathbf{J}$	$\mathbf{M}$
$\mathbf{A}$	$\mathbf{F}$
ϵ	μ
μ	ϵ
k	k
η	$1/\eta$
$1/\eta$	η

According to the duality theorem, from (1.3.25) and (1.3.26) we may find the solution for the z-directed infinitesimal magnetic dipole $I_m l$ as

$$H_\theta \approx -j\omega F_\theta = je^{-jkr}\frac{1}{\eta}\frac{kI_m l}{4\pi r}\sin\theta \tag{1.3.35}$$

$$H_r \approx H_\Phi = E_r = E_\theta = 0 \tag{1.3.36}$$

$$E_\Phi \approx -\eta H_\theta \tag{1.3.37}$$

For the finite length magnetic dipole, based on the duality and (1.3.33)

$$F_2 = \int_{-l/2}^{l/2} I_m(x', y', z')e^{jkz'\cos\theta}dz' \tag{1.3.38}$$

If the current is uniformly distributed, with unit amplitude, that is $I_m = 1/l$, then

$$F_2 = \frac{\sin(kl\cos\theta/2)}{kl\cos\theta/2} \tag{1.3.39}$$

$$F = F_1 F_2 = \sin\theta \frac{\sin(kl\cos\theta/2)}{kl\cos\theta/2} \tag{1.3.40}$$

Now let us consider a magnetic current loop with unit amplitude and $\cos n\phi'$ current distribution which is shown in Figure 1.4. Unlike the linear antenna, for a loop antenna, the radiation pattern of the infinitesimal dipoles is a function of source point coordinate ϕ'. Therefore in the field solution, it is impossible to separate the element factor and the space factor. Following Table 1.1, (1.3.28) and the duality principle, we have

$$\mathbf{H}_F = -j\omega(F_\theta\hat{\mathbf{a}}_\theta + F_\phi\hat{\mathbf{a}}_\phi)$$

$$\mathbf{E}_F = \eta\mathbf{H}_F \times \hat{\mathbf{a}}_r = j\omega\eta(-F_\phi\hat{\mathbf{a}}_\theta + F_\theta\hat{\mathbf{a}}_\phi)$$

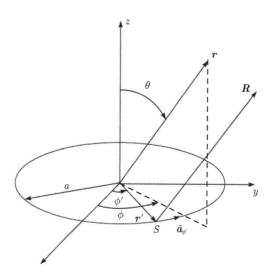

Figure 1.4   Magnetic current loop geometry (S denotes the source point).

From (1.3.19), the duality principle and the far field approximation similar to (1.3.33), we have

$$\mathbf{F} = \frac{\epsilon e^{-jkr}}{4\pi}\int_C \hat{\mathbf{a}}_{\phi'}\cos n\phi' \frac{e^{jkd}}{r}dl' = \frac{\epsilon e^{-jkr}}{4\pi r}\int_0^{2\pi}\hat{\mathbf{a}}_{\phi'}\cos n\phi' e^{jkd}ad\phi'$$

where $\hat{\mathbf{a}}_{\phi'}$ is the unit tangential vector of the loop at S, $d = a\sin\theta\cos(\phi'-\phi)$ is the projection of the source vector $\mathbf{r}'$ on $\mathbf{r}$. Based on the coordinate transformation

$$\hat{\mathbf{a}}_{\phi'} = \sin\theta\sin(\phi-\phi')\hat{\mathbf{a}}_r + \cos\theta\sin(\phi-\phi')\hat{\mathbf{a}}_\theta + \cos(\phi-\phi')\hat{\mathbf{a}}_\phi$$

we have

$$E_{\theta,n} = -j\omega\eta F_\phi = -j\frac{ae^{-jkr}}{2\lambda r}\int_0^{2\pi}\cos n\phi'\cos(\phi'-\phi)e^{jka\sin\theta\cos(\phi'-\phi)}d\phi' \tag{1.3.41}$$

$$E_{\phi,n} = j\omega\eta F_\theta = -j\frac{ae^{-jkr}}{2\lambda r}\int_0^{2\pi}\cos n\phi'\cos\theta\sin(\phi'-\phi)e^{jka\sin\theta\cos(\phi'-\phi)}d\phi' \tag{1.3.42}$$

The integrals may be carried out by the integral representation of Bessel function

$$
\int_0^{2\pi} e^{jx\cos(\phi'-\phi)} \begin{bmatrix} \cos(\phi'-\phi) \\ \sin(\phi'-\phi) \end{bmatrix} \cos n\phi' d\phi'
$$
$$
= j^{n+1}\pi[\mathrm{J}_{n+1}(x) \mp \mathrm{J}_{n-1}(x)] \begin{bmatrix} \cos n\phi \\ \sin n\phi \end{bmatrix}
\tag{1.3.43}
$$

where $\mathrm{J}_{n+1}(x)$ and $\mathrm{J}_{n-1}(x)$ are the first kind Bessel functions of order $n+1$ and $n-1$ respectively. Then, (1.3.41) and (1.3.42) reduce to

$$
E_{\theta,n} = j^n \frac{a\pi e^{-jkr}}{2\lambda r} \cos n\phi \left[\mathrm{J}_{n+1}(ka\sin\theta) - \mathrm{J}_{n-1}(ka\sin\theta)\right]
\tag{1.3.44}
$$

$$
E_{\phi,n} = j^n \frac{a\pi e^{-jkr}}{2\lambda r} \sin n\phi \cos\theta \left[\mathrm{J}_{n+1}(ka\sin\theta) + \mathrm{J}_{n-1}(ka\sin\theta)\right]
\tag{1.3.45}
$$

If the magnetic current loop is placed on an infinite ground plane, formulas (1.3.44) and (1.3.45) should be multiplied by factor 2. These new formulas give the radiation fields of a circular microstrip patch antenna with the cavity mode approximation that will be discussed in Chapter 3.

1.3.3   Radiation Fields Generated by Huygens' Planar Element

Consider an aperture on the xoy plane shown in Figure 1.5. If there are both a tangential electric field and a tangential magnetic field, to find the radiation field according to the equivalence principle, we should use both the electric current $\mathbf{J}_s = \hat{\mathbf{n}} \times \mathbf{H}$ and the magnetic current $\mathbf{M}_s = \mathbf{E} \times \hat{\mathbf{n}}$ as the sources. The solution may be obtained by combining the results in Sections 1.3.1 and 1.3.2. Consider the differential element dx, dy and the tangential field on the aperture to be E_x, H_y. The relationship between E_x and H_y is assumed to be $H_y = E_x/\eta$. The surface current densities are $J_x = -H_y = -E_x/\eta$ and $M_y = -E_x$ respectively. They form the infinitesimal electric dipole $(J_x dy)dx$ and infinitesimal magnetic dipole $(M_y dx)dy$. This differential element is called the Huygens' planar element. The radiation fields of this element are generated by a pair of orthogonal electric and magnetic dipoles. The total field $d\mathbf{E}$, $d\mathbf{H}$ generated by the differential element will be the superposition of two dipoles

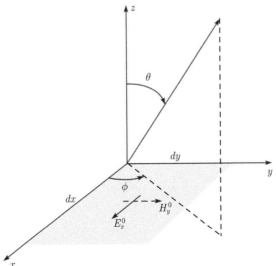

Figure 1.5   Huygens' planar element.

$$d\mathbf{E} = d\mathbf{E}_A + d\mathbf{E}_F, \quad d\mathbf{H} = d\mathbf{H}_A + d\mathbf{H}_F$$

In our case

$$dE_\theta = dE_{\theta,A} + dE_{\theta,F}, \quad dE_\phi = dE_{\phi,A} + dE_{\phi,F}$$

From (1.3.28) and the duality principle

$$dE_{\theta,A} = -j\omega A_\theta = -j\omega A_x \cos\phi \cos\theta$$
$$dH_{\theta,F} = -j\omega F_\theta = -j\omega F_y \sin\phi \cos\theta$$
$$dE_{\phi,A} = -j\omega A_\phi = j\omega A_x \sin\phi$$
$$dH_{\phi,F} = -j\omega F_\phi = -j\omega F_y \cos\phi$$

From (1.3.20) and the duality principle

$$A_x = \frac{\mu(J_x dy)dx}{4\pi r} e^{-jkr} = -\frac{\mu E_x^o dx dy e^{-jkr}}{\eta 4\pi r}$$

$$F_y = \frac{\epsilon(M_y dx)dy}{4\pi r} e^{-jkr} = -\frac{\epsilon E_x^o dx dy e^{-jkr}}{4\pi r}$$

Finally we have

$$dE_\theta = dE_{\theta,A} + dE_{\theta,F} = dE_{\theta,A} + \eta dH_{\phi,F}$$
$$= \frac{j E_x^o dx dy e^{-jkr}}{2\lambda r}(\cos\phi\cos\theta + \cos\phi) \tag{1.3.46}$$

$$dE_\phi = dE_{\phi,A} + dE_{\phi,F} = dE_{\phi,A} - \eta dH_{\theta,F}$$
$$= \frac{-j E_x^o dx dy e^{-jkr}}{2\lambda r}(\sin\phi + \sin\phi\cos\theta) \tag{1.3.47}$$

$$dH_\phi = dE_\theta/\eta, \quad dH_\theta = -dE_\phi/\eta \tag{1.3.48}$$

It is interesting to note that, compared with the radiation fields generated by infinitesimal electric or magnetic dipole alone, the radiation fields of Huygens's planar element are no longer omnidirectional but unidirectional. This is a result of the Huygens equivalence principle. The total field from an aperture E_θ will be

$$E_{\theta \ or \ \phi} = \int dE_{\theta or \phi} = A \cdot F_1 F_2 \tag{1.3.49-a}$$

$$A = \frac{j e^{-jkr}}{2\lambda r} \tag{1.3.49-b}$$

$$F_1 = (\cos\phi\cos\theta + \cos\phi), \qquad for \ E_\theta$$
$$F_1 = -(\sin\phi + \sin\phi\cos\theta), \qquad for \ E_\phi \tag{1.3.50-a}$$

$$F_2 = \iint_S E_x^0(x', y')e^{jkr' \cos\psi} ds' \tag{1.3.50-b}$$

Similar to (1.3.33), (1.3.38), $r'\cos\psi$ is the projection of $\mathbf{r}'$ onto $\mathbf{r}$, ψ is the angle between $\mathbf{r}'$ and $\mathbf{r}$ as shown in Figure 1.6. The path length difference $r'\cos\psi$ may be calculated as

$$r'\cos\psi = \mathbf{r}' \cdot \hat{\mathbf{a}}_r = x'\sin\theta\cos\phi + y'\sin\theta\sin\phi$$

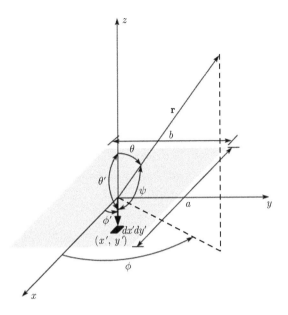

Figure 1.6   Rectangular aperture.

As an example, let us consider a rectangular aperture with both the equivalent electric and magnetic current distribution

$$\mathbf{M}_s = -\hat{\mathbf{a}}_y E_x^0, \quad \mathbf{J}_s = -\hat{\mathbf{a}}_x E_x^0/\eta, \quad \begin{cases} -a/2 \leqslant x' \leqslant a/2 \\ -b/2 \leqslant y' \leqslant b/2 \end{cases}$$
$$\mathbf{M}_s = 0, \quad \mathbf{J}_s = 0, \qquad\qquad elsewhere$$

Based on (1.3.49-a)–(1.3.50-b)

$$E_\theta = A \cos\phi (1 + \cos\theta) \cdot F_2 \tag{1.3.51}$$

$$E_\phi = -A \sin\phi (1 + \cos\theta) \cdot F_2 \tag{1.3.52}$$

$$F_2 = \int_{-b/2}^{b/2} \int_{-a/2}^{a/2} E_x^0 e^{jkx' \sin\theta \cos\phi} e^{jky' \sin\theta \sin\phi} dx' dy' \tag{1.3.53}$$

$$= E_x^0 \cdot ab \cdot \frac{\sin X}{X} \cdot \frac{\sin Y}{Y} \tag{1.3.54}$$

where $X = (ka \sin\theta \cos\phi)/2$, $Y = (kb \sin\theta \sin\phi)/2$.

In some cases, the equivalent currents may be only one current by making a judicious choice of the equivalent model. For example, if in Figure 1.6, the rectangular aperture is on an infinite electric ground plane, the equivalent model may be chosen as equivalent electric current to be zero everywhere, the magnetic current to be $\mathbf{M}_s = -\hat{\mathbf{a}}_x 2E_x^0$ and zero elsewhere. Consequently, in (1.3.46), (1.3.47), we should delete the first term. For the case of electric current only, we should delete the second term.

In the above derivation, we assumed the relationship between E_x^0 and H_y^0 on the aperture simply to be $E_x^0 = \eta H_y^0$. In some practical problems, it is necessary to find a more precise result. For example when concerned with the fields on the open waveguide aperture, if we only consider the dominant mode in waveguide, according to the waveguide theory

$$E_x^0 = \frac{\omega \mu_0}{\gamma} \frac{1 + \Gamma}{1 - \Gamma} H_y^0 = \xi \eta H_y^0$$

where

$$\xi = \frac{k}{\gamma}\frac{1+\Gamma}{1-\Gamma}$$

$k = 2\pi/\lambda$, $\gamma = 2\pi/\lambda_g$, λ_g is the wavelength in the waveguide and Γ is the reflection coefficient of the aperture, usually it may be measured. In that case, we should place the coefficient ξ before the second term in (1.3.46), (1.3.49-b).

The linear source may be considered as the limit case of the aperture source. For example, for y-directed linear source, that is, $ka \to 0$, (1.3.52) reduces to $F_2 = E_x^0 \sin Y/Y$.

1.4 Fundamental Parameters of Antennas

1.4.1 Radiation Pattern

Various parts of a radiation pattern are referred to as lobes, which may be subclassified into major or main, minor, side, and back lobes.

Figure 1.7(a) demonstrates a symmetrical three-dimensional polar pattern with a number of radiation lobes. Some are of greater intensity than others, but all are classified as lobes.

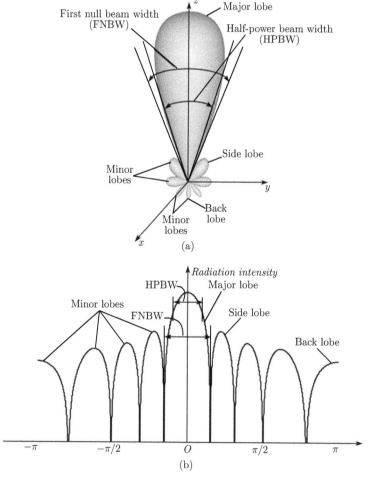

Figure 1.7 (a) Radiation lobes and beamwidths of an antenna pattern, (b) Linear plot of power pattern and its associated lobes and beamwidths. (After Balanis [7], © 1997 Wiley)

Figure 1.7(b) shows a linear two-dimensional plot of the power pattern [one plane of Figure 1.7(a)] where the same pattern characteristics are indicated.

A major lobe, also called main beam, is defined as "the radiation lobe containing the direction of maximum radiation." In Figure 1.7 the major lobe is pointing at the $\theta = 0$ direction. A minor lobe is any lobe except the major lobe. A side lobe is "a radiation lobe in any direction other than the intended lobe." (Usually a side lobe is adjacent to the main lobe and occupies the hemisphere in the direction of the main beam. A back lobe is "a radiation lobe whose axis makes an angle of approximately 180° with respect to the main beam of an antenna." It refers to a minor lobe that occupies the hemisphere in a direction opposite to that of the major (main) lobe.

Minor lobes usually represent radiation in undesired directions, and they should be minimized. Side lobes are normally the largest of the minor lobes. The level of minor lobes is usually expressed as a ratio of the power density to that of the major lobe. This ratio is often termed the side lobe ratio or side lobe level (SLL). Side lobe levels of −20dB or lower are desirable in most applications. Half-power beam width (HPBW) and first null beam width (FNBW) are illustrated in Figure 1.7 and usually they are given in degrees. Attainment of a side lobe level lower than −30dB usually requires very careful design and construction. A side lobe level lower than −40dB is considered to be an ultra low side lobe level and is difficult to achieve. In most systems, a low side lobe level is very important, to minimize false target indications through the side lobes.

In most cases, the radiation pattern is determined in the far-field region (Fraunhofer). In the radiating near-field region (Frensnel), the radiation pattern is different due to the square phase error as shown in Figure 1.8.

1.4.2   Directivity and Gain

The directivity of an antenna is defined as "the ratio of the radiation intensity U in a given direction from the antenna to the radiation intensity averaged over all directions. The average radiation intensity is equal to the total power P_{rad} radiated by the antenna divided by 4π. If the direction is not specified, the direction of maximum radiation intensity is implied." Stated more simply, the directivity of a nonisotropic source is equal to the ratio of its radiation intensity U in a given direction over the radiation intensity U_0 of an isotropic source. In mathematical form, it can be written as

$$D = \frac{U}{U_0} \tag{1.4.1}$$

According to the definition, $P_{rad} = \oiint_\Omega U d\Omega$. For an isotropic source, $U = U_0$ is independent of θ, ϕ, thus $P_{rad} = \oiint_\Omega U_0 d\Omega = 4\pi U_0$, and

$$U_0 = \frac{P_{rad}}{4\pi} \tag{1.4.2-a}$$

U may be computed by

$$U = \frac{4\pi r^2 (|\mathbf{E}|^2 / 2\eta)}{4\pi} = \frac{r^2}{2\eta}|\mathbf{E}|^2 = \frac{r^2 |\mathbf{E}|^2}{2 \times 120\pi} \tag{1.4.2-b}$$

where $\mathbf{E}$ represents the peak value of the far-zone electric field of the antenna. Substituting (1.4.2-a). (1.4.2-b) into (1.4.1) results in

$$D = \frac{U}{U_0} = \frac{4\pi U}{P_{rad}} = \frac{r^2 |\mathbf{E}|^2}{60 P_{rad}} \tag{1.4.3}$$

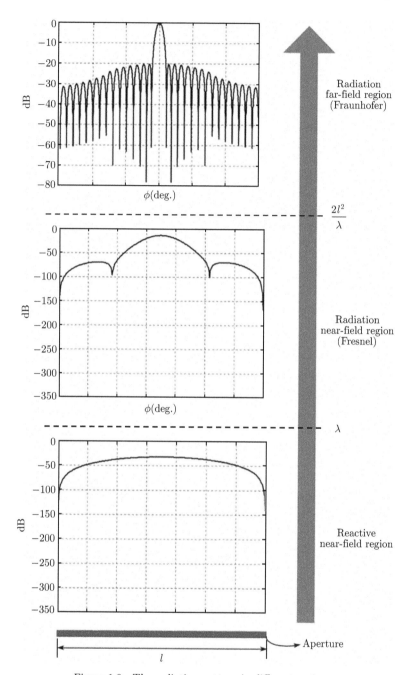

Figure 1.8   The radiation pattern in different regions.

Alternatively, let the radiation intensity of an antenna be of the form

$$U = B_0 f^2(\theta, \phi) \tag{1.4.4}$$

where B_0 is a constant. The total radiated power is found using

$$P_{rad} = \oiint_{\Omega} B_0 f^2(\theta, \phi) d\Omega \tag{1.4.5}$$

We now write the general expression for the directivity and maximum directivity using (1.4.4) and (1.4.5), respectively as

$$D(\theta, \phi) = 4\pi \frac{F(\theta, \phi)}{\int_0^{2\pi} \int_0^{\pi} F(\theta, \phi) \sin\theta d\theta d\phi} \qquad (1.4.6)$$

$$D_{\max} = D_0 = 4\pi \frac{F(\theta, \phi)|_{\max}}{\int_0^{2\pi} \int_0^{\pi} F(\theta, \phi) \sin\theta d\theta d\phi} = \frac{4\pi}{\Omega_A} \qquad (1.4.7)$$

where $F(\theta, \phi) = f^2(\theta, \phi)$, Ω_A is the beam solid angle and is defined as the solid angle through which all the power of the antenna would flow if its radiation intensity is constant and equal to the maximum value of U for all angles within Ω_A.

Instead of using the exact expression of (1.4.7), it is often convenient to derive simpler approximate expressions to compute the directivity. For antennas with a narrow major lobe and negligible minor lobes, the beam solid angle is approximately equal to the product of the half-power beam widths in two perpendicular planes. With this approximation, (1.4.7) can be approximated by

$$D_0 = \frac{4\pi}{\Omega_A} \approx \frac{4\pi}{\theta_{1r}\theta_{2r}}$$

where
θ_{1r}=half-power beamwidth in one plane (rad)
θ_{2r}=half-power beamwidth in a plane at a right angle to the other (rad)

If the beamwidths are known in degrees, then

$$D_0 \approx \frac{4\pi(180/\pi)^2}{\theta_{1d}\theta_{2d}} = \frac{41253}{\theta_{1d}\theta_{2d}} = \frac{C}{\theta_{1d}\theta_{2d}} \qquad (1.4.8)$$

C may be different for different types of antennas. The suggested values are[10]:

Antenna Type	$C(deg^2)$
Uniform rectangular aperture	32,383
Cosine-uniform rectangular aperture such as open-ended waveguide	35,230
Uniform circular aperture	33,709
Circular aperture with a parabolic-on–a-12-dB-pedestal distribution	38,933
General use for practical antennas	26,000

For an aperture antenna, for example a rectangular aperture with in-phase field distribution $E_x^0 = E_x$, we may write an alternative expression, which involves not the radiation pattern but the fields on the direction of maximum radiation, $\phi = \theta = 0°$, and then

$$U_{\max} = \frac{r^2}{2\eta}|\mathbf{E}|_{\max}^2 \qquad (1.4.9\text{-a})$$

From (1.3.51)–(1.3.54) where E_x is assumed to be real

$$|\mathbf{E}|_{\max}^2 = \left(\frac{1}{\lambda r}\iint_S E_x ds\right)^2 = \left(\frac{1}{\lambda r}\iint_S |E_x| ds\right)^2 \qquad (1.4.9\text{-b})$$

P_{rad} may be considered as the total radiation power on the aperture

$$P_{rad} = \frac{1}{2\eta} \iint_S |E_x|^2 ds \qquad (1.4.10)$$

Formulas (1.4.3), (1.4.9-a), (1.4.9-b) and (1.4.10) yield an alternative expression of the directivity

$$D_0 = \frac{4\pi \left(\iint_S |E_x| ds \right)^2}{\lambda^2 \iint_S |E_x|^2 ds} \qquad (1.4.11)$$

For uniform distribution, that is when $|E_x|$ is a constant, we have

$$D_0 = \frac{4\pi S}{\lambda^2}$$

with S being the physical aperture area. For other distributions

$$D_0 = \epsilon_{ap} \frac{4\pi S}{\lambda^2} \qquad (1.4.12)$$

where $\epsilon_{ap} \leqslant 1$ is called the aperture efficiency and is related to C in (1.4.8).

Losses are represented by radiation efficiency e_r ($0 \leqslant e_r \leqslant 1$). Gain, G, is directivity reduced by losses on the antenna structure

$$G = e_r D \qquad (1.4.13)$$

For electrically small antennas, e_r can be very small. For others, such as horn antennas, e_r approaches 1. Exceptions are antennas that include lossy device, such as radomes, cable/waveguide runs, filters, etc. Losses due to impedance mismatch are not included in the definition of gain, but are often unavoidable. The term "realized gain" is used when mismatch effects are included. Usually, such effects can be estimated and removed to obtain gain. In many wireless applications, gain is expressed in units of dBd. Gain in dBd equals gain in dB reduced by the 2.15dB, gain of the half-wave dipole reference antenna.

1.4.3   Polarization[7]

Polarization of a radiated wave is defined as "that property of an electromagnetic wave describing the time varying direction and relative magnitude of the electric-field vector; specially, the figure traced as a function of time by the extremity of the vector at a fixed location in space, and the sense in which it is traced, as observed along the direction of propagation." Polarization then is the curve traced by the end point of the arrow representing the instantaneous electric field. The field must be observed along the direction of propagation. A typical trace as a function of time is shown in Figure 1.9(a) and (b). The polarization of a wave received by an antenna is defined as the polarization of a plane wave, incident from a given direction and having a given power flux density, which results in maximum available power at the antenna terminals.

Polarization may be classified as linear, circular, or elliptical. If the vector that describes the electric field at a point in space as a function of time is always directed along a line, the field is said to be linearly polarized. If the figure that the electric field traces is an ellipse,

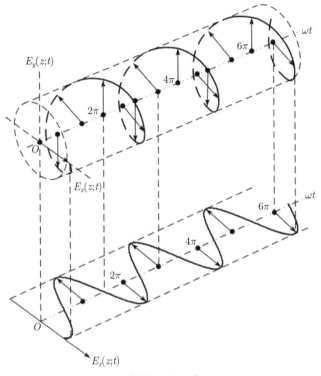

(a) Rotation of wave

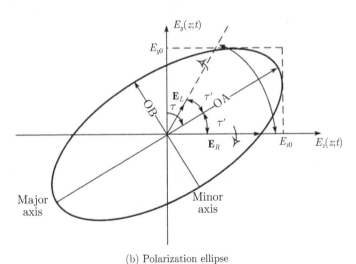

(b) Polarization ellipse

Figure 1.9   Rotation of a plane electromagnetic wave and its polarization ellipse
at $z = 0$ as function of time.

and the field is said to be elliptically polarized. Linear and circular polarizations are special cases of elliptical, and they can be obtained when the ellipse becomes a straight line
or a circle, respectively. When the electric field is traced in a clockwise (CW) direction,
that is, the rotation of the electric field vector and the direction of the wave propagation
form a right-handed screw, it is called right-hand polarization, while in a counterclockwise
(CCW), left-hand polarization. The polarization is usually resolved into a pair of orthogonal
polarizations, the co-polarization and cross polarization.

The instantaneous field of a plane wave, travelling in the negative z direction shown in Figure 1.9, can be written as

$$\mathbf{E}(z;t) = \hat{\mathbf{a}}_x E_x(z;t) + \hat{\mathbf{a}}_y E_y(z;t) \tag{1.4.14}$$

The instantaneous components are related to their complex counterparts by

$$\begin{aligned} E_x(z;t) &= \mathrm{Re}\left[E_{xo}e^{j(\omega t + kz + \phi_x)}\right] \\ &= E_{x0}\cos(\omega t + kz + \phi_x) \end{aligned} \tag{1.4.15}$$

$$\begin{aligned} E_y(z;t) &= \mathrm{Re}\left[E_{yo}e^{j(\omega t + kz + \phi_y)}\right] \\ &= E_{y0}\cos(\omega t + kz + \phi_y) \end{aligned} \tag{1.4.16}$$

where E_{x0} and E_{y0} are, respectively, the maximum magnitudes of the x and y components.

Linear Polarization

For the wave to have linear polarization, the time-phase difference between the two components must be

$$\Delta\phi = \phi_y - \phi_x = n\pi, n = 0, 1, 2, 3, \cdots \tag{1.4.17}$$

Circular Polarization

Circular polarization can be achieved only when the magnitudes of the two components are the same and the time-phase difference between them is odd multiples of $\pi/2$. That is,

$$|E_x| = |E_y| \Rightarrow E_{x0} = E_{y0} \tag{1.4.18}$$

$$\Delta\phi = \phi_y - \phi_x = +(1/2 + 2n)\pi, \quad n = 0, 1, 2, \cdots, \quad for\ CW \tag{1.4.19}$$
$$\Delta\phi = \phi_y - \phi_x = -(1/2 + 2n)\pi, \quad n = 0, 1, 2, \cdots, \quad for\ CCW \tag{1.4.20}$$

If the direction of wave propagation is reversed (i.e., $+z$ direction), the phase in (1.4.19) and (1.4.20) for CW and CCW rotation must be interchanged.

Elliptical Polarization

Elliptical polarization can be obtained when the time-phase difference between the two components is odd multiples of $\pi/2$ and their magnitudes are not the same or when the time-phase difference between the two components is not equal to multiples of $\pi/2$ (irrespective of their magnitudes). That is, when

$$\Delta\phi = \phi_y - \phi_x = +\left(\frac{1}{2} + 2n\right)\pi, \quad n = 0, 1, 2, \cdots, \quad for\ CW \tag{1.4.21}$$

$$\Delta\phi = \phi_y - \phi_x = -\left(\frac{1}{2} + 2n\right)\pi, \quad n = 0, 1, 2, \cdots, \quad for\ CCW \tag{1.4.22}$$

$$|E_x| \neq |E_y| \Rightarrow E_{x0} \neq E_{y0}$$

or

$$\Delta\phi = \phi_y - \phi_x \neq \pm\frac{n}{2}\pi > 0, \quad n = 0, 1, 2, \cdots, \quad for\ CW \tag{1.4.23}$$
$$\Delta\phi = \phi_y - \phi_x \neq \pm\frac{n}{2}\pi < 0, \quad n = 0, 1, 2, \cdots, \quad for\ CCW \tag{1.4.24}$$

For elliptical polarization, the curve traced at a given position as a function of time is a tilted ellipse, as shown in Figure 1.9(b). The ratio of the major axis to the minor axis is referred to as the axial ratio (AR), and it is equal to

$$\gamma = \frac{\text{major axis}}{\text{minor axis}} = \frac{OA}{OB}, \quad 1 \leqslant \gamma \leqslant \infty \tag{1.4.25}$$

where γ denotes the axial ratio (AR),

$$OA = \left\{ \frac{1}{2} \left[E_{x0}^2 + E_{y0}^2 + \left(E_{x0}^4 + E_{y0}^4 + 2E_{x0}^2 E_{y0}^2 \cos(2\Delta\phi) \right)^{1/2} \right] \right\}^{1/2} \tag{1.4.26}$$

$$OB = \left\{ \frac{1}{2} \left[E_{x0}^2 + E_{y0}^2 - \left(E_{x0}^4 + E_{y0}^4 + 2E_{x0}^2 E_{y0}^2 \cos(2\Delta\phi) \right)^{1/2} \right] \right\}^{1/2} \tag{1.4.27}$$

The tilt of the ellipse, relative to the y axis, is represented by the angle τ given by

$$\tau = \frac{\pi}{2} - \frac{1}{2} \tan^{-1} \left[\frac{2E_{x0}E_{y0}}{E_{x0}^2 - E_{y0}^2} \cos(\Delta\phi) \right] \tag{1.4.28}$$

When the ellipse is aligned with the principal axes $[\tau=n\pi/2, n=0,1,2,\cdots]$, the major (minor) axis is equal to $E_{x0}(E_{y0})$ or $E_{y0}(E_{x0})$ and the axial ratio is equal to E_{x0}/E_{y0} or E_{y0}/E_{x0}.

Figure 1.10 gives an example of the realization of circular polarization. In front of the aperture of the horn, there is a polarizer. On aperture A, E_x may be divided into two orthogonal components E_u and E_v. They are in phase and in the same amplitude. The polarizer may provide a different phase constant to E_u and E_v. With proper design, on aperture B, E_u and E_v will be in 90° phase difference and the same amplitude. Therefore the wave radiated from aperture B will be a circular polarization wave.

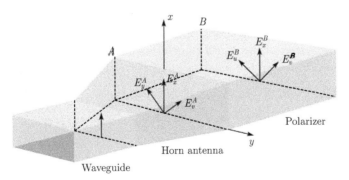

Figure 1.10   An example of the realization of circular polarization.

Alternatively, $\mathbf{E}(z,t)$ in (1.4.14) may be expressed by the superposition of circular polarization orthogonal vectors

$$\mathbf{E}(z,t) = \text{Re} \left[(\hat{\mathbf{a}}_R E_R + \hat{\mathbf{a}}_L E_L) e^{j(\omega t + kz)} \right] \tag{1.4.29}$$

where

$$\hat{\mathbf{a}}_R = \frac{1}{\sqrt{2}} (\hat{\mathbf{a}}_x + j\hat{\mathbf{a}}_y), \quad \hat{\mathbf{a}}_L = \frac{1}{\sqrt{2}} (\hat{\mathbf{a}}_x - j\hat{\mathbf{a}}_y) \tag{1.4.30}$$

E_R, E_L are complex scalars. Based on (1.4.30) and considering only the relative phase between E_R and E_L, assuming that E_R is with zero phase and E_L is with the phase of $-\theta$, (1.4.29) yields

$$\mathbf{E}(z,t) = \frac{1}{\sqrt{2}}\,\mathrm{Re}\left\{[(E_L + E_R)\,\hat{\mathbf{a}}_x + j\,(E_R - E_L)\,\hat{\mathbf{a}}_y]\,e^{j(\omega t + kz)}\right\} \qquad (1.4.31)$$

where $E_R = |E_R|$ and $E_L = |E_L|\exp(j\theta)$. From (1.4.31), we have both X and Y components that are functions of ωt, $X = f(\omega t)$ and $Y = g(\omega t)$. After eliminating ωt, we have the ellipse equation. The alternative representation of $\mathbf{E}(z,t)$ gives a clearer picture of the polarization properties. For example, it is easy to see that the axial ratio γ will be

$$\gamma = \frac{|E_R| + |E_L|}{|E_R| - |E_L|} = \frac{1 + |\Gamma|}{1 - |\Gamma|}, \quad |E_R| > |E_L| \qquad (1.4.32\text{-a})$$

where

$$\Gamma = \frac{E_L}{E_R} = \frac{|E_L|}{|E_R|}e^{j\theta} = \frac{|E_L|}{|E_R|}e^{j2\tau'} \qquad (1.4.32\text{-b})$$

$\tau' = \theta/2$ is the tilt of the ellipse, relative to the x axis measured counterclockwise, and $\tau = (\pi/2) - \tau'$ is the tilt of the ellipse, relative to the y axis measured clockwise as shown in Figure 1.9. From (1.4.14)–(1.4.16) and (1.4.31) we have

$$\xi = \frac{E_R - E_L}{E_R + E_L} = \frac{1 - \Gamma}{1 + \Gamma}$$
$$= -j\frac{E_{y0}}{E_{x0}}e^{j\Delta\phi} = \frac{E_{y0}}{E_{x0}}e^{j(\Delta\phi - \frac{\pi}{2})} \qquad (1.4.33)$$

From (1.4.32-a,b)–(1.4.33), it is seen that γ, Γ and ξ correspond to the standing wave ratio, reflection coefficient and normalized input admittance of a transmission line. Therefore the relationship between state of orthogonal source $\xi = E_{y0}\exp\left(j(\Delta\phi - \pi/2)\right)/E_{x0}$ and that of polarization $\Gamma = E_L/E_R$ is a bilinear transformation as shown in (1.4.33). In applications, it is desirable to know how to adjust the antenna in order to satisfy the required polarization. Through measurement, we may obtain the state of polarization Γ and based on (1.4.33), we get the state of source ξ. Sometimes, we know the state of orthogonal source ξ and wish to evaluate the state of polarization. The polarization chart may serve as a convenient tool for these purposes. The construction of this chart is based on the same mathematical relationship (1.4.33) as that of the Smith chart used for transmission lines.

Finally, we discuss the loss of the polarization. Let $z = 0$, $t = 0$ and define

$$\mathbf{E}'_1 = (E_{R1} + E_{L1})\,\hat{\mathbf{a}}_x + j\,(E_{R1} - E_{L1})\,\hat{\mathbf{a}}_y$$
$$= |E_{L1}|\,e^{j(\psi_{L1}+\psi_{R1})/2}\frac{2}{1 - \gamma_1}\left[(-\gamma_1\cos\tau'_1 + j\sin\tau'_1)\hat{\mathbf{a}}_x\right.$$
$$\left. + (-\gamma_1\sin\tau'_1 - j\cos\tau'_1)\hat{\mathbf{a}}_y\right] \qquad (1.4.34)$$

where subscript 1 denotes transmitting antenna No.1, γ_1 is the axial ratio defined in (1.4.25) and (1.4.32-a), $\tau'_1 = (\psi_{L1} - \psi_{R1})/2$, $E_{L1} = |E_{L1}|e^{j\psi_{L1}}$ and $E_{R1} = |E_{R1}|e^{j\psi_{R1}}$. If the receiving antenna No.2 is facing the main beam of transmitting antenna No.1 with the same polarization, in (1.4.34), then taking antenna No.1 as the reference, γ should be with plus sign

$$\mathbf{E}'_2 = |E_{L2}|e^{j(\psi_{L2}+\psi_{R2})/2}\frac{2}{1 + \gamma_2}\left[(\gamma_2\cos\tau'_2 + j\sin\tau'_2)\hat{\mathbf{a}}_x\right.$$
$$\left. + (\gamma_2\sin\tau'_2 - j\cos\tau'_2)\hat{\mathbf{a}}_y\right] \qquad (1.4.35)$$

The received field is proportional to $|\mathbf{E}'_1 \cdot \mathbf{E}'_2|$ and the received power P is proportional to $|\mathbf{E}'_1 \cdot \mathbf{E}'_2|^2$. From (1.4.34), (1.4.35) we have

$$
\begin{aligned}
p &\propto |\mathbf{E}'_1 \cdot \mathbf{E}'_2|^2 \\
&= A\left[(1+\gamma_1\gamma_2)^2 \cos^2(\tau'_1 - \tau'_2) + (\gamma_1 + \gamma_2)^2 \sin^2(\tau'_1 - \tau'_2)\right] \\
&= \frac{A}{2}\left[(1+\gamma_1^2)(1+\gamma_2^2) + 4\gamma_1\gamma_2 + (1-\gamma_1^2)(1-\gamma_2^2)\cos 2(\tau'_1 - \tau'_2)\right] \\
&= A(1+\gamma_1^2)(1+\gamma_2^2)\left[\frac{1}{2} + \frac{2\gamma_1\gamma_2}{(1+\gamma_1^2)(1+\gamma_2^2)} + \frac{(1-\gamma_1^2)(1-\gamma_2^2)}{2(1+\gamma_1^2)(1+\gamma_2^2)}\right. \\
&\quad \left. \cdot \cos 2(\tau'_1 - \tau'_2)\right]
\end{aligned} \tag{1.4.36}
$$

where A is a coefficient and will be eliminated by normalization. The normalizing of (1.4.36) relative to the perfect matching yields the polarization loss factor (PLF) for co-polarization

$$
P_{co} = \frac{1}{2} + \frac{2\gamma_1\gamma_2}{(1+\gamma_1^2)(1+\gamma_2^2)} + \frac{1}{2}\frac{(1-\gamma_1^2)(1-\gamma_2^2)}{(1+\gamma_1^2)(1+\gamma_2^2)}\cos 2(\tau'_1 - \tau'_2) \tag{1.4.37}
$$

It is observed for example, that for perfect circular co-polarization, $\gamma_1 = \gamma_2 = 1$, then $P_{co} = 1$. If the transmitting antenna and the receiving antenna are with orthogonal polarizations, the PLF for cross-polarization is

$$
P_{cross} = \frac{1}{2} - \frac{2\gamma_1\gamma_2}{(1+\gamma_1^2)(1+\gamma_2^2)} + \frac{1}{2}\frac{(1-\gamma_1^2)(1-\gamma_2^2)}{(1+\gamma_1^2)(1+\gamma_2^2)}\cos 2(\tau'_1 - \tau'_2) \tag{1.4.38}
$$

where subscript cross means cross-polarization.

Example 1.2:
   The electric field of a left-hand polarization electromagnetic wave with $\gamma_1 = 4$, $\tau'_1 = 15°$ is incident upon a right-hand polarization antenna with $\gamma_2 = 2$, $\tau'_2 = 45°$. Find the polarization loss factor (PLF) P_{cross}

Solution:
   From (1.4.38),

$$
P_{cross} = \frac{1}{2} - \frac{2(2)(4)}{(1+4)(1+16)} + \frac{1}{2}\cdot\frac{(1-4)(1-16)}{(1+4)(1+16)}\cos 2(45° - 15°) = 0.44.
$$

It means that the received power reduces to 44% of the received power in the case of perfect polarization matching.

Example 1.3:
   The measured parameters of the state of polarization are $\gamma = 1.4$, $\tau' = 37°$, determine the state of source.

Solution:
   According to (1.4.32-a) and (1.4.32-b), $\Gamma = |\Gamma|e^{j2\tau'} = 0.17e^{j(2\times 37°)}$. From (1.4.33),

$$
\frac{E_y}{E_x} = j\frac{1-\Gamma}{1+\Gamma} = j\frac{1 - 0.17(\cos 74^0 + j\sin 74^0)}{1 + 0.17(\cos 74^0 + j\sin 74^0)} = 0.91e^{j72^0}.
$$

This information is useful for adjusting the state of source, if the pure circular polarization is desired.

1.4.4   Characteristics and Parameters of an Antenna in Receiving Mode[7, 11, 12]

An antenna can also be used to capture (collect) electromagnetic waves and to extract power from them when it is in the receiving mode. The main characteristics and the parameters are receiving pattern, inner impedance, gain and effective receiving area.

An antenna pattern in receiving mode viewed as either a mathematical function or a graphical representation of the receiving properties of the antenna is a function of space coordinates. For a receiving antenna, when it is excited it looks like a generator with an inner impedance. This inner impedance is the same as the input impedance in transmission mode. These characteristics are useful in establishing the relationship between antenna parameters of an antenna in transmitting and in receiving modes, which will be discussed later. They are also useful in some other cases. Measuring the radiation pattern is usually not very convenient because of the heavy weight of the transmitter. As an alternative, we may measure the pattern when the antenna is in the receiving mode. In order to enhance the ratio of signal to interference in an adaptive antenna array, it is possible to get the nulls of the receiving pattern at the directions of the interferences through adaptively weighting the amplitude and phase for each antenna element. As we will see later, it is difficult to obtain the pattern in receiving mode and as an alternative we may find the weights for the desired pattern in transmitting mode.

Consider a half-wavelength dipole ($2l = \lambda/2$) shown in Figure 1.11. The current distribution is a cosine function. When a plane wave is incident upon the receiving antenna at angle θ as is shown in Figure 1.11(a) and assuming that $|\mathbf{E}| = 1$, the induced voltages on the two differential elements dz_1 and dz_2 are $du_{z1} = dz \sin\theta e^{j\tau' z \cos\theta}$ and $du_{z2} = dz \sin\theta e^{-j\tau' z \cos\theta}$ respectively. The resultant induced voltage will be

$$
\begin{aligned}
du_z &= dz \sin\theta (e^{j\tau' z \cos\theta} + e^{-j\tau' z \cos\theta}) \\
&= 2 \sin\theta \cos(\tau' z \cos\theta) dz
\end{aligned}
\tag{1.4.39}
$$

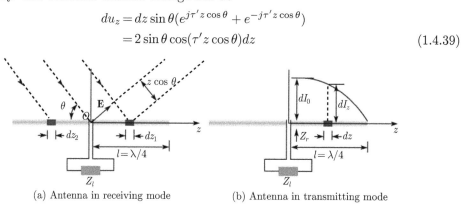

(a) Antenna in receiving mode   (b) Antenna in transmitting mode

Figure 1.11   Finding the pattern and inner impedance of receiving antenna using dual reciprocity.

The resultant current through the load impedance $Z_l = R_l + jX_l$ is the sum of the currents produced by the induced voltages du_z. R_l is the load resistance and X_l is the load reactance. These currents are easily found if the reciprocity theorem is applied. Assume that the induced voltages on dz_1 and dz_2 are du_{z1} and du_{z2} respectively, the port currents through the load Z_l produced by these voltages are dI_l. If we add voltage du_z on the port, according to the reciprocity theorem, it will produce the same current dI_z on dz_1 or dz_2, that is, $dI_l = dI_z$. Based on Figure 1.11(b), we may find dI_z and then dI_l

$$
\begin{aligned}
dI_l = dI_z &= dI_0 \cos\beta z = \frac{du_z \cos\beta z}{Z_r + Z_l} \\
&= \frac{2 \sin\theta \cos(\beta z \cos\theta) \cos\beta z dz}{Z_r + Z_l}
\end{aligned}
\tag{1.4.40}
$$

where the internal impedance $Z_r = (R_r+R_d)+jX_r$ is the input impedance when the antenna is in transmitting mode. R_r, R_d are radiation resistance and loss resistance respectively, X_r is the antenna reactance. The induced voltage dV at the output of the receiving antenna is

$$dV = dI_l(Z_r + Z_l) = 2\sin\theta \cos(\beta z \cos\theta)\cos\beta z dz \tag{1.4.41}$$

The total induced voltage is given as

$$V = \int_0^{\frac{\lambda}{4}} dV = \int_0^{\frac{\lambda}{4}} 2\sin\theta \cos\beta z \cos(\beta z \cos\theta)dz$$
$$= \frac{2}{\beta}\frac{\cos(\pi \cos\theta/2)}{\sin\theta} = F(\theta) \tag{1.4.42}$$

The current I_l through the load or the current I within the loop is

$$I = I_l = \frac{u_A}{Z_r + Z_l} = \frac{F(\theta)}{Z_r + Z_l} \tag{1.4.43}$$

From the above derivation, it is seen that the receiving pattern is the same as the radiation pattern obtained in Section 1.3.1; the inner impedance for the receiving mode is the same as the input impedance for the transmitting mode. Although the derivation is for a specific antenna, it can be proved that the conclusion holds for antennas of any shape or size.

Understanding the physical insight of the above results is helpful. The current distribution on the transmitting antenna follows that of a pure standing wave. The current distribution on the receiving antenna depends on both the excitation condition and the loading condition and is different with that on the transmitting antenna[16]. In transmitting mode, the field at the infinity is the interference (superposition) of all the differential elements on the antenna with certain amplitude distribution, uniform electric phase in the case of half-wavelength dipole and progressive space phase difference caused by the path difference in space. In receiving mode, the excitation is a plane wave with uniform amplitude and progressive phase difference caused by the path difference between the wavefront in space and the antenna. Therefore the electric phase is different at the different point of the receiving antenna. The induced current through the load is the superposition of the differential current produced by the differential elements with certain amplitude and electric phase distributions. Although the meaning of superposition is different for transmitting and receiving modes, the final integral expression is the same, resulting in the same pattern.

To use the reciprocity theorem, we put the two ports on the same antenna. However, it could be used in other ways, for example, the other port could be put at infinity or at the output/input of the other antenna so long as the reciprocity is satisfied, that is, two ports are separated by a linear and isotropic (but not necessarily homogeneous) medium.

The equivalent circuit of Figure 1.11(a) is shown in Figure 1.12. The captured power P_{cap} is

$$P_{cap} = \frac{1}{2}\text{Re}(VI^*) = \frac{1}{2}\text{Re}\,V\left[\frac{V}{(R_d + R_r + R_l) + j(X_r + X_l)}\right]^* \tag{1.4.44}$$

The maximum captured power occurs when we have conjugate matching; that is when $R_d + R_r = R_l$ and $X_r = -X_l$. For this case

$$P_{cap} = \frac{1}{4}\frac{|V|^2(R_d + R_r)}{(R_d + R_r)^2} = \frac{1}{4}\frac{|V|^2}{R_d + R_r} \tag{1.4.45}$$

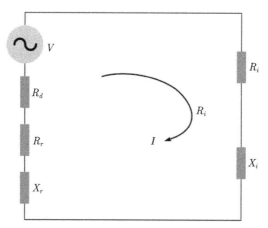

Figure 1.12   Equivalent circuit for an antenna in receiving mode.

The power delivered to the load P_l, the reradiated (scattered) power P_r, and the dissipated power P_d are then respectively given by

$$P_l = \frac{1}{8} \frac{|V|^2 R_l}{(R_d + R_r)^2} = \frac{1}{8} \frac{|V|^2}{R_l} \tag{1.4.46}$$

$$P_r = \frac{1}{8} \frac{|V|^2 R_r}{(R_d + R_r)^2} \tag{1.4.47}$$

$$P_d = \frac{1}{8} \frac{|V|^2 R_d}{(R_d + R_r)^2} \tag{1.4.48}$$

From (1.4.45)–(1.4.48), it is seen that

$$P_{cap} = P_l + P_r + P_d \tag{1.4.49}$$

The power may be expressed by the effective area if we define

$$A = \frac{P}{W} \tag{1.4.50}$$

where W is the power density of incident wave. Accordingly we have

$$A_{cap} = A_e + A_s + A_d \tag{1.4.51}$$

where A_{cap} is the receiving area which is related to the captured power, A_e is the effective area which is related to the power delivered to the load, A_s is the scattering area which is related to the scattered or reradiated power and A_d is the loss area which is related to the power dissipated as heat through R_d. It is seen that not all of the power that is captured by an antenna is delivered to the load. Under conjugate matching condition actually only half of the captured power is delivered to the load; the other half is scattered and dissipated as heat[1]. This conclusion is not rigorous and should be carefully examined in some cases. Because further investigation shows that the internal resistance in receiving mode depends on the excitation and loading, it is not always equal to the input resistance in transmitting mode[16].

There is a relationship between effective area and the directivity previously defined for an antenna in transmitting mode. Consider transmitting antenna A and receiving antenna B; the power delivered to the load of antenna B due to antenna A is P_{lB}. Based on (1.4.46), noticing that the induced voltage $|V_B|$ is proportional to the amplitude of the current $|I_{0A}|$ at the input of antenna A, we have

$$P_{lB} = \frac{|V_B|^2}{8R_{lB}} = \frac{|I_{0A}|^2 \xi}{8R_{lB}} \qquad (1.4.52)$$

The radiation power of antenna A is

$$P_{tA} = \frac{1}{2}|I_{0A}|^2 R_{rA} \qquad (1.4.53)$$

From (1.4.52) and (1.4.53), we have

$$\frac{P_{lB}}{P_{tA}} = \frac{\xi}{4R_{lB}R_{rA}} \qquad (1.4.54)$$

If the distance between antenna A and B is r, according to the definition of the power density, W is written as

$$W = \frac{P_{tA}D_{0A}}{4\pi r^2} \qquad (1.4.55)$$

From (1.4.50) and (1.4.55), we have

$$\frac{P_{lB}}{P_{tA}} = \frac{A_{eB}D_{0A}}{4\pi r^2} \qquad (1.4.56)$$

where D_{0A} is the maximum directivity of antenna A and A_{eB} is the maximum effective area of antenna B. Equating (1.4.54) and (1.4.56) leads to

$$\xi = \frac{R_{lB}R_{rA}A_{eB}D_{0A}}{\pi r^2} \qquad (1.4.57)$$

When the transmitting and receiving antennas A and B are interchanged, we have another formula due to the reciprocity

$$\xi = \frac{R_{lA}R_{rB}A_{eA}D_{0B}}{\pi r^2} \qquad (1.4.58)$$

Making use of the above two expressions and assuming that $R_r = R_l$ then we obtain the following identity

$$\frac{D_{0A}}{A_{eA}} = \frac{D_{0B}}{A_{eB}} \qquad (1.4.59)$$

This identity states that the ratio of maximum directivity and maximum effective area is a constant for any kind of antenna. To find this constant, we may choose the simplest antenna, the fundamental dipole. For this dipole, $F(\theta,\phi)$ in (1.4.7) is $\sin^2\theta$ and D_0 may be easily found to be $3/2$. To find A_e, use (1.4.50), assuming a uniform plane wave with the amplitude E is incident upon the dipole. The incident power density W is

$$W = \frac{E^2}{2\eta} \qquad (1.4.60)$$

where $\eta = 120\pi$ is the intrinsic impedance in free space. The received power in (1.4.50) is determined by (1.4.46). In this formula, the induced voltage V for a fundamental dipole with length l is El. We may find the radiation resistance through (1.4.53) when the dipole is used as the transmitting antenna. Using the formula for the radiation electric field given in (1.3.24), it is easy to prove that the radiation resistance is

$$R_l = R_r = \frac{2P_t}{I_0^2}$$

$$= \frac{2 \int_0^{2\pi} \int_0^{\pi} \left(\frac{60\pi I_0 l}{\lambda r} \right)^2 \frac{1}{2\eta} \sin^3\theta \, r^2 d\theta d\phi}{I_0^2}$$

$$= 80\pi^2 \left(\frac{l}{\lambda} \right)^2 \tag{1.4.61}$$

The combination of (1.4.46), (1.4.50), (1.4.60) and (1.4.61) yields

$$A_e = \frac{3\lambda^2}{8\pi} \tag{1.4.62}$$

Finally, we have

$$A_e = \left(\frac{\lambda^2}{4\pi} \right) D_0 \tag{1.4.63}$$

The maximum value of (1.4.63) is achieved when the antenna has no losses and is matched to the load, and the incoming wave is polarization-matched to the antenna. Otherwise, all of these effects should be taken into account. That is, (1.4.63) should be multiplied by a certain loss coefficient to take the antenna loss into account, the factor $(1 - |\Gamma|^2)$ to take the reflection loss into account and the factor $P_{H,co}$ or $P_{H,cross}$ given in Section 1.4.3 to take the polarization loss into account.

1.4.5   Radar Equation and Friis Transmission Formula[13–15]

The basic concept of radar is very simple. A radar operates by radiation electromagnetic wave and detecting the echo returned from reflecting targets. The echo signal provides information about the target such as the range to the target, the angular location.

With the definition of effective area A_e in Section 1.4.4, it is easy to give the radar equation which gives the range of a radar in terms of the radar characteristics. One form of this equation gives the received signal power P_r as

$$P_r = \left(\frac{P_t G_t}{4\pi R^2} \right) \left(\frac{\sigma}{4\pi R^2} \right) (e_r A_e) \tag{1.4.64}$$

The right-hand side is written as the product of three factors to represent the physical processes taking place. The first factor is the power density at a distance R meters from a radar that radiates a power of P_t watts from an antenna of gain G_t. The numerator of the second factor is the radar cross Section (RCS) σ in square meter. The denominator accounts for the divergence on the return path of the wave with range and is the same as the denominator of the first factor, which accounts for the divergence on the outward path. The product of the first two terms represents the power per square meter returned to the radar. The antenna of effective aperture area A_e intercepts a portion of this power in an amount given by the product of the three factors. The efficiency coefficient e_r is to include the losses. If the maximum radar range $R_{\max}$ is defined as that which results in the received power P_r being equal to the receiver minimum detectable signal $S_{\min}$, the radar equation may be written as

$$R^4_{\max} = \frac{P_t G_t (e_r A_e) \sigma}{(4\pi)^2 S_{\min}} \tag{1.4.65}$$

When the same antenna is used for both transmitting and receiving, the transmitting gain G_t and A_e are related by $G_t = 4\pi e_r A_e / \lambda^2$ based on (1.4.13) and (1.4.63). Substituting it into (1.4.65) gives the radar equation:

$$R^4_{\max} = \frac{P_t G_t^2 \lambda^2 \sigma}{(4\pi)^3 S_{\min}} \tag{1.4.66}$$

The example of the radar equation given above is useful for rough computations of range performance but is overly simplified and does not give realistic results. Actually, the RCS and $S_{\min}$ are statistical in nature, thus the specification of the range must be made in statistical terms. In addition, the propagation loss and various other losses (some of them, such as the reflection loss, polarization loss, we have mentioned previously) should be taken into account.

Similar derivation of the formula for the communication system may also be carried out. In that case, the received signal power P_r is given as

$$P_r = \left(\frac{P_t G_t}{4\pi R^2} \right) (e_r A_e) \tag{1.4.67}$$

The meaning of the first and second factors is the same as that of the first and third factors in (1.4.64), and R is the distance between the transmit antenna and the receiver antenna. If the gain of receiver antenna is G_r, using the formula $G_r = 4\pi e_r A_e / \lambda^2$ again, we have

$$\frac{P_r}{P_t} = \frac{G_t G_r \lambda^2}{(4\pi R)^2} \tag{1.4.68}$$

which is called the Friis transmission formula. This formula may also be used in the measurement of antenna gain.

Bibliography

[1] C. A. Balanis, "Antenna theory: a review," *Proceedings of the IEEE*, vol. 80, no. 1, Jan. 1992, pp. 7–23.

[2] J. D. Kraus, "Antennas since Hertz and Marconi," *ibid*, pp. 131–136.

[3] D. M. Pozar, "Microstrip antennas," *ibid*, pp. 79–91.

[4] N. Fourikis, *Phased Array-Based Systems and Applications*, John Wiley & Sons, Inc., 1997.

[5] J. C. Liberti, JR. T. S. Rappaport, *Smart Antennas for Wireless Communications: IS-95 and Third Generation CDMA Applications*, Prentice Hall PTR, 1999.

[6] J. Litva and T. K. Lo, *Digital Beamforming in Wireless Communications*, Artech House, 1996.

[7] C. A. Balanis, *Antenna Theory, Analysis and Design* (second edition), John Wiley & sons, Inc., 1997.

[8] N. H. Fang, *Introduction to Electromagnetic Theory*, Science Press, 1986.

[9] E. V. Jull, *Aperture Antenna and Diffraction Theory*, Peter Peregrinus, 1981.

[10] W. L. Stutzman, "Estimating directivity and gain of antennas," *IEEE Antennas and Propagation Magazine*, vol. 40, no. 4 pp. 7–11, Aug., 1998.

[11] F. Y. Zhang, *Antennas and Their Feeding System*, Beijing Science Education Press, 1961.

[12] D. Q. Zhang, *Fundamentals of Microwave Antenna*, Beijing Industry Institute Press, 1985.

[13] M. Skolnik, *Radar Handbook* (second edition), McGraw-Hill Publishing Company, 1990.

[14] W. L. Stutzman and G.A.Thiele, *Antenna Theory and Design*, John Wiley & Sons, Inc., 1981.

[15] J. D. Kraus and R. J. Marhefka, *Antennas: For All Applications* (third edition), McGraw-Hill Companies, Inc., 2002.

[16] C. C. Su, "On the equivalent generator voltage and generator internal impedance for receiving antennas," *IEEE Trans. on Antennas Propagat.*, vol. 51, no. 2, pp. 279–285, 2003.

Problems

1.1 For a linear antenna directed along z axis with length l

    1. when $I_e = \cos(\pi z'/l)$ $l \neq \lambda/2$, prove the radiation pattern is

$$F(\theta) = \frac{2l \cos(\pi l \cos \theta/\lambda)}{\pi \left[1 - (2/\pi)^2 (\pi l/\lambda)^2 \cos^2 \theta\right]} \sin \theta$$

    2. when

$$I_e = \begin{cases} \sin k(\frac{l}{2} + z'), & -\frac{l}{2} \leqslant z' \leqslant 0 \\[2mm] \sin k(\frac{l}{2} - z'), & 0 \leqslant z' \leqslant \frac{l}{2} \end{cases}$$

    prove the radiation pattern is

$$F(\theta) = \frac{2}{k} \frac{\cos(\pi l \cos \theta/\lambda) - \cos(kl/2)}{\sin \theta}$$

    3. verify when $l = \lambda/2$, these two expressions will be the same.

1.2 In Figure 1.5, if $\mathbf{E}^0 = \hat{\mathbf{a}}_y E_y^0$, $\mathbf{H}^0 = -\hat{\mathbf{a}}_x E_y^0/\eta$, show that $F_1 = \sin \phi(1 + \cos \theta)$ for E_θ, and $F_1 = \cos \phi(1 + \cos \theta)$ for E_ϕ. If $\mathbf{H}^0 = 0$, show that $F_1 = \sin \phi$ for E_θ, and $F_1 = \cos \phi \cos \theta$ for E_ϕ.

1.3 In Figure 1.6, if the aperture is located in xoz plane, that is $x \to z$, $y \to x$, $z \to y$, $\mathbf{E}^0 = E_x^0 \hat{\mathbf{a}}_x$, $\mathbf{H}^0 = 0$, and the definition of θ and ϕ is the same as that in Figure 1.6, prove that for the far field $E_r = E_\theta = 0$, $E_\phi = A F_1 F_2$, $A = je^{-jkr}/2\lambda r$ as shown in (1.3.49-b), $F_1 = -\sin \theta$, $F_2 = E_x^0 ab(\sin X/X)(\sin Z/Z)$, $X = (kb \sin \theta \cos \phi)/2$, $Z = (ka \cos \theta)/2$. Hint:

$$\begin{pmatrix} \hat{\mathbf{a}}_r \\ \hat{\mathbf{a}}_\theta \\ \hat{\mathbf{a}}_\phi \end{pmatrix} = \begin{pmatrix} \sin \theta \cos \phi & \sin \theta \sin \phi & \cos \theta \\ \cos \theta \cos \phi & \cos \theta \sin \phi & -\sin \theta \\ -\sin \phi & \cos \phi & 0 \end{pmatrix} \begin{pmatrix} \hat{\mathbf{a}}_x \\ \hat{\mathbf{a}}_y \\ \hat{\mathbf{a}}_z \end{pmatrix}$$

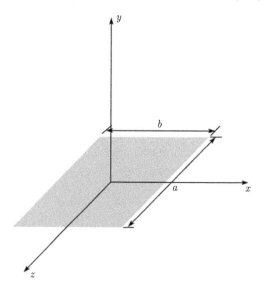

1.4 For a rectangular and uniformly distributed aperture, show that

1. half-power beamwidth (HPBW) for two principal planes are $50.8/(b/\lambda)$ and $50.8/(a/\lambda)$ respectively.

2. find D_0 using (1.4.8).

3. find D_0 using (1.4.12).

4. compare the result from 2,3.

1.5 Prove (1.3.28) using the far field approximation.

1.6 A uniform plane wave, of a form similar to (1.4.14), is traveling in the positive z-axis. Find the polarization (linear, circular, or elliptical), sense of rotation (CW or CCW), axial ratio (AR), and tilt angle τ (in degree) when

(a) $E_x = E_y$, $\Delta\phi = \phi_y - \phi_x = 0$ (b) $E_x \neq E_y$, $\Delta\phi = \phi_y - \phi_x = 0$
(c) $E_x = E_y$, $\Delta\phi = \phi_y - \phi_x = \pi/2$ (d) $E_x = E_y$, $\Delta\phi = \phi_y - \phi_x = -\pi/2$
(e) $E_x = E_y$, $\Delta\phi = \phi_y - \phi_x = \pi/4$ (f) $E_x = E_y$, $\Delta\phi = \phi_y - \phi_x = -\pi/4$
(g) $E_x = 0.5E_y$, $\Delta\phi = \phi_y - \phi_x = \pi/2$ (h) $E_x = 0.5E_y$, $\Delta\phi = \phi_y - \phi_x = -\pi/2$

1.7 Derive (1.4.26)–(1.4.28).

1.8 Prove (1.4.37) and (1.4.38). When using the horn antenna with circular polarization shown in Figure 1.10 to transmit a circular polarization wave, prove that it is impossible to receive the back scattering from a perfect conducting sphere.

1.9 For an aperture of circular waveguide with TM_{0m} mode, prove that

$$E_{\theta,0m} = j2\pi C(1 + \cos\theta)\frac{a}{(\xi_{0m})^2 - \eta^2}\left[\eta\, \mathrm{J}_0\,(a\eta)\,\mathrm{J}_1\,(a\xi_{0m}) - \xi_{0m}\,\mathrm{J}_0\,(a\xi_{0m})\,\mathrm{J}_1\,(a\eta)\right]$$

where

$$C = B\frac{\beta a}{\mu_{0m}}\frac{je^{-jkr}}{\lambda r}$$

the aperture distribution of radial component $E_{\rho,0m}$ of TM_{0m} mode on $z = 0$ plane is assumed to be

$$E_{\rho,0m} = B\frac{\beta a}{\mu_{0m}}\mathrm{J}_1\left(\frac{\mu_{0m}}{a}\rho\right)$$

B is a constant related to the excitation, β is the propagation constant in waveguide of the relevant mode, a is the radius of the waveguide, J_1 is the first order Bessel function, μ_{0m} is the root of zero order Bessel function J_0, $\xi_{0m} = \mu_{0m}/a$, $\eta = k\sin\theta$.

Hints: (1) Due to the symmetry of $E_{\rho,0m}$ in ϕ, in (1.3.51), let $\phi = 0$ and

$$F_2 = \int_0^{2\pi}\int_0^a E_\rho(\rho',\phi')e^{jk\rho'\sin\theta\cos\phi'}\rho'\,d\rho'\,d\phi'$$

(2) Use the identity

$$e^{jt\cos\theta} = \mathrm{J}_0(t) + 2\sum_{n=1}^{\infty}j^n\,\mathrm{J}_n(t)\cos n\theta$$

and the integral formula

$$\int_0^a \mathrm{J}_m(\xi\rho)\,\mathrm{J}_m(\eta\rho)\rho\,d\rho = \frac{a}{\xi^2 - \eta^2}\left[\eta\,\mathrm{J}_{m-1}(\eta a)\,\mathrm{J}_m(\xi a) - \xi\,\mathrm{J}_{m-1}(\xi a)\,\mathrm{J}_m(\eta a)\right]$$

$$(\xi^2 - \eta^2 \neq 0)$$

1.10   For two linearly polarized waves propagating along $\hat{\mathbf{a}}_x$ and $\hat{\mathbf{a}}_u$, where unit vector $\hat{\mathbf{a}}_u$ is at an angle θ with respect to the unit vector $\hat{\mathbf{a}}_x$.   If $E_x = \cos\theta e^{j\varphi_x}$, $E_y = \sqrt{(1+\sin\theta\cos\varphi_x)^2 + (\sin\theta\sin\varphi_x)^2}\, e^{j\varphi_y}$, where φ_x and φ_y are the time-phases of E_x and E_y respectively, $\varphi_y = \tan^{-1}\left[\sin\theta\sin\varphi_x / (1 + \sin\theta\cos\varphi_x)\right]$, prove that when $\varphi_x = \pi/2 + \theta$, we have $|E_x| = |E_y| = \cos\theta$, $\varphi_y = \theta$ and $\varphi_x - \varphi_y = \pi/2$. (It is seen that these two linearly polarized waves form the circularly polarized wave.)

CHAPTER 2

Arrays and Array Synthesis

2.1 Introduction

In Chapter 1, the linear antenna shown in Figure 1.3 may be considered as formed by continuously distributed infinitesimal electric dipoles. This is actually an example of continuous linear array. The rectangular aperture in Figure 1.6 may be considered as a continuous planar array. The radiation pattern of both examples may be obtained by (1.3.34), where the element factor F_1 is the radiation pattern of differential element and the space factor F_2 is called the array factor AF.

In the above examples, all the elements forming the array are identical and have the same orientation. Therefore, in computing the far field $\mathbf{E}$, the element factor F_1 may be taken out of the integral. Consequently, the total radiation pattern F is equal to the product of the element factor F_1 and the space (array) factor F_2, as given in (1.3.34). This is referred to as pattern multiplication for both the continuously distributed and the discretely distributed sources. It is seen that F_2 is actually the radiation pattern of a point-source array formed by omnidirectional elements. The array factor is a function of the elements, their geometrical arrangement, their relative magnitudes, their relative phases, and their spacing. Through knowledge of all these controlled parameters, it is possible to obtain a required radiation pattern of the antenna.

2.2 N-Element Linear Array: Uniform Amplitude and Spacing

Referring to the geometry shown in Figure 2.1, assume that all the elements have identical amplitude but each succeeding element has a progressive phase lead β current excitation relative to the preceding one.

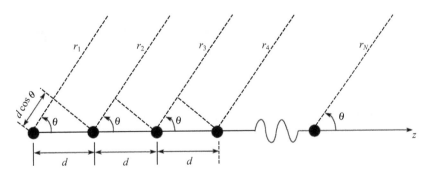

Figure 2.1   Far-field geometry of an N-element array of isotropic sources along the z-axis.

An array of identical elements with same magnitude and a progressive phase shift is referred to as a uniform array. According to the geometry in Figure 2.1, the array factor AF is given by:

$$AF = F_2 = \sum_{n=1}^{N} e^{j(n-1)(kd\cos\theta+\beta)} = \sum_{n=1}^{N} e^{j(n-1)\psi} \qquad (2.2.1)$$

where $\psi = kd\cos\theta + \beta$, and N is the number of the elements. Based on the formula of summation of geometrical series, (2.2.1) may be written as

$$AF = \frac{1 - e^{jN\psi}}{1 - e^{j\psi}} = e^{j(N-1)\psi/2}\frac{\sin\left(\dfrac{N}{2}\psi\right)}{\sin\left(\dfrac{1}{2}\psi\right)} \tag{2.2.2-a}$$

If the reference point is the physical center of the array, the AF in (2.2.2-a) reduces to

$$AF = \frac{\sin\left(\dfrac{N}{2}\psi\right)}{\sin\left(\dfrac{1}{2}\psi\right)} \tag{2.2.2-b}$$

For small values of ψ, the above expression can be approximated by

$$AF \approx \frac{\sin\left(\dfrac{N}{2}\psi\right)}{\dfrac{\psi}{2}} \tag{2.2.2-c}$$

The normalized array factor $(AF)_n$ is

$$(AF)_n = \frac{1}{N}(AF) \tag{2.2.2-d}$$

All the properties of F, i.e., the nulls, -3dB point, side lobe level (SLL) of the first side lobe and so on, may be found from (2.2.2-d). For example, from (2.2.2-c), the first side lobe occurs approximately when $N\psi/2 \approx \pm(3\pi/2)$. From (2.2.2-d)

$$(AF)_n \approx \left|\frac{\sin\left(N\psi/2\right)}{N\psi/2}\right|_{\frac{N}{2}\psi=\pm\frac{3\pi}{2}} = \frac{2}{3\pi} = 0.212 \tag{2.2.3}$$

Which in dB is equal to

$$AF_n = 20\lg\left(\frac{2}{3\pi}\right) = -13.46\text{dB}$$

From (2.2.2-a), it is seen that, when $\beta=0$, the maximum radiation of an array will be directed normal to the axis of the array at $\theta = \pm 90°$. This array is called the broadside array. To direct the maximum toward $\theta=0°$, we should have

$$\psi = kd\cos 0° + \beta = 0, \quad \text{that is }, \quad \beta = -kd \tag{2.2.4-a}$$

If the maximum is desired toward $\theta = 180°$, then

$$\psi = kd\cos 180° + \beta = 0, \quad \text{that is,} \quad \beta = kd \tag{2.2.4-b}$$

When β satisfies (2.2.4-a), the array is called the end-fire array.

2.3   Phased (Scanning) Array, Grating Lobe and Sub-Array

In many applications, it is desired to have the beam pointed to a new direction in microseconds. It may be realized through electrical control of the feeding phase β between the elements. If the desired direction is at $\theta = \theta_0$, β must be adjusted so that

$$\psi = kd\cos\theta_0 + \beta = 0, \quad \text{that is }, \quad \beta = -kd\cos\theta_0 \tag{2.3.1}$$

Thus by controlling the progressive phase shift between the elements, the maximum radiation can be steered to any desired direction to form a scanning array. This is the basic principle of electronic scanning phased array operation. If in phased array technology the scanning should be continuous, then the system should be capable of continuously varying the progressive phaseshift between the elements. In practice, this is accomplished electronically by the use of ferrite, ferroelectric or diode phase shifters.

A grating lobe is defined as "a lobe, other than the main lobe, produced by an array antenna when the inter-element spacing is sufficiently large to permit the in-phase addition of radiated fields in more than one direction in real space." From (2.2.2-b), the maximum of main lobe occurs when $\psi = \pm 2\pi$. Consider the general case when the beam is steered to $\theta = \theta_0$, so that, $\beta = -kd \cos \theta_0$ as given in (2.3.1), then we have

$$\psi = kd \cos \theta - kd \cos \theta_0 = \pm 2\pi$$

or

$$\cos \theta = \cos \theta_0 \pm \frac{\lambda}{d}$$

From the condition $|\cos \theta| \leqslant 1$, we have

$$-1 \leqslant \cos \theta_0 \pm \frac{\lambda}{d} \leqslant 1 \tag{2.3.2}$$

From (2.3.2), we may obtain a condition to avoid the grating lobes. Notice that $d > 0$ and its value should be chosen so that

$$d < \frac{\lambda}{1 + |\cos \theta_0|} \tag{2.3.3}$$

The array factor $(AF)_n$ in (2.2.2-a) is expressed in terms by ψ. A polar diagram of such function can easily be obtained by means of the geometrical construction shown in Figure 2.2[1]. The function $(AF)_n$ is plotted as a function of ψ. A circle of radius kd is then drawn with its center at a distance β from the line $\psi = 0$. With the aid of this circle, ψ may be found for a given value of θ, and the corresponding value of $|(AF)_n|$ can be determined from the curve. This length is then laid out from the center of the circle in the direction θ to give a point on the desired diagram. Now $0 \leqslant \theta \leqslant \pi$ and the interval for ψ is of length $2kd$, situated symmetrically with respect to $\psi = \beta$. This interval corresponds to real values of the angle θ and is called the visible range. The part of the curve $(AF)_n$ within the visible range is called the visible part of the curve. In the visible range, $|\cos \theta| \leqslant 1$, θ and the propagation constant are real.

When $|\cos \theta| \geqslant 1$, θ and the propagation constant will be complex. In this case, the evanescent waves which decay rapidly when propagate will appear. These waves are recognized as invisible. It is important to note that the length of the visible range is determined by the spacing d and position of the visible range is determined by the feeding phase β. In Figure 2.2, it is seen that the grating lobe appears just when $kd + \beta = kd + kd |\cos \theta_0| = 2\pi$. This confirms the condition in (2.3.3). Usually, the radius kd should be smaller than that in Figure 2.2 to avoid the grating lobe.

An antenna array may be divided into sub-arrays. For example, a 1×8 array may be considered as 1×2 sub-arrays, with each sub-array being a 1×4 array. The array factor $F_{2,n} = (AF)_n$ may be obtained from (2.2.2-a) as

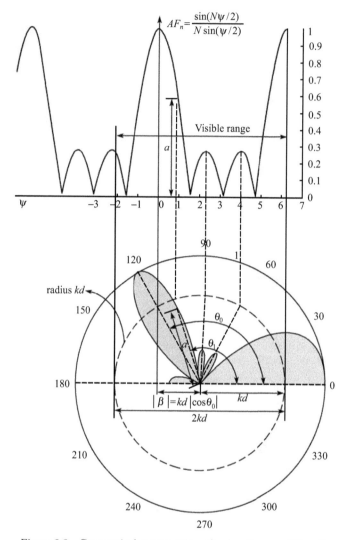

Figure 2.2   Geometrical construction of polar diagram ($N = 4$).

$$F_{2n,1\times8,d} = \frac{\sin\left[\dfrac{8}{2}kd\cos\theta\right]}{8\sin\left[\dfrac{1}{2}kd\cos\theta\right]} = \frac{\sin\left[\dfrac{4}{2}kd\cos\theta\right]}{4\sin\left[\dfrac{1}{2}kd\cos\theta\right]} \times \frac{\sin\left[\dfrac{2}{2}k4d\cos\theta\right]}{2\sin\left[\dfrac{1}{2}k4d\cos\theta\right]}$$

$$= F_{2n,1\times4,d} \times F_{2n,1\times2,4d} \tag{2.3.4}$$

where the second subscript denotes number of elements, the third one denotes the spacing between elements. Assume $d = 0.5\lambda$, we have

$$F_{2n,1\times8,d} = \frac{\sin\left(4\pi\cos\theta\right)}{8\sin(\dfrac{\pi}{2}\cos\theta)} \tag{2.3.5}$$

$$F_{2n,1\times4,d} = \frac{\sin\left(2\pi\cos\theta\right)}{4\sin\left(\dfrac{\pi}{2}\cos\theta\right)} \tag{2.3.6}$$

$$F_{2n,1\times2,4d} = \frac{\sin(4\pi\cos\theta)}{2\sin(2\pi\cos\theta)} \tag{2.3.7}$$

From the geometrical construction of the polar diagram, we can find the radiation pattern, that is, the angle dependence of the array factor given in (2.3.5)–(2.3.7). They may also be drawn in Cartesian coordinate system as shown in Figure 2.3. Only half of the figure is given due to the symmetry. It is seen that it does not make any difference whether you consider it as a 1×8 array or a 1×2 array formed by two 1×4 arrays. Even if the array is scanned, the equivalence still keeps because all the array factors have the same phase shift β. From (2.3.4)–(2.3.7)

$$F_{2n,1\times8,d} = \frac{\sin[4(\pi\cos\theta+\beta)]}{8\sin[(\pi\cos\theta+\beta)/2]}$$

$$= \frac{\sin[2(\pi\cos\theta+\beta)]}{4\sin[(\pi\cos\theta+\beta)/2]} \times \frac{\sin[4(\pi\cos\theta+\beta)]}{2\sin[2(\pi\cos\theta+\beta)]}$$

$$= F_{2n,1\times4,d} \times F_{2n,1\times2,4d} \tag{2.3.8}$$

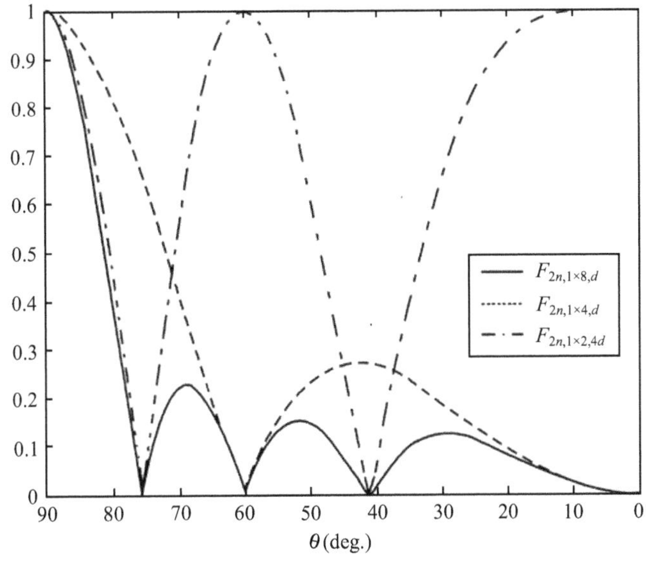

Figure 2.3   Angle dependence of array factors.

However, if only the 1×2 array is scanned, the grating lobe of $F_{2n,1\times2,4d}$ no longer appears at the zero point ($60°$) of $F_{2n,1\times4,d}$. As a result, the side lobe level will be increased due to the grating lobe. This may happen in a phased array.

In application of a phased array, the phantom-bit technique is used to reduce the number of bits in digital phase shifters. In this technique, several lower bits of the digital phase shifter are removed, but in the calculation of the feeding phase the total bits are used. In the example of (2.3.8), if the last two bits are removed, it is equivalent to set $2^2 = 4$ elements in the sub-array to be the same phase and the quantization error is increased and consequently, the sub-array $F_{2n,1\times4,d}$ has no capability to scan. When $F_{2n,1\times2,d}$ is scanned, the side lobe level is increased. The increased side lobe level due to the quantization is called the quantization lobe. The randomization of the feeding phase may solve this problem. The main principle may be illustrated as follows. Suppose the phase step is Δ, then instead of being in phase, the phase on each element will be superimposed by a random phase 0 or Δ.

In doing so, the radiation pattern by sub-array factor $F_{2n,1\times4,d}$ will shift randomly around the scanned array factor $F_{2n,1\times2,4d}$ to keep the zero point of the sub-array mean array factor appearing at the position of grating lobe.[24] A combined approach to disrupt the periodicity in the array was proposed to reduce the grating lobe levels. They are: (1) the optimized amplitude weighting at the sub-array ports; (2) using the random sub-array; and (3) the random staggering of the rows. This approach can be used for a phased array with limited scanning and for the digital beamforming antenna array with adaptive nulling[25].

2.4   N-Element Linear Array: Uniform Spacing, Nonuniform Amplitude

The uniform amplitude distribution yields the smallest half-power beam-width but the highest side lobe level. To reduce the side lobe level, non-uniform amplitude distribution is required. It has been shown analytically that for a given side lobe level, the Dolph-Tschebyscheff (D-T) distribution produces the smallest beam-width between the first nulls. Conversely, for a given beam-width between the first nulls, the D-T distribution leads to the smallest possible side lobe level. If the array factor is given, the field and phase distribution may be found.

2.4.1   Schelkunoff's Unit Circle Representation (SUCR)[2]

The array factor for non-uniform amplitude distribution may be obtained by modifying (2.2.1) as

$$AF = F_2 = \sum_{n=1}^{N} \frac{I_n}{I_1} e^{j(n-1)(kd\cos\theta+\beta)} = \frac{I_N}{I_1} \sum_{n=1}^{N} \frac{I_n}{I_N} W^{n-1} \qquad (2.4.1)$$

where $W = \exp(j\psi)$ and $\psi = kd\cos\theta + \beta$. I_n is the amplitude excitation of each element, the subscripts denote the number of elements. N is the total number of elements. Formula (2.4.1) may also be written as

$$|AF| = \left|\frac{I_N}{I_1}\right| \times \left|W^{N-1} + \left(\frac{I_{N-1}}{I_N}\right) W^{N-2} + \cdots + \frac{I_1}{I_N}\right| = \left|\frac{I_N}{I_1}\right| |f(W)| \qquad (2.4.2)$$

and

$$|AF_n| = |f(W)| = |W - W_1| \cdot |W - W_2| \cdots |W - W_{N-1}| \qquad (2.4.3)$$

where $W_1, W_2, \cdots, W_{N-1}$ are the roots of $f(W)$.

The SUCR provides an alternative approach by using complex W-plane to find the relationship between array factor and θ which is different from the approach shown in Figure 2.2. Observe that as θ varies in real space from 0 to π, ψ varies from $\psi_s = kd + \beta$ to $\psi_f = -kd + \beta$ and that W traces out a path along a unit circle in the complex plane shown in Figure 2.4. The total excursion of W is from $\exp(j\psi_s)$ to $\exp(j\psi_f)$, proceeding clockwise as θ goes from 0 to π. Thus ψ_s(read ψ-start) and ψ_f(read ψ-finish) mark the initial and terminal points of the W-excursion and the angular extent is $2kd$ radians.

Further inspection of (2.4.3) reveals that if the roots W_m are placed on the unit circle in the range of W, then $|f(W_m)| = 0$, and a pattern with N nulls will result. It turns out that the SUCR may be used to synthesize arrays whose patterns possess nulls in desired directions.

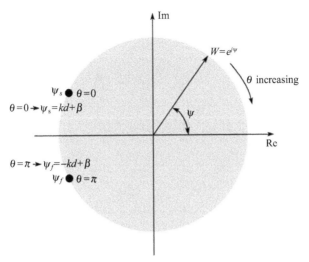

Figure 2.4   Schelkunoff's Unit Circle.

As an illustrative example, consider a five-element array with uniform spacing $d = \lambda/2$ and $\beta = 0$. For uniform excitation, according to (2.2.2-a), $|AF|$ has roots at the positions $\exp(jN\psi_m) = \exp(j2m\pi)$, that is, $\psi_m = 2m\pi/N$, $m = 1, 2, \cdots, N-1$. Here, at $N = 5$, the roots will be $\pm 72°$, $\pm 144°$ as shown in Figure 2.5(a). Suppose the new positions $\pm 87°$ and $\pm 149°$ are tried as shown in Figure 2.5(b); this places the roots $62°$ apart, instead of $72°$. When a large unit circle is constructed with roots placed in these positions, the product of distance measurements gives the innermost side lobe at -18.5dB and the other side lobes at -21.3dB. This suggests that the roots W_2 and W_3 have been shifted too close together, but that the shift in W_1 and W_4 might be just about right. One could continue to try revised roots positions, perhaps $\pm 89°$ and $\pm 145.5°$, thus converging to the positions that will give the desired -20dB level for all side lobes.

With the correct root positions known, one can return to (2.4.3) and write

$$f(W) = \left(W - e^{j1.55}\right)\left(W - e^{-j1.55}\right)\left(W - e^{j2.54}\right)\left(W - e^{-j2.54}\right)$$
$$= \left(W^2 - 2W\cos 89° + 1\right)\left(W^2 - 2W\cos 145.5° + 1\right)$$
$$= W^4 + 1.60W^3 + 1.95W^2 + 1.60W + 1$$

The relative amplitude distribution is

$$1 \qquad 1.60 \qquad 1.95 \qquad 1.60 \qquad 1$$

This distribution that results in the uniform side lobe is called the Dolph-Tschebyscheff distribution and will be discussed in next section.

2.4.2   Dolph-Tschebyscheff (DT) Distribution

The Tschebyschelf polynomial is defined as

$$T_m(z) = \cos[m\cos^{-1}(z)], \qquad -1 \leqslant z \leqslant 1 \qquad (2.4.4\text{-a})$$

$$T_m(z) = \cosh[m\cosh^{-1}(z)], \quad |z| > 1 \qquad (2.4.4\text{-b})$$

The examples of $T_3(z)$ and $T_4(z)$ are shown in Figure 2.6 and may be written as

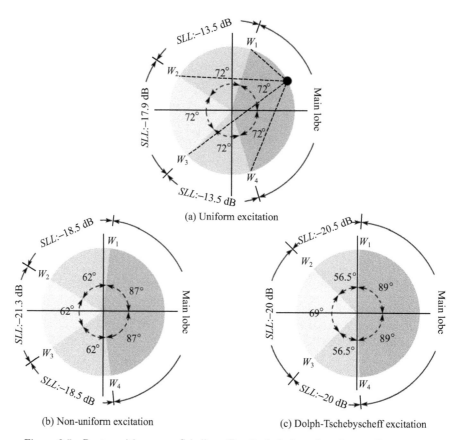

(a) Uniform excitation

(b) Non-uniform excitation   (c) Dolph-Tschebyscheff excitation

Figure 2.5   Root positions on a Schelkunoff unit circle for a five-element linear array.

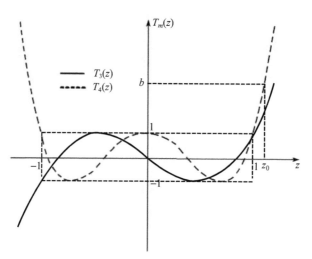

Figure 2.6   Tschebyscheff polynomials of order 3 and 4.

$$T_3(z) = 4z^3 - 3z \qquad\qquad\qquad (2.4.5\text{-a})$$

$$T_4(z) = 8z^4 - 8z^2 + 1 \qquad\qquad (2.4.5\text{-b})$$

This polynomial possesses interesting properties: all polynomials, of any order, pass through the point $(1,1)$; within the range $-1 \leqslant z \leqslant 1$, the polynomials have values within -1 to 1; all roots occur within $-1 \leqslant z \leqslant 1$, and all maxima and minima have values of 1 and -1. These properties are useful in designing antenna arrays and filters for some applications where equal side lobe level is desired. Based on physical insight, it is understandable that for given width of the main lobe, the equal side lobe level may result in the minima of the largest side lobe level.

Since the array factor of an even or odd number of elements can be expressed by a similar polynomial as well, the unknown coefficients of the array factor can be determined by equating the two polynomials[3]. We shall explain the principle through an example with an odd number of elements. For an even number case, see Prob. 2.5. If we choose the phase reference point at the center of the array, similar to Prob. 2.1, the array factor may be expressed by a series involving cosine functions. For the uniformly spaced five-element array with progressive phaseshift β, taking the phase reference point as O, the array factor AF is

$$AF = \left| 2I_1 + I_2 e^{j(\alpha d \cos \theta + \beta)} + I_2 e^{-j(\alpha d \cos \theta + \beta)} \right.$$

$$\left. + I_3 e^{j2(\alpha d \cos \theta + \beta)} + I_3 e^{-j2(\alpha d \cos \theta + \beta)} \right|$$

$$= 2I_1 + 2I_2 \cos \psi + 2I_3 \cos 2\psi$$

$$= 2I_1 + 2I_2 \left(2\cos^2 \frac{\psi}{2} - 1 \right) + 2I_3 \left(8\cos^4 \frac{\psi}{2} - 8\cos^2 \frac{\psi}{2} + 1 \right)$$

$$= 2I_1 - 2I_2 + 2I_3 + (4I_2 - 16I_3) \cos^2 \frac{\psi}{2} + 16I_3 \cos^4 \frac{\psi}{2} \qquad (2.4.6)$$

Comparing the polynomials of order 4 ($m = 4$) in (2.4.5-b) and (2.4.6), we see that the correspondence of variable is $z \to \cos(\psi/2)$. If we specify the side lobe level to be R, then $20 \lg b = R$ and $b = 10^{R/20}$. Figure 2.6 shows the corresponding value of z to be z_0. At this point, $\cos(\psi/2) = 1$. To accommodate the range of variation, it is natural to let

$$z = z_0 \cos \frac{\psi}{2} \qquad (2.4.7)$$

Supposing $R = 20$dB, then we have $b = 10$. From (2.4.4-b), $10 = \cosh \left[4 \cosh^{-1} z_0 \right]$, we have $z_0 = 1.2933$. Finally we have the following identity:

$$(2I_1 - 2I_2 + 2I_3) + (4I_2 - 16I_3) \cos^2 \frac{\psi}{2} + 16I_3 \cos^4 \frac{\psi}{2}$$

$$= 1 - 8z_0^2 \cos^2 \frac{\psi}{2} + 8z_0^4 \cos^4 \frac{\psi}{2} \qquad (2.4.8)$$

The coefficients $2I_1, I_2, I_3$ may be obtained by matching the terms with the same order, giving the result: $I_3 = 1.4$, $I_2 = 2.26$, $2I_1 = 2.67$. After normalizing to I_0, we have the current distribution: $1.00, 1.60, 1.94, 1.60, 1$. The above result is quite close to that in Section 2.4.1.

The DT design procedure may also be implemented by using root matching[3]. From (2.4.4-a), the roots of $T_m(z)$ can be determined readily and are given by

$$z_p = \cos \left[(2p - 1) \frac{\pi}{2m} \right], \qquad p = 1, 2, \cdots, m \qquad (2.4.9)$$

Based on (2.4.7)

$$\psi_p = 2\cos^{-1}\left(\frac{z_p}{z_0}\right) \tag{2.4.10}$$

Consider the same five-element array given above, where $m = 4$, (2.4.9) gives $z_p = \cos \pi/8$, $\cos 3\pi/8$, $\cos 5\pi/8$, $\cos 7\pi/8$. Substituting z_p and $z_0 = 1.2933$ into (2.4.10) yields $\psi_p = \pm 88.82°, \pm 145.16°$. From (2.4.3) we have

$$f(w) = w^4 + 1.60w^3 + 1.93w^2 + 1.60w + 1$$

The result is again quite close to that in Section 2.4.1.

The polar diagram of the pattern may be obtained similarly to what is done in Figure 2.2. The only difference is we need extra transformation $z = z_0 \cos(\psi/2)$. The determination of AF should trace $\theta \to \psi \to z \to AF = |T_m(z)|$. Again we take the same example of a five-element array shown in Figure 2.7 to illustrate the geometrical construction of the array pattern for a DT array which is shown in Figure 2.8.

In practical design, formulas are available to determine the excitation coefficients as follows.

For odd $2N + 1$ elements, such as in Figure 2.7, $N = 2$, the coefficients can be obtained using the following formula[3]

$$I_n = \sum_{q=n}^{N+1} (-1)^{N-q+1}(z_0)^{2(q-1)} \frac{(q+N-2)!N}{\epsilon_n(q-n)!(q+n-2)!(N-q+1)!}$$

$$n = 1, 2, \cdots, N+1$$

$$\epsilon_n = \begin{cases} 2, & n = 1 \\ 1, & n \neq 1 \end{cases} \tag{2.4.11-a}$$

with

$$AF = \sum_{n=1}^{N+1} 2I_n \cos(n-1)\psi \tag{2.4.11-b}$$

For even $2N$ elements, such as shown in Prob. 2.5 $N = 2$, the coefficients can be obtained using the following formula[3]

$$I_n = \sum_{q=n}^{N} (-1)^{N-q}(z_0)^{2q-1} \frac{(q+N-2)!(2N-1)}{2(q-n)!(q+n-1)!(N-q)!}, \quad n = 1, 2, \cdots, N \tag{2.4.11-c}$$

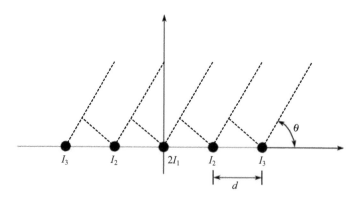

Figure 2.7   Uniformly spaced five-element array.

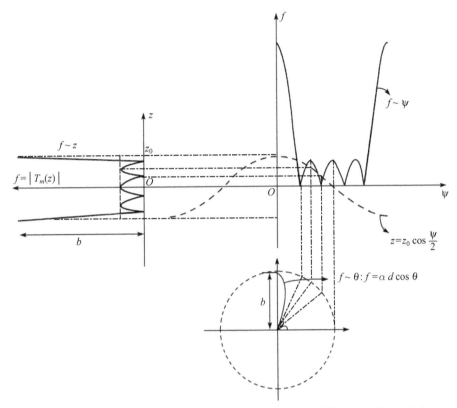

Figure 2.8   Geometrical construction of array pattern for DT array with $d = \lambda/2$.

with

$$AF = \sum_{n=1}^{N} 2I_n \cos \frac{(2n-1)\psi}{2} \qquad (2.4.11\text{-d})$$

Formulas (2.4.11-a) and (2.4.11-b) are suitable for computer calculations and there are also computer program available in [3].

2.4.3   Taylor Distribution[4]

In some cases, the current distribution of DT array varies very sharply near the edge and the side lobe level is very sensitive to the errors of the currents. Consequently, it will cause feeding difficulties. The Taylor distribution usually is quite smooth. This distribution leads to a pattern whose first few minor lobes (closest to the main lobe) are maintained at an equal and specified level, and the remaining lobes decay monotonically.

Taylor distribution was developed for the continuous sources. However it may also be used for discrete array through discrete sampling of the continuous distribution or the root matching. These two methods will be described later.

Taylor succeeded in extending his technique to circular planar apertures. These Taylor circular distributions can also be sampled, and thus can give the excitation coefficients for discrete planar arrays with a circular boundary. Since there is no similar extension of Dolph's technique to circular planar apertures, it is important for the antenna designer to be familiar with the principal features of Taylor's procedure.

Taylor chose to start his analysis by considering the general array factor as given in (1.3.33) with the source distributed along the z direction.

$$F_2(\theta) = \int_{-\frac{l}{2}}^{\frac{l}{2}} I_e(z')e^{jkz'\cos\theta}dz'$$

If $I_e(z') = K$, then

$$F_2(\theta) = Kl\frac{\sin\left[kl\cos\theta/2\right]}{kl\cos\theta/2} \qquad (2.4.12)$$

Letting $kl\cos\theta/2 = \pi l\cos\theta/\lambda = u$, the normalized array factor $F_{2,n}(\theta)$ will be

$$F_{2,n}(\theta) = F_{2,n}(u) = \frac{\sin u}{u} \qquad (2.4.13)$$

This array factor has symmetrical side lobes whose intensity falls off as u^{-1} with the pair of closest side lobes being down 13.5dB. Based on (2.4.13), Taylor suggested an array factor comprised of a product of factors whose roots are the nulls of the pattern. This factor is given by

$$F_T(u, A, \overline{n}) = \frac{\sin u}{u}\frac{\displaystyle\prod_{n=1}^{\overline{n}-1}\left[1-(u/u_n)^2\right]}{\displaystyle\prod_{n=1}^{\overline{n}-1}\left[1-(u/(n\pi))^2\right]} \qquad (2.4.14)$$

Which removes the innermost $\bar{n}-1$ pairs of nulls from the original $\sin(u)/u$ pattern and replaces them with new pairs at modified positions $\pm u_n$ which are a slight modification of the ideal Tschebyscheff space factor. The determination of u_n will be given in (2.4.22).

In Section 2.4.2, if the number of elements of the DT array are allowed to become infinite, the space factor suitable for continuous source can be derived[5]. For Tschebyscheff polynomial $T_m(z)$, a change of variable is now made to achieve two objects: to allow the side lobe region on either side of the main lobe to be represented by T_m, and to join together the two large amplitude regions of the T_m in such a way that they form the main beam with zero slope at $z = 0$ as shown in Figure 2.9.

Taylor called this polynomial $W_{2m}(z)$. It is defined by

$$W_{2m}(z) = T_m(z_0 - a^2z^2) \qquad (2.4.15)$$

where a is a constant. The expression for T_m valid in the oscillatory region is

$$T_m(z) = \cos\left(m\cos^{-1}z\right)$$

the zeros of W_{2m} are found from (2.4.4-a) and (2.4.15)

$$\cos\left[m\cos^{-1}(z_0 - a^2z_n^2)\right] = 0$$

or

$$z_n = \pm\frac{1}{a}\left[z_0 - \cos(\frac{n\pi}{m} - \frac{\pi}{2m})\right]^{1/2} \qquad (2.4.16)$$

assuming for brevity that

$$\cosh^{-1}b = \pi A \qquad (2.4.17)$$

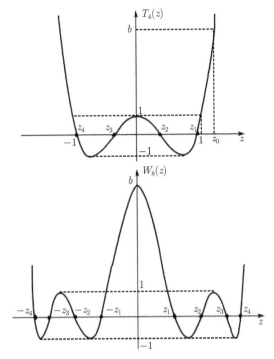

Figure 2.9 $T_4(z)$ and $W_8(z)$.

The constant z_0 may be expressed in terms of the main beam to side lobe ratio b according to (2.4.4-b)

$$z_0 = \cosh\left(\frac{1}{m}\cosh^{-1}b\right) = \cosh\frac{\pi A}{m} \tag{2.4.18}$$

Now letting the order m tend to infinity, and at the same time changing the argument scale of W_{2m} with the choice of the constant $a = \pi/m\sqrt{2}$, gives the following simple expression for z_n

$$\begin{aligned}
z_n^2 &= \lim_{\substack{a\to 0 \\ m\to\infty}} \frac{1}{a^2}\left\{\left[\cosh a\sqrt{2}A\right] - \cos\left[\frac{a\sqrt{2}}{2}(2n-1)\right]\right\} \\
&\approx \lim_{\substack{a\to 0 \\ m\to\infty}} \frac{1}{a^2}\left\{\left[1 + \frac{1}{2}\left(a\sqrt{2}A\right)^2\right] - \left[1 - \frac{a^2}{4}(2n-1)^2\right]\right\} \\
&= A^2 + \left(n - \frac{1}{2}\right)^2, \quad n = 1, 2, \cdots
\end{aligned}$$

Therefore

$$z_n = \sqrt{A^2 + \left(n - \frac{1}{2}\right)^2} \tag{2.4.19}$$

The corresponding space factor has unity amplitude side lobes and main beam amplitude b. Based on the zeros in (2.4.19), the ideal space factor $F(z, A)$ may be constructed as

$$F(z, A) = \lim_{m\to\infty} W_{2m}(z) = C\prod_{n=1}^{\infty}(z_n^2 - z^2)z_n^{-2}$$

$$= C \prod_{n=1}^{\infty} \left[A^2 + (n - \frac{1}{2})^2 - z^2 \right] z_n^{-2}$$

$$= C \prod_{n=1}^{\infty} (n - \frac{1}{2})^2 z_n^{-2} \prod_{n=1}^{\infty} \left[1 - \frac{z^2 - A^2}{(n - \frac{1}{2})^2} \right]$$

The constant C may be chosen so as to obtain

$$F(z, A) = \prod_{n=1}^{\infty} \left[1 - \frac{z^2 - A^2}{(n - 1/2)^2} \right] = \cos \left[\pi \sqrt{z^2 - A^2} \right] \tag{2.4.20}$$

Letting $z = u/\pi$, (2.4.20) may serve as the radiation pattern of the line source

$$F(u, A) = \cos \sqrt{u^2 - (\pi A)^2} \tag{2.4.21}$$

The space factor $F(u, A)$ given in (2.4.21) is called the ideal space factor. It is unrealizable because the remote side lobes do not decay and the corresponding line source has a singularity at each end. A clever solution was found by Taylor where the z scale is stretched slightly by a factor of σ, closely approximating the ideal space factor zeros in close. At some point, say $\bar{n}$, a zero occurs at this integer due to the stretching. From this transition point on, making the zeros occur at $\pm n$, the "approximate" space factor then will have zeros at:

$$u_n = \frac{\pi l}{\lambda} \cos \theta_n = \begin{cases} \pm \pi \sigma \sqrt{A^2 + (n - 1/2)^2}, & 1 \leqslant n \leqslant \bar{n} \\ \pm n\pi, & \bar{n} \leqslant n \leqslant \infty \end{cases} \tag{2.4.22}$$

where σ is also called the stretchout factor of the beamwidth and may be determined by

$$\pi \sigma \sqrt{A^2 + (\bar{n} - 1/2)^2} = \bar{n}\pi$$

giving

$$\sigma = \frac{\bar{n}}{\sqrt{A^2 + (\bar{n} - 1/2)^2}} \tag{2.4.23}$$

As $\bar{n} \to \infty$, the approximate space factor approaches the ideal space factor which has $\sigma = 1$. From $\partial \sigma / \partial \bar{n} = 0$, it is seen that when $\bar{n} = 2A^2 + (1/2)$, σ reaches the maximum. To keep the monotonical decrease of σ, $\bar{n}$ should be $\geqslant 2A^2 + (1/2)$.

The corresponding approximate space factor is given by (2.4.14), where u_n is modified by (2.4.22).

The formula (2.4.14) may be reformulated as follows[6]. The denominator in (2.4.14) is rewritten as

$$\prod_{n=1}^{\bar{n}-1} \left[1 - \left(\frac{u}{n\pi} \right)^2 \right] = \left(\prod_{n=1}^{\bar{n}-1} n^2 \right)^{-1} \left[\prod_{n=1}^{\bar{n}-1} \left(n + \frac{u}{\pi} \right) \left(n - \frac{u}{\pi} \right) \right] \tag{2.4.24}$$

Making use of the formulas of Γ function

$$\Gamma \left(1 + \frac{u}{\pi} \right) \Gamma \left(1 - \frac{u}{\pi} \right) = \frac{u}{\sin u} \tag{2.4.25}$$

$$\prod_{n=1}^{\bar{n}-1} n^2 = [\Gamma(\bar{n})]^2 \tag{2.4.26}$$

$$\prod_{n=1}^{\bar{n}-1} (n \pm \frac{u}{\pi}) = \frac{\Gamma(\bar{n} \pm u/\pi)}{\Gamma(1 \pm u/\pi)} \tag{2.4.27}$$

the combination of (2.4.24) and (2.4.25)–(2.4.27) yields

$$\prod_{n=1}^{\bar{n}-1}\left[1-\left(\frac{u}{n\pi}\right)^2\right]=\frac{\sin u}{u}\frac{\Gamma\left(\bar{n}+u/\pi\right)\Gamma\left(\bar{n}-u/\pi\right)}{\left[\Gamma\left(\bar{n}\right)\right]^2}\tag{2.4.28}$$

Substitution of (2.4.28) into (2.4.14) results in

$$F_T\left(p,A,\bar{n}\right)=\begin{cases}\dfrac{\left[(\bar{n}-1)!\right]^2}{(\bar{n}-1+p)!\,(\bar{n}-1-p)!}\displaystyle\prod_{n=1}^{\bar{n}-1}\left[1-\left(\dfrac{\pi p}{u_n}\right)^2\right],&|p|<\bar{n}\\[12pt]0,&|p|\geqslant\bar{n}\end{cases}\tag{2.4.29}$$

where u_n is given by (2.4.22), $p=u/\pi$.

With the Taylor pattern defined by (2.4.14), it becomes a simple matter to find the corresponding aperture distribution from (1.3.33). If we let $I_e\left(z'\right)$ be represented by Fourier series

$$I_e\left(z'\right)=\sum_{m=0}^{\infty}B_m\cos\frac{2m\pi z'}{l}\tag{2.4.30}$$

then substitution in (1.3.33) gives

$$F_2(\theta)=F_T(u)=\sum_{m=0}^{\infty}B_m\int_{-l/2}^{l/2}\cos\frac{2m\pi z'}{l}e^{j2p\pi z'/l}dz'\tag{2.4.31}$$

The odd part of the integrand of (2.4.31) can be discarded, which leaves

$$F_T(p)=\sum_{m=0}^{\infty}B_m\int_{-l/2}^{l/2}\cos\frac{2m\pi z'}{l}\cos\frac{2p\pi z'}{l}dz'\tag{2.4.32}$$

If p is an integer, the integral in (2.4.32) is zero unless $m=p$; as a consequence,

$$lB_0=F_T\left(0\right),\quad\frac{1}{2}B_m=F_T\left(m\right),m=1,2,\cdots\tag{2.4.33}$$

However, (2.4.14) indicates that $F_T\left(m\right)=0$, when $m\geqslant\bar{n}$, so this Fourier series truncates, and thus the continuous aperture distribution is given by

$$I_e(z')=\frac{1}{l}\left[F_T(0)+2\sum_{p=1}^{\bar{n}-1}F_T\left(p,A,\bar{n}\right)\cos\frac{2p\pi z'}{l}\right]\tag{2.4.34}$$

$F_T(p,A,\bar{n})$ is given in (2.4.29).

The half-power beamwidth is given approximately by[4]

$$\theta_0\approx2\sin^{-1}\left\{\frac{\lambda\sigma}{\pi l}\left[\left(\cosh^{-1}b\right)^2-\left(\cosh^{-1}\frac{b}{\sqrt{2}}\right)^2\right]^{1/2}\right\}\tag{2.4.35}$$

When $l\gg\lambda$, (2.4.35) may be approximated as

$$\theta_0\approx\frac{2\lambda\sigma}{\pi l}\left[\left(\cosh^{-1}b\right)^2-\left(\cosh^{-1}\frac{b}{\sqrt{2}}\right)^2\right]^{1/2}\tag{2.4.36}$$

Formulas (2.4.17), (2.4.23) and (2.4.36) may be tabulated as shown in Table 2.1.

Table 2.1   Taylor line source design parameters.

Side lobe level R(dB)	Voltage ratio	$\theta_0 l$ $\sigma\lambda(°)$	A^2	$\bar{n}=2$	$\bar{n}=3$	$\bar{n}=4$	σ $\bar{n}=5$	$\bar{n}=6$	$\bar{n}=7$	$\bar{n}=8$
0	1.00000	28.65	0.00000	1.33333	1.20000	1.14286	1.11111	1.09091	1.07692	1.06667
5	1.77828	34.49	0.14067	1.29351	1.18672	1.13635	1.10727	1.08838	1.07514	1.06534
10	3.16228	40.33	0.33504	1.24393	1.16908	1.12754	1.10203	1.08493	1.07268	1.06350
15	5.62341	45.93	0.58950	1.18689	1.14712	1.11631	1.09528	1.08034	1.06949	1.06112
20	10.00000	51.17	0.90777	1.12549	1.12133	1.10273	1.08701	1.07490	1.06554	1.05816
25	17.7828	56.04	1.29177	——	1.09241	1.08698	1.07728	1.06834	1.06083	1.05463
30	31.6228	60.55	1.74229	——	——	1.06934	1.06619	1.06079	1.05538	1.05052
35	56.2341	64.78	2.25976	——	——	——	1.05386	1.05231	1.04932	1.04587
40	100.0000	68.76	2.84428	——	——	——	——	1.04298	1.04241	1.04068

A continuous line source distribution can be sampled at equispaced values of z' to determine the excitation of a linear array. Obviously, if the number of the elements is large enough, the sampling interval will be so small that all the fine detail in the continuous aperture distribution will be captured. Under this circumstance, the pattern from the discrete array will differ but little from pattern due to the continuous aperture distribution. However, in many practical applications, this number is not so large that the sampling results in an excitation, which gives a badly degraded pattern. It is possible to circumvent this difficulty by working directly with the desired pattern, rather than its continuous aperture distribution by using the root matching method[2]. This procedure is similar to that introduced in Section 2.4.2.

The design steps are summed up as follows:

1. For given side lobe level (SLL) to find b. For example, $R = 20$, from $20\lg b = R$, $b = 10^{R/20} = 10$.

2. Use (2.4.17) to find A. For example, $A = (\cosh^{-1} 10)/\pi = 0.95277$.

3. For given $\bar{n}$, use (2.4.23) to find σ. For example, $\bar{n} = 5$ ($\bar{n} \geqslant 2A^2 + 0.5 = 2.32$), $\sigma = 1.087$.

4. For given l, or number of elements N and spacing d between two elements, for example, $l = 7\lambda$, or $N = 15$ (if $d = \lambda/2$), to find the nulls by (2.4.22) or

$$u_n = \pm 1.17\pi, \ \pm 1.93\pi, \ \pm 2.91\pi, \ \pm 3.94\pi, \ \pm 5.00\pi, \ \pm 6.00\pi, \ \pm 7.00\pi.$$

5. From (2.4.34), (2.4.29) find the amplitude distribution. For a discrete array, use the equal-space sampling to find the amplitude distribution of the elements by the following formula:

$$I_n(m) = 1 + 2\sum_{p=1}^{\bar{n}-1} F_T(p)\cos(mp)$$

where

$$m = \frac{2\pi}{l}z', \quad l = (N-1)d$$

$$z' = \begin{cases} nd, & n = 0, \cdots, \dfrac{N-1}{2}, & N \text{ odd} \\[2ex] \dfrac{2n+1}{2}d, & n = 0, \cdots, \dfrac{N}{2}-1, & N \text{ even} \end{cases}$$

6. If the root matching method is used for discrete array, define the effective aperture $l_e = Nd$, for broadside array, $u_n = \pi l_e \cos\theta_n/\lambda$ and $\psi_n = kd\cos\theta_n$, we may have

$$\psi_n = \frac{2\pi d}{\lambda}\frac{\lambda}{\pi l_e}u_n = \frac{2u_n}{N}$$

The amplitude distribution may be found by

$$|f(W)| = \left|(W - e^{j\psi_1})(W - e^{j\psi_2})\cdots(W - e^{j\psi_n})\right|$$

2.4.4   Woodward-Lawson (WL) Method

Some antenna applications require pattern without nulls. An example is the airport beacon antenna, which must radiate uniformly in azimuth ϕ to be able to communicate with aircraft arriving from all directions. It must also radiate without nulls in elevation θ if it is to maintain contact with incoming aircraft, which fly at constant height, and thus appear at a constantly changing angle θ with respect to the antenna. To choose the cosecant function as the radiation pattern is a particularly practical selection because it would ensure that an airplane flying at a constant height would continue to receive a constant level signal from the beacon as its range changes[2]. Unlike the synthesis of the narrow pencil-beam patterns such as the DT synthesis and Taylor synthesis, this type of synthesis is the shaped beam synthesis. The method is introduced for the continuous source; however, it can be used for the discrete array through discretization as shown previously.

Assuming a continuous source $I(z')$ within $(-l/2, l/2)$ and zero elsewhere, the relationship between current distribution $I(z')$ and the radiation pattern F_2 given in (1.3.33) may be written in the form of Fourier transform after extending the limit to infinity

$$F_2 = \int_{-l/2}^{l/2} I_e(z')e^{jkz'\cos\theta}dz' = \int_{-\infty}^{\infty} I_e(z')e^{j\xi z'}dz' \tag{2.4.37-a}$$

F_2 may be a function of θ, ξ, u or p with the relationship $k\cos\theta = \xi = 2u/l = 2\pi p/l$ as defined here and previously. The current distribution $I_e(z')$ may be determined through the inverse Fourier transform of (2.4.37-a)

$$I_e(z') = \frac{1}{2\pi}\int_{-\infty}^{\infty} F_2(\theta)e^{-jz'\xi}d\xi \tag{2.4.37-b}$$

Eqn. (2.4.37-b) indicates that if $F_2(\theta)$ represents the desired pattern, the excitation distribution $I(z')$ that will yield the exact desired pattern must exist for all values of z' $(-\infty \leqslant z' \leqslant \infty)$. Since physically only sources of finite dimensions are realizable, the excitation distribution of (2.4.37-b) is truncated at $z' = \pm l/2$ (beyond $z' = \pm l/2$ it is set to zero). The approximate current distribution is denoted as $I_{e,a}(z')$. This distribution yields approximate pattern $F_{2,a}(\theta)$

$$F_2(\theta) \approx F_{2,a}(\theta) = \int_{-l/2}^{l/2} I_{e,a}(z')e^{j\xi z'}dz' \tag{2.4.38}$$

The above procedure of determining current distribution $I_{e,a}(z')$ directly through the Fourier transform of the radiation pattern is actually the Fourier transform method. Alternatively, one may expand the current distribution $I_e(z')$ first

$$I_e(z') = \frac{1}{l}\sum_{m=-M}^{M} b_m e^{-j\beta_m z'} \tag{2.4.39}$$

where $\beta_m = k \cos \theta_m$. Substituting (2.4.39) into (2.4.37-a) yields

$$
\begin{aligned}
F_2(\theta) &= \sum_{m=-M}^{M} \frac{b_m}{l} \int_{-l/2}^{l/2} e^{-j\beta_m z'} e^{j\xi z'} dz' \\
&= \sum_{m=-M}^{M} b_m \frac{\sin \left[kl(\cos \theta - \cos \theta_m)/2\right]}{kl \left(\cos \theta - \cos \theta_m\right)/2} \\
&= \sum_{m=-M}^{M} f_{2m}(\theta)
\end{aligned}
\tag{2.4.40}
$$

It is seen that the maximum of each individual term in (2.4.40) occurs when $\theta = \theta_m$, and it is equal to $F_2(\theta_m)$. In addition, if we choose $\cos \theta = m(\lambda/l)$, all other terms of (2.4.40) which are associated with the other samples are zero at $\theta = \theta_m$. In other words, all sampling terms $f_{2m}(\theta)$ (composing functions) of (2.4.40) are zero at all sampling points other than at their own. Thus at each sampling point the total field is equal to that of the sample. Consequently

$$
b_m = F_2(\theta_m)
\tag{2.4.41}
$$

$$
\beta_m = k \cos \theta_m = \frac{km\lambda}{l}, \quad m = 0, \pm 1, \pm 2, \cdots, \pm M
\tag{2.4.42}
$$

In order to ensure real values of θ (visible region), M should be the closest integer to $M \leqslant l/\lambda$.

The physical meaning of (2.4.40) is that the radiation pattern of an aperture with the current distribution $I_e(z')$ may be considered as the sum of the radiation patterns with different uniform aperture distribution b_m and linear phase shift β_m. Moreover, (2.4.40) is also recognized as the Shannon sampling theorem. The composing function is the spline function. This method may be generalized to two-dimensional cases, for example, the circular aperture. In that case, the radiation pattern from a uniformly distributed circular aperture is expressed by Bessel function of order 1, that is, $J_1(\xi)/\xi$. This function should be the spline function.

For discrete linear arrays with spacing d, l should be replaced by Nd in all the related formulas.

Example:

Use W-L method to determine the current distribution and the approximate radiation pattern of a line-source $l = 5\lambda$ placed along the z-axis whose desired radiation pattern is symmetrical about $\theta = \pi/2$, and is given by

$$
F_2(\theta) = \begin{cases} 1, & \pi/4 \leqslant \theta \leqslant 3\pi/4 \\ 0, & elsewhere \end{cases}
$$

This is referred to as a sectorial pattern.

Solution:

Since $l = 5\lambda$, $M = 5$, the sampling separation is 0.2. The total number of sampling points is 11. The angles where the sampling is performed are given, according to (2.4.42), by

$$
\theta_m = \cos^{-1}\left(m\frac{\lambda}{l}\right) = \cos^{-1}(0.2m), \quad m = 0, \pm 1, \cdots, \pm 5
$$

The angles and the excitation coefficients at the sample points are listed in Table 2.2.

Table 2.2 Values of θ_m and b_m.

m	$\theta_m(°)$	$b_m = F_2(\theta_m)$	m	$\theta_m(°)$	$b_m = F_2(\theta_m)$
0	90	1			
1	78.46	1	−1	101.54	1
2	66.42	1	−2	113.58	1
3	53.13	1	−3	126.87	1
4	36.87	0	−4	143.13	0
5	0	0	−5	180	0

The computed pattern is shown in Figure 2.10(a) where it is compared with the desired pattern and a good reconstruction is indicated. The side lobe level, relative to the value of the pattern at $\theta = 90°$, is 0.160 (−15.9dB).

To demonstrate the synthesis of the pattern using the sampling concept, all seven nonzero composing functions $f_{2m}(\theta)$ used for the reconstruction of the $l = 5\lambda$ line-source pattern are also given in Figure 2.10(b). It can be seen that the value of the desired pattern at each sampling point is determined solely by the maximum value of a single composing (spline) function.

For an array of $N=10$ elements (number of sampling points $N_s=11$ in this example) with an element spacing of $d = \lambda/2$, the excitation coefficients of the array at the sampling points are the same as those of the line source. Using the values of b_m listed above, the computed array factor pattern using (2.4.40) by setting $l = Nd$ is shown in Figure 2.10(a). A good synthesis of the desired pattern is displayed. The side lobe level is 0.221(−13.1dB). The excitation coefficient of each array element is given by

$$ I_n(z_n') = \frac{1}{N_s} \sum_{m=-M}^{M} b_m e^{-jkz_n' \cos\theta_m} $$

where z_n' indicates the position of the nth element symmetrically located about the geometrical center of the array. This position is in the middle, between two sampling points. The excitation coefficients along with their symmetrical position are listed in Table 2.3. To achieve the normalized amplitude pattern of unity at $\theta = 90°$ in Figure 2.10(a), the array factor must be multiplied by $1/\sum I_n = 1/0.4545463$.

There are some comments on the Fourier transform method and the W-L method by C. A. Balanis in [3]: "The Fourier transform method is best suited for reconstruction of desired patterns which are analytically simple and which allow the integrations to be performed in closed form. Today, with the advent of high-speed computers, this is not a major restriction since integration can be performed (with high accuracy) numerically. In contrast, the W-L method is more flexible, and it can be used to synthesize any desired pattern. In fact, it can even be used to reconstruct pattern which, because of their complicated nature, cannot be expressed analytically. Measured patterns, either of analog or digital form, can also be synthesized using the W-L method."

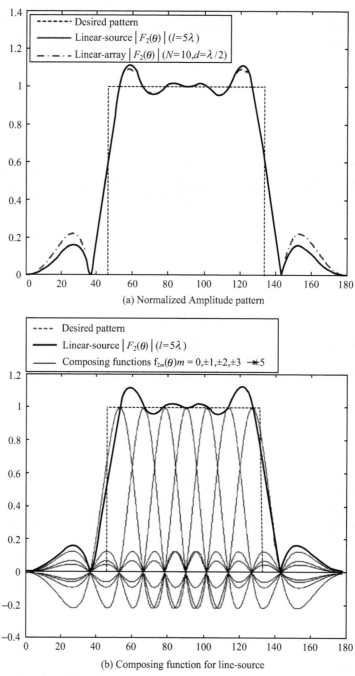

(a) Normalized Amplitude pattern

(b) Composing function for line-source

Figure 2.10   Desired and synthesized patterns, and composing functions for Woodward-Lawson designs.
(After Balanis [3], © 1997 Wiley)

Table 2.3   Values of I_n. (After Balanis [3], © 1997 Wiley)

Element Number n	Element Position z'_n	Excitation Coefficient I_n
± 1	$\pm 0.25\lambda$	0.5177930
± 2	$\pm 0.75\lambda$	-0.0313252
± 3	$\pm 1.25\lambda$	-0.0909091
± 4	$\pm 1.75\lambda$	0.1007740
± 5	$\pm 2.25\lambda$	-0.0417864

2.4.5 Supergain Arrays[2]

As is seen in Section 2.4.1, the total excursion of W along the unit circle is $2kd$. The intriguing possibility arises that one could make the inter-element spacing d smaller and smaller, simultaneously repositioning the roots on the unit circle so that they always remain within the range of W, and in such a way that the pattern in real space is unaltered. In this manner, a sum pattern with a main beam of a prescribed beamwidth and side lobes of prescribed heights could be generated by an equispaced linear array of a specified number of elements, but with the total length of the array arbitrarily small.

As a specific illustration of this possibility of a reduced-length array, consider once again the five-element equispaced linear array ($d = \lambda/2$, $\beta = 0$) given in Section 2.4.1. According to (2.4.3), the array factor is

$$AF = |f(W)| = \left|\left(W - e^{j\psi_1}\right)\left(W - e^{-j\psi_1}\right)\left(W - e^{j\psi_2}\right)\left(W - e^{-j\psi_2}\right)\right|$$
$$= W^4 - 2W^3\left(\cos\psi_1 + \cos\psi_2\right) + 2W^2\left(1 + 2\cos\psi_1\cos\psi_2\right)$$
$$-2W\left(\cos\psi_1 + \cos\psi_2\right) + 1 \qquad (2.4.43)$$

Assuming $\psi_2 = 2\psi_1$, the current distribution is then given by

$$1 \quad -2\left(\cos\psi_1 + \cos 2\psi_1\right) \quad 2\left(1 + 2\cos\psi_1\cos 2\psi_1\right) \quad -2\left(\cos\psi_1 + \cos 2\psi_1\right) \quad 1 \qquad (2.4.44)$$

When $\psi_1 = 72°$, $\psi_2 = 2\psi_1 = 144°$ as given in Figure 2.5(a), all of these currents are in unity and $\left[(1)^2 + (1)^2 + (1)^2 + (1)^2 + (1)^2\right] R$ denotes the ohmic losses, with R as some appropriate ohmic representation of the resistivity and shape of an element. The field strength at the peak of the main beam is measured by the sum of the currents and therefore the total radiated power can be represented by $(5)^2 k$, with k a factor that depends on pattern shape. Assume that, with $\psi_1 = 72°$, the ohmic losses are 1% of the power radiated. Then $5^2 k = 100(5R)$, or $k = 20R$.

Now assume $\psi_1 = 1°$, that is, there has been a 72-fold contraction in the length of the array and in the root replacement on the unit circle. For this case, (2.4.44) gives, for the current distribution, 1 −3.998477 5.996954 −3.998477 1. Compared with the antenna of the same length with uniform current distribution, the antenna with this current distribution has much larger gain and is known as the supergain antenna. In this case, $\sum(I)^2 = 2 \times 1^2 + 2 \times (3.998477)^2 + (5.996954)^2 = 70$.

$$\sum I = 2 + 2(1 + 2\cos\psi_1\cos 2\psi_1) - 4(\cos\psi_1 + \cos 2\psi_1)$$
$$= 8\left[(\cos^2\psi_1 - 1)(\cos\psi_1 - 1)\right]$$
$$\approx 16\left[1 - \psi_1^2/2 - 1\right]^2 = 4\psi_1^4 = 0.371 \times 10^{-6}$$

The ohmic losses have become $\sum(I^2)R = 70R$, and the radiation power $(\sum I)^2 k = 0.13764 \times 10^{-12}k$. The ratio of the power radiated to the ohmic losses is

$$\frac{0.13764 \times 10^{-12}k}{70R} = 0.39326 \times 10^{-12}$$

The ohmic losses, which were assumed to be only 1% of the radiated power at $d = \lambda/2$ spacing, are in contrast a trillion times as large as the radiated power at $d = \lambda/(2 \times 72) = \lambda/144$ spacing. Even with a modest reduction in spacing to $d = \lambda/4$ the ohmic losses are four times as large as that of the radiated power. This simple example serves to illustrate that the drastic penalty is the loss of efficiency if reduction of length is contemplated for linear arrays. Further study shows that mechanical and electrical tolerances become severe and frequency bandwidth is sharply narrowed as the inter-element spacing is contracted. Before these problems are properly solved, the application of supergain is actually impractical. The supergain phenomenon should also be considered in the research of electrically small antennas.

2.5   N-Element Linear Array: Uniform Amplitude, Nonuniform Spacing

The element spacings provide another parameter, in addition to the amplitude and phase of the element current, which can be used to control the radiation pattern. In practice, array elements cannot be located much closer than a half wavelength as indicated in Section 2.4.5. Furthermore, the sizes of practical antenna elements are of the order of a half-wavelength dimension, and it would be difficult to make elements much smaller without loss of efficiency.

An aperture that contains N elements equally spaced at half-wavelength intervals contains more elements than if the spacings are made unequal and if the minimum spacing is a half-wavelength. Since the unequally spaced array contains fewer elements than the conventional array occupying the same aperture, it is said to be "thinned." The conventional array with half-wavelength spacing is called a "filled" array.

Unequally spaced arrays may be used to obtain radiation patterns with low peak side lobes without the need for an amplitude taper (although the amplitude taper is still possible). This might be of importance in applications where it is not convenient to individually adjust the amplitude of the current on the elements. Moreover, fewer elements result in low cost and easy maintenance.

Grating lobes, as explained in Section 2.3, are equal in magnitude to the main lobe and are formed in equally spaced arrays when the electrical spacing between elements is wide enough to cause phase differences between adjacent elements of more than 2π radians. The phantom-bit technique described in Section 2.3 provides a solution to this problem for phased array by randomly shifting the phase center to destroy the periodicity. The unequally spaced array provides an alternative, to permit the antenna to scan over a wide angle or to operate over a wide frequency range without the formation of grating lobes that could appear with an equally spaced array.

2.5.1   Density Taper-Deterministic[1]

When the element spacings are of the order of one-half wavelength, the radiation pattern of an equally spaced array is a close approximation to that from a continuous aperture of the same size and illumination function. With this as a guide it is of interest to consider the design of an unequally spaced array by attempting to approximate the continuous-aperture current density with equal-amplitude samples spaced nonuniformly. In so doing, the density of the equally excited, unequally spaced radiating elements as a function of location within the aperture will be of the same form as the continuous current density function of the conventional antenna used as the model. This design procedure is called density tapering to distinguish it from the more usual amplitude tapering. It has also been known as space tapering. The continuous aperture illumination from which the density taper is derived is called the model function and will be denoted by $i_0(x)$.

The quality of the approximation to a continuous aperture current density function by an array depends on the number of elements. The thinner the array, the poorer the approximation. It is found that the main lobe of the density-tapered array and the main lobe of the continuous aperture used as the amplitude-tapered model are close approximations of one another. Also, the near-in side lobes are generally similar. However, the far-out side lobes can deviate considerably and generally rise to relatively large values.

One method of selecting the element locations of the density-tapered array is trial and error. Although this can often produce satisfactory results, it can be a tedious procedure, especially when the number of elements is large. This section describes a systematic procedure for finding the element locations by a method other than trial and error. It is based on the equal-area approximation to the aperture illumination function and uses the cumulative current distribution (integral of the density) rather than the current density itself. It is called a deterministic method to distinguish it from the statistical method discussed in the next section.

To employ density tapering, an amplitude-tapered illumination function is first selected as a model. One criterion for its selection is that, when used with a continuous aperture, its radiation pattern should be similar to that desired of the density-tapered array. The illumination function of the amplitude-tapered model might be as shown in Figure 2.11 (a). To locate the positions of the M elements the area under the curve is divided into M equal parts and an element is placed at center of each of the intervals defined by the equal areas, as illustrated in Figure 2.11 (b). The density of the equally excited, unequally spaced discrete current depicted in (b) is seen to approximate the continuous current density function of (a).

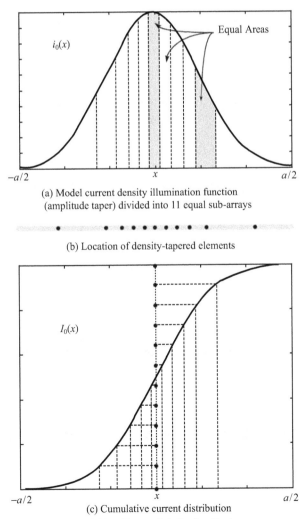

(a) Model current density illumination function
(amplitude taper) divided into 11 equal sub-arrays

(b) Location of density-tapered elements

(c) Cumulative current distribution

Figure 2.11   Deterministic density taper.

The element locations in a linear array may be determined with the equal-area approximation applied to the cumulative distribution $I_0(x)$ of the model aperture illumination, rather than the current density $i_0(x)$. The relationship between the two is given by

$$I_0(x) = \int_{-a/2}^{x} i_0(x)dx \tag{2.5.1}$$

which is the integral of the amplitude-tapered current density $i_0(x)$ taken over the limits $(-a/2, x)$, or $(-\infty, x)$, since $i_0(x) = 0$ for $x < -a/2$. The cumulative distribution is plotted in Figure 2.11 (c). The equal areas may be found by dividing the ordinate into M equal increments and projecting these points onto the x axis, as shown in Figure 2.11 (c). The elements are then located within the center of each interval. The procedure is similar to the trapezoidal rule for approximating an integral.

2.5.2   Density Taper-Statistical[1]

In this approach to the design of the density-tapered array, the model amplitude-tapered illumination function is employed to determine, on a probabilistic basis, whether or not an element should be located at a particular point within the aperture. The model illumination function serves a role analogous to that of the probability density function of probability theory, although it does not necessarily conform to the strict definition of a probability density function. The elements are located randomly (actually pseudorandomly) rather than in some definite manner as in the deterministic density taper of the preceding section. However, the elements are not uniformly random across the aperture, but their average density, computed statistically, follows the form of the model amplitude-taper illumination function. This method is called a statistical density taper, since the radiation pattern of a particular class of design can be specified beforehand only in statistical terms.

The procedure for designing a statistical, density-tapered array may be illustrated with the aid of Figure 2.12. The curve represents the amplitude taper $i_0(x)$ of the model aperture illumination. The scale of the ordinate is so adjusted that the maximum of the model amplitude taper is equal to k, where $0 \leqslant k \leqslant 1$. In the present discussion and in Figure 2.12, $k=1$. Along the abscissa are N possible element-pair locations. (The array consists of pairs of elements, since symmetry is assumed throughout this chapter when considering linear arrays.) The possible element locations are equally spaced (generally one-half wavelength).

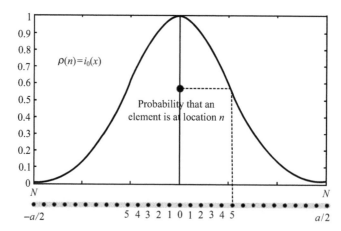

Figure 2.12   Model aperture illumination used to determine placement of elements in statistical density-taper, method of designing unequally spaced arrays.

The model amplitude taper specifies the probability that an element pair will be located at the nth equally spaced position of the aperture.

The design of the statistical density taper begins by selecting a continuous amplitude taper $i_0(x)$ whose pattern is to be approximated by the density-tapered array. From the continuous distribution $i_0(x)$, a discrete approximation is obtained. This is denoted A_n and is the aperture illumination of an amplitude-tapered filled array used as the model for the density taper, where $0 \leqslant A_n \leqslant 1$. Each possible element location is examined in turn to determine whether or not an element is to be located there. An element is placed at a particular point if the value of the amplitude taper A_n at that location is less than a number, chosen at random, between the values of 0 and 1. Let us say that the amplitude taper A_n at some element location n has a value of 0.7. A random-number generator or table is consulted, and a number between zero and one is selected. If it is less than A_n (in this case 0.7), the element remains. If the random number is greater, it is removed. If this is repeated in many designs, the element will remain at that particular location a fraction of times equal to the value of the model amplitude taper at that point. Thus in this example, an element will be allowed to remain seven times out of ten on the average. With enough elements, the pattern of the statistically designed density-tapered array will approximate that of the amplitude-tapered filled array used as the model. This is a relatively simple design procedure and can be readily implemented with a digital computer.

The degree of thinning achieved will depend upon the value of k and the shape of amplitude taper. With typical amplitude tapers, the degree of thinning with $k = 1$ is about 40 to 60 percent for circular planar apertures. When $k = 1$, the design is said to be naturally thinned. Greater thinning is obtained with values of k less than unity.

Finally, there are some features of the thinned array that should be pointed out:

1. The gain of a thinned array of isotropic elements each radiating equal power is approximately equal to the number of elements within the aperture.

2. The beamwidth is of the order of λ_0/D, where λ_0 is the wavelength and D is the aperture dimension.

3. Removing elements in the thinned array results in reduced gain compared with a filled array with the same beamwidth. In a receiving array, there is a decrease in signal-to-noise ratio.

4. The average side lobe level of a highly thinned array relative to that of the main beam approaches a value equal to the reciprocal of the number of elements remaining in the array. For the thinned array, the energy is not assigned to broaden the main beam. As a result, the side lobe level of especially the far-out side lobe level will be increased. If the thinning is not too severe (of the order of half the elements removed) the peak side lobe level can be kept to a reasonable value and can be made competitive with that of a conventional design.

2.6 Signal Processing Antenna Array

Signal processing is used in an antenna array to form the desired beam. This process is also called beamforming (BF) and is often referred to as spatial filtering. The outputs from elements in the array can be subjected to various forms of signal processing, where phase or amplitude adjustments are made to produce outputs that can provide concurrent angular information for signals arriving in several different directions in space.

If the beamforming is carried out at radio frequency (RF), the analog beamforming network usually consists of devices that change the phase and power of the signals. It is sometimes desirable to form multiple beams that are offset by finite angles from each other. A multiple-beamforming network is known as a beamforming matrix. The best-known example is given by the Butler matrix[7] and will be introduced in Section 2.6.1.

In some applications, it is necessary to electronically scan the beam of an antenna. This can be accomplished by changing the phase of the signals at the antenna elements. If only the phases are changed, with the amplitude weights remaining fixed as the beam is steered, the array is commonly known as a phased array, which was introduced in Section 2.3.

Beamforming can be carried out at intermediate frequencies (IF). The beamforming network can be implemented using resistors, hybrid circuits, and trapped delay lines, which are constructed using lumped circuits. The beamforming may be more convenient in many respects, since it may be performed after amplification has taken place so that the losses in the beamforming network are less important. However , it requires that each element must have its own RF-to-IF receiver.

Alternatively, the beamforming may be carried out at element level through digitalization, which is recognized as digital beamforming (DBF)[8]. The DBF antenna system is based on capturing the radio frequency (RF) signals at each of the antenna elements and converting them into two streams of binary baseband signals (i.e, in-phase (I) and quadrature-phase (Q) channels). Included within the digital baseband signals are the amplitudes and phases of signals received at each element of the array. The beamforming is carried out by weighting these digital signals, thereby adjusting their amplitudes and phases such that when added together they form the desired beam. The main advantage gained from DBF is greatly added flexibility without any attendant degradation in signal-to-noise ratio (SNR). In many ways it can be considered to be the ultimate antenna, in that all of the information arriving at the antenna aperture is captured in the digital streams that flow from this face. The price paid to obtain the advantages of this kind of array is cost and complexity.

In one channel receiver application, it is still possible to capture all the information arriving at the antenna aperture through time sequence phase or angle weighting. After recovering the amplitude and phase distribution on the aperture, the angular superresolution may be realized by using the signal processing technique. This kind of array will be discussed in Section 2.6.2.

The adaptive beamforming is used to optimize the array pattern by adjusting the element control weights until a prescribed objective function is satisfied. Adaptive beamforming technology is also referred to as smart antenna technology in the sense that the information on the aperture of the antenna is used adaptively and completely.

Although the antenna synthesis described in previous sections may be considered as antenna beamforming in general, in the following some specific topics will be addressed.

The comparison between array signal processing and time signal processing is helpful, which is shown in Table 2.4.

Table 2.4   Comparison between array signal processing and time signal processing.

	Sampling	Variable	Spectrum	System Function	Filtering
Time Signal Processing	$X(n)$	Time	Frequency spectra	Transfer Function	Enhance or Suppress Signals of Different Frequencies
Array Signal Processing	X_n	Space	Spatial Spectra	Radiation Pattern	Enhance or Suppress Signals of Different Directions

2.6.1   Multi-Beam Antenna Array (Analog Beamforming)

In a beamforming matrix, an array of hybrid junctions and fixed-phase shifters are used to achieve the desired results. For example, a Butler beamforming matrix for a four-element array is shown in Figure 2.13(a). This matrix uses four 90° phase-lag hybrid junctions with the transmission properties shown and two 45° fixed-phase shifters. In transmitting, by tracing the signal from the four ports to the array elements or by using (2.6.1), one should be able to find the aperture relative phase distribution corresponding to the individual ports of a four-port Butler matrix. Consequently the beam direction may easily be found.

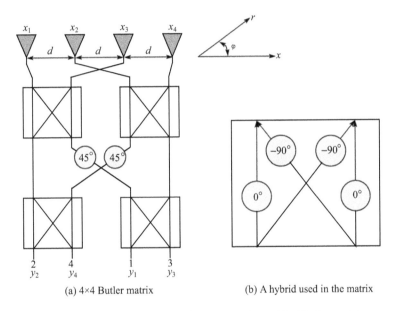

(a) 4×4 Butler matrix   (b) A hybrid used in the matrix

Figure 2.13   Principle of multi-beam array formed by Butler matrix.

Butler matrix (network) beamforming is similar to the fast Fourier transform (FFT) process. In fact, they have a 1 to 1 equivalence. Surprisingly, the Butler matrix was developed before the FFT. However, there is an important difference between them: a Butler matrix processes signals in the analog domain, whereas the FFT processes signals in the digital domain.

The multi-beam antenna array has many applications. As an example, Figure 2.14 shows a switched beam smart antenna system. Base transceiver stations (BTS) in which antenna arrays employ narrow beams pointing to each user rather than omni or directional antenna covering a large number of users or areas can provide better network performance. It is the theoretical basis of smart antenna system (SAS) designed for a modern cellular network, which is much challenged by network capacity and service quality due to the increased internet traffic volume. The application of SASs with BTSs reduces the co-channel interference level in a cellular network on both the uplink and the downlink directions due to the narrower antenna beamwidth. It in turn improves the bit energy-to-interference power spectral density ratio of received signals and consequently enhances the network capacity and/or improves the service quality.

After having the received signal-strength indicator (RSSI) and the direction of arrival (DOA) information, the selecting and switching of the beam to the desired target will be carried out. According to the structure shown in Figure 2.13(a), the transform matrix of the Butler network is given as

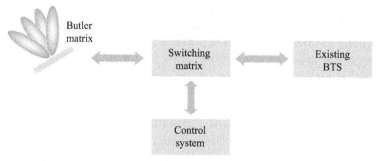

Switched beam SAS

Figure 2.14   Switched beam smart antenna system.

$$\begin{pmatrix} x_1 \\ x_2 \\ x_3 \\ x_4 \end{pmatrix} = \begin{pmatrix} e^{-j45} & e^{-j180} & e^{j45} & e^{-j90} \\ e^{j0} & e^{-j45} & e^{-j90} & e^{-j135} \\ e^{-j135} & e^{-j90} & e^{-j45} & e^{j0} \\ e^{-j90} & e^{j45} & e^{-j180} & e^{-j45} \end{pmatrix}^{-1} \begin{pmatrix} y_1 \\ y_2 \\ y_3 \\ y_4 \end{pmatrix} \tag{2.6.1}$$

where the number in matrix is in degrees. If we denote the matrix as $\mathbf{B}$, then we have

$$\mathbf{X} = \mathbf{B}^{-1}\mathbf{Y} = \mathbf{B}^T\mathbf{Y} \tag{2.6.2}$$

where $\mathbf{X} = [x_1, x_2, x_3, x_4]^T$ and $\mathbf{Y} = [y_1, y_2, y_3, y_4]^T$. The second identity is due to the orthogonality of matrix $\mathbf{B}$. In receiving, $\mathbf{X}$ is known and we may use (2.6.2) to find $\mathbf{Y}$. Super-resolution algorithms are applied to estimate the DOA from $\mathbf{X}$. In case of K mobile users in a sector, the vector $\mathbf{X}$ can be written as

$$\mathbf{X}(t) = \sum_{i=1}^{K} \mathbf{a}(\theta_i)S_i(t) + \mathbf{n}(t) = \mathbf{A}(\theta)\mathbf{S}(t) + \mathbf{n}(t) \tag{2.6.3}$$

and

$$\mathbf{A}(\theta) = [\mathbf{a}(\theta_1), \mathbf{a}(\theta_2), \cdots, \mathbf{a}(\theta_k)] \tag{2.6.4}$$

where $\mathbf{S}(t) = [S_1(t), S_2(t), \cdots, S_k(t)]^T$ is the incoming signal, and $\mathbf{n}(t)$ is a complex noise vector. The kth direction vector $\mathbf{a}(\theta_i)$ is

$$\mathbf{a}(\theta_i) = \left[1, e^{jkd\sin\theta_i}, e^{jk2d\sin\theta_i}, e^{jk3d\sin\theta_i}\right]^T \tag{2.6.5}$$

where d denotes the inter-element spacing.

The angle resolution depends on the first null beamwidth (FNBW). If two targets are inside the FNBW, they will not be distinguished. The FNBW $2\theta'_0$ may be derived from $N\psi/2 = \pi$ for a uniform array (see formula (2.2.2-a)) and it gives $\cos\theta_0 = \sin\theta'_0 = \lambda/Nd$, where $\theta'_0 = 90° - \theta_0$. Therefore

$$2\sin\theta' \approx 2\theta' = \frac{2\lambda}{Nd} \approx \frac{2\lambda}{D} \tag{2.6.6}$$

where D is the dimension of the aperture. The limitation of the angle resolution subject to (2.6.6) is known as the Rayleigh criterion. The resolution surpassing the Rayleigh criterion is called the super-resolution. The super-resolution algorithm, RELAX and MUSIC may be applied to (2.6.3) to get the super-resolution. The improvement factor of resolution against Rayleigh criterion is listed in Table 2.5[8–10].

Table 2.5 Comparison of resolution capability of **RELAX** and **MUSIC** in terms of the resolution improvement factor against Rayleigh criterion.

SNR (dB)	Improvement Factor of Resolution	
	RELAX	MUSIC
∞	6.7	40.0
20	5.0	2.7
10	2.9	2.0
0	1.7	0.7

It is seen that higher resolution DOA estimation can be achieved as super-resolution algorithms are applied to the switch beam smart antenna. The resolution capability of the RELAX algorithm for SNR is better than that of the MUSIC algorithm. However, the MUSIC method provides better resolution for higher SNR and takes less computer time. For lower SNR, the RELAX algorithm is a better candidate.

2.6.2 Angular Super-Resolution for Phased Antenna Array through Phase Weighting[11]

In the example of angular super-resolution given in Section 2.6.1, both amplitude and phase on each element should be acquired. In most of the conventional phased array, there is only one output channel. To realize the angular super-resolution, the amplitude and phase distributions on the aperture still need to be acquired. In this case, this distribution may be obtained by time sequence phase weighting. The principle of this technique may be illustrated through the four-element phased array shown in Figure 2.15. The phase weighting is realized by variable phase shifter. The power divider is a 3dB hybrid coupler. For example, when x_1, x_2 are with equal amplitude and in-phase, the output will be the sum of them and zero output to the matched load. If out-of-phase, the output will be zero and the signal will be absorbed by the matched load.

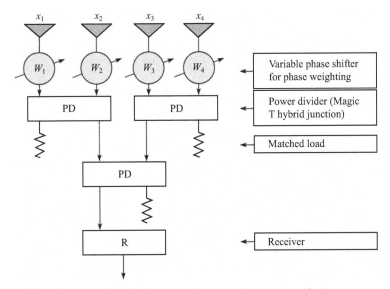

Figure 2.15   Configuration of phase weighting angular super-resolution antenna array.

If the signal on the aperture is $\mathbf{X'} = [x_1, \cdots, x_N]^T$, where N is the number of the elements, and in Figure 2.15 it is 4. The output $\mathbf{Y}$ to the receiver is

$$\mathbf{Y} = \frac{1}{\sqrt{N}}\mathbf{W}\mathbf{X}'$$

(2.6.7)

$\mathbf{W}$ is the weighting matrix. Define $\mathbf{X}'/\sqrt{N} = \mathbf{X}$, then we have

$$\mathbf{Y} = \mathbf{W}\mathbf{X}, \quad \mathbf{X} = \mathbf{W}^{-1}\mathbf{Y}$$

(2.6.8)

The weighting matrix is chosen to be the Hadamard matrix $\mathbf{H}_N$, that is $\mathbf{W} = \mathbf{H}_N$. According to the property of Hadamard matrix

$$\mathbf{W} = \mathbf{H}_N = \mathbf{H}_2 \otimes \mathbf{H}_{N/2}, \quad \mathbf{W}^{-1} = \frac{1}{N}\mathbf{H}_N, \quad \mathbf{H}_2 = \begin{bmatrix} 1 & 1 \\ 1 & -1 \end{bmatrix}$$

(2.6.9)

where $\otimes$ denotes Kronecker product. If $N=4$

$$\mathbf{H}_4 = \mathbf{H}_2 \otimes \mathbf{H}_2 = \begin{bmatrix} 1 & 1 & 1 & 1 \\ 1 & -1 & 1 & -1 \\ 1 & 1 & -1 & -1 \\ 1 & -1 & -1 & 1 \end{bmatrix}$$

(2.6.10)

From (2.6.10), it is seen that the weighting matrix relates to a $0 - \pi$ phase modulation. It is easy both for real time calculation and for the hardware realization. Actually, $\mathbf{X}$ and $\mathbf{Y}$ are a Walsh-Hadamard transform pair

$$\mathbf{Y} = \mathbf{H}_N\mathbf{X}, \quad \mathbf{X} = \frac{1}{N}\mathbf{H}_N\mathbf{Y}$$

(2.6.11)

Therefore, for an N element array, through n times phase weighting, the amplitude and phase distribution may be recovered. This technique may also be considered as using pattern diversity to obtain the space information. After the aperture distribution is recovered, the non-linear spectrum estimation algorithm, such as RELAX, is used to obtain the super-resolution.

According to the Parseval's theorem, the following identities hold

$$\sum_{i=1}^{N} |x_i|^2 = \frac{1}{N}\sum_{k=1}^{N} |y_k|^2$$

(2.6.12)

or

$$\sum_{i=1}^{N} |x_i'|^2 = \sum_{k=1}^{N} |y_k|^2$$

(2.6.13)

Formulas (2.6.12) and (2.6.13) show that the energy in i domain (space or element domain) is equal to that in k domain (time domain). That is, the Walsh-Hadamard transformation is subject to the energy conservation law. The total power P received by the receiver for N times is equal to that received by N elements once. However, the total received power by the array n times should be NP; obviously the power $(N-1)P/N$ is absorbed by the matched loads in the system when the Wilkinson power divider is used and is scattered when the conventional power divider is used. This technique was used in a one-dimensional phased antenna array with 139 antenna elements. The angle resolution is improved by a factor of 2 under 15dB signal-to-noise ratio(SNR).

2.6.3 Angular Super-Resolution for Conventional Antenna through Angle Weighting[12, 13]

If linear phase weighting is used in the phase weighting technique, the pattern diversity is actually the beam scan. In conventional antennas this scan can be realized through mechanical rotation. Therefore, based on the similar idea as that in phase weighting technique, we may use the outputs from different angles to recover the aperture distribution of a virtual uniform linear array antenna. After it is done, the nonlinear spectrum estimation algorithm is applied to achieve the super-resolution.

Consider a one-dimensional linear antenna array with M elements. The spacing between array elements is d. We assume that the weights of each element are $H(1), H(2), \cdots, H(M)$. The far field pattern of this array can be written as:

$$h(\theta) = \sum_{m=1}^{M} H(m)e^{j2\pi \frac{[m-(M+1)/2]d\sin\theta}{\lambda}}, \quad -90^0 \leqslant \theta \leqslant 90^0 \qquad (2.6.14)$$

Sampling the far field pattern with interval $\Delta\theta$, then the discrete far field pattern is:

$$h(n) = \sum_{m=1}^{M} H(m)e^{j2\pi \frac{[m-(M+1)/2]d\sin(n\Delta\theta)}{\lambda}}, \quad -\text{round}(90^0/\Delta\theta) \leqslant n \leqslant \text{round}(90^0/\Delta\theta) \qquad (2.6.15)$$

$h(n)$ and $H(m)$ are discrete Fourier transform pair, so the weights of the antenna elements can be obtained from $h(n)$ as:

$$H(m) = \sum_{n=-\text{round}(90^\circ/\Delta\theta)}^{\text{round}(90^\circ/\Delta\theta)} h(n)e^{-j2\pi \frac{[m-(M+1)/2]d\sin(n\Delta\theta)}{\lambda}}, \quad m=1,2,\cdots,M \qquad (2.6.16)$$

For a real aperture antenna with same aperture size $D = (M-1)d$, an equivalent linear array antenna can always be found from its discrete far field pattern $h(n)$.

When the real aperture antenna is scanned in an angle, the output of the antenna is the convolution of the target distribution to the antenna pattern. So the output of the real aperture antenna sampled with interval $\Delta\theta$ as the antenna is stepped in angle can be written as follows:

$$s(n) = x(n) \otimes h(n) \qquad (2.6.17)$$

where $x(n)$ is the angular distribution of radar targets, and $\otimes$ denotes the convolution operator. The discrete Fourier transform of (2.6.17) is as follows:

$$S(m) = X(m) \cdot H(m) \qquad (2.6.18)$$

where $X(m)$ are complex response values of the equivalent linear array antenna elements related to the scattered wave from targets while the normal direction of the equivalent antenna points to $\theta = 0$, and can be obtained by:

$$X(m) = \frac{S(m)}{H(m)}, \quad m=1,2,\cdots,M \qquad (2.6.19)$$

$X(m)$ can also be expressed as

$$\mathbf{X} = \mathbf{AS} + \mathbf{N} \qquad (2.6.20)$$

where $\mathbf{X} = [X(1), X(2), \cdots, X(M)]^T$, $\mathbf{S}$ is a $p \times 1$ vector of scattering strength of targets, p is the number of targets, $\mathbf{A}$ is an $m \times p$ transfer matrix whose columns are steering vectors of targets, and $\mathbf{N}$ is an $m \times 1$ vector of independent random complex noise. Then, the nonlinear super-resolution algorithm is adopted to get the super-resolution angular distribution from $X(m)$.

The experimental results in the laboratory and from a conventional array antenna all show that the angle resolutions are improved by a factor of 2 under the reasonable SNR.

2.6.4   Adaptive Beamforming Antenna Array

An adaptive array differs from a conventional array in that the complex weights on the antenna elements are not determined by the designer in advance. Instead, feedback control signals are obtained from within the system for optimization of the element weights. The control signals may be obtained from samples of the radiation field received by the antenna elements. Alternatively, they may be obtained from the system output by comparison of the output with some known or desired output. There are $2N$ degrees of freedom in an N-element array, one for amplitude and one for phase of each complex weight. When only the shape of the radiation pattern is to be controlled, $2(N-1)$ suffice because the amplitude and phase of one of the complex weights may be chosen arbitrarily. The element with the arbitrary weight is called the reference element. Its amplitude represents an arbitrary gain setting for the array; its phase represents an arbitrary fixed phase shift across the array. Neither affects the normalized radiation pattern.

The idea of interference suppression is illustrated through a two-element array shown in Figure 2.16(a)[8]. The desired signal, $s(t)$, arrives from the foresight direction ($\theta_s = 0$) and the interference signal, $I(t)$, arrives from the angle ($\theta_I = \pi/6$) radians. The incident interference signal arrives at element 2 with a phase lead with respect to element 1 of value $(2\pi/\lambda_0)d\sin(\pi/6) = \pi/2$. After the weighting, the outputs for desired signal and interference are $y_d = A\exp(j2\pi f_0 t)(W_1 + W_2)$ and $y_I = N\exp(j2\pi f_0 t)(W_1 + \exp(j\pi/2)W_2)$ respectively. The solution for $W_1 + W_2 = 1$, $W_1 + \exp(j\pi/2)W_2 = 0$ is found to be $W_1 = (\sqrt{2}/2)\exp(-j\pi/4)$ and $W_2 = (\sqrt{2}/2)\exp(j\pi/4)$ respectively. The radiation pattern is

$$F(\theta) = (\sqrt{2}/2)\left[e^{-j(\pi/4 + \pi\sin\theta/2)} + e^{j(\pi/4 + \pi\sin\theta/2)}\right] = \sqrt{2}\cos(\pi/4 + \pi\sin\theta/2)$$

It indeed produces zero radiation at $\theta_s = \pi/6$ radians.

Even if the directions of the interference are unknowns, it is still possible to find the weighting coefficients which may produce nulls at the direction of interference (See [23] and Prob. 2.13).

The general configuration to illustrate the principle is shown in Figure 2.16(b). Suppose the received signal is $\mathbf{y}(t) = [y_1(t), \cdots, y_N(t)]^T$, the weighting vector $\mathbf{w} = [w_1, \cdots, w_N]^T$, the output then is $\mathbf{w}^H\mathbf{y}(t)$. If the desired (reference) signal is $r(t)$, the error signal is $e(t) = r(t) - \mathbf{w}^H\mathbf{y}(t)$. The time-average value of $|e(t)|^2$ is[14]

$$\overline{|e(t)|^2} = \overline{|r(t)|^2} - 2\mathbf{w}^H\mathbf{S}_{yr} + \mathbf{w}^H\mathbf{R}_{yy}\mathbf{w} \tag{2.6.21}$$

where $\mathbf{S}_{yr} = \overline{\mathbf{y}(t)r^*(t)}$, $\mathbf{R}_{yy} = \overline{\mathbf{y}(t)\mathbf{y}^H(t)}$. (2.6.21) is a quadratic function of the weights. This means that there is a unique set of values w_i such that $\overline{|e(t)|^2}$ is a minimum. Now the mean square error (2.6.21) when differentiated with respect to w_i leads to

$$\nabla(\overline{|e(t)|^2}) = -2\mathbf{S}_{yr} + 2\mathbf{R}_{yy}\mathbf{W} \tag{2.6.22}$$

Letting the gradient $\nabla(\overline{|e(t)|^2}) = 0$, one may have the optimum solution to $\mathbf{W}$ as

$$\mathbf{W}_{opt} = \mathbf{R}_{yy}^{-1}\mathbf{S}_{yr} \tag{2.6.23}$$

According to the received signal, the system may adaptively produce the optimized weights.

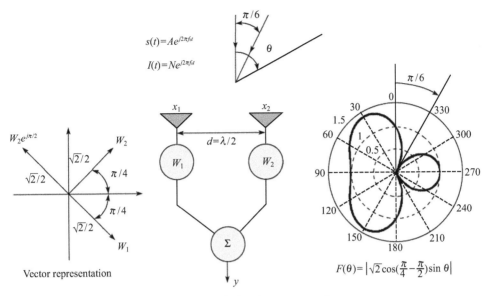

(a) Two-element array for interference suppression

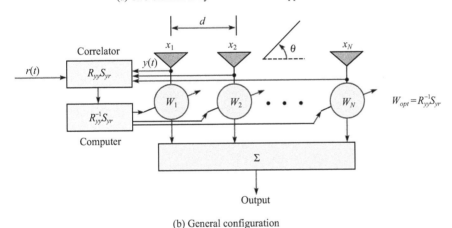

(b) General configuration

Figure 2.16   Principle of adaptive array configuration.

An explicit reference signal may not always exist. For this particular reason, a blind adaptive beamforming concept has been conceived and developed[8]. The adaptive techniques used in blind adaptive beamforming are essentially the same as those used in conventional adaptive beamforming. The key difference is that blind adaptive beamforming does not require an explicit reference signal. Instead, a blind adaptive beamforming system generates its own reference signal based on the implicit characteristics of the wanted signal. These hidden characteristics include the constant modulus property, information from the decision process, cyclostationarity and other similar communications signal features. The constant modulus algorithm (CMA)[15] has gained a lot of interest. Some efforts to improve this algorithm, including the combination of DOA (direction of arrival), have been made[16, 17].

Finally, the following points should be emphasized:

- A digital beamforming (DBF) adaptive antenna system can be considered to be the ultimate or perfect smart antenna system in the sense that it is possible to capture and to adaptively make use of all the information that falls on the antenna aperture.

It can provide independently controllable simultaneous beams. The DBF antenna is truly the product of a marriage between electromagnetics and digital signal processing.

- Adaptability and signal processing are separate attributes. An array may have just one or both. A nonadaptive example of signal processing is synthetic aperture radar(SAR); an example of an adaptive system without signal processing is the monopulse radar. The use of analog or digital techniques for the adaptive circuits or the signal processing circuits is a matter of designer's choice.

- The smart antenna system (SAS) is composed of three parts: antenna array (linear, sector, circular, $\cdots$), beam-forming network and DOA estimator (with reference or blind). SAS is significant in improving the performance of communication and radar system, such as greater capacity, higher signal interference noise ratio (SINR), less base station requirement and so on. The limitations of the application are the real time signal processing ability and the cost barrier.

- In the case of a single channel, the time sequence phase weighting can be used to obtain the aperture distribution without DBF. The price paid for is the time. To enhance the angle resolution, the aperture may be equivalently enlarged by using SAR technique or/and the super-resolution algorithm.

2.7 Planar Arrays

In addition to placing elements along a line (to form a linear array) individual radiators can be positioned along a rectangular grid to form a rectangular or planar array. Planar arrays provide an additional variable, which can be used to control and shape the pattern of the array. Planar arrays are more versatile and can provide more symmetrical patterns with lower side lobes. In addition, they can be used to scan the main beam of the antenna toward any point in space[3].

2.7.1 Array Factor

If M by N elements are initially placed along the x-axis and y-axis, as shown in Figure 2.17, the array factor of it can be written similar to (2.4.1) as

$$AF = \sum_{m=1}^{M} \sum_{n=1}^{N} \frac{I_{mn}}{I_{11}} e^{j(m-1)(kd_x \sin\theta \cos\phi + \beta_x)} e^{j(n-1)(kd_y \sin\theta \sin\phi + \beta_y)} \quad (2.7.1)$$

where I_{mn} is the excitation coefficient of each element. The spacing and progressive phase shift between the elements along the x-axis and y-axis are represented, respectively, by d_x, d_y and β_x, β_y. If each row has the same current distribution, even though the current levels are different in different rows, that is, $I_{mn}/I_{m1} = I_{1n}/I_{11}$, then $I_{mn}/I_{11} = (I_{m1}/I_{11})(I_{1n}/I_{11})$. In that case, the current distribution is said to be separable and the array factor can be expressed in the form

$$AF = S_{xM} S_{yN} \quad (2.7.2)$$

where

$$S_{xM} = \sum_{m=1}^{M} \frac{I_{m1}}{I_{11}} e^{j(m-1)(kd_x \sin\theta \cos\phi + \beta_x)} \quad (2.7.3\text{-a})$$

$$S_{yN} = \sum_{n=1}^{N} \frac{I_{1n}}{I_{11}} e^{j(n-1)(kd_y \sin\theta \sin\phi + \beta_y)} \quad (2.7.3\text{-b})$$

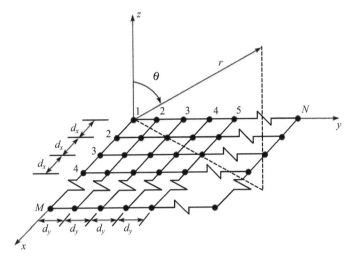

Figure 2.17 Planar array geometries.

Eqn. (2.7.2) indicates that the pattern of a rectangular array is the product of the array factors of the arrays in the x and y directions.

If $I_{mn}/I_{m1} = I_{1n}/I_{11} = I_0$, (2.7.2) can be expressed as

$$AF = I_0 \sum_{m=1}^{M} e^{j(m-1)(kd_x \sin\theta \cos\phi + \beta_x)} \sum_{n=1}^{N} e^{j(n-1)(kd_y \sin\theta \sin\phi + \beta_y)} \qquad (2.7.4)$$

According to (2.7.2), the normalized form of (2.7.4) can be written as

$$AF_n(\theta,\phi) = \left\{ \frac{1}{M} \frac{\sin\left(\frac{M}{2}\psi_x\right)}{\sin\left(\frac{\psi_x}{2}\right)} \right\} \left\{ \frac{1}{N} \frac{\sin\left(\frac{N}{2}\psi_y\right)}{\sin\left(\frac{\psi_y}{2}\right)} \right\} \qquad (2.7.5)$$

where $\psi_x = kd_x \sin\theta \cos\phi + \beta_x$, $\psi_y = kd_y \sin\theta \sin\phi + \beta_y$.

When the spacing between the elements is equal or greater than λ, multiple maxima of equal magnitude are formed. The principle maximum is referred to as the major lobe and the remaining as the grating lobe as discussed in Section 2.3.

For a rectangular array, the major lobe and grating lobes of S_{xM} and S_{yN} in (2.7.3-a) and (2.7.3-b) are located at

$$\begin{aligned} kd_x \sin\theta \cos\phi + \beta_x = \pm 2m\pi, \quad & m = 0, 1, 2, \cdots \\ kd_y \sin\theta \sin\phi + \beta_y = \pm 2n\pi, \quad & n = 0, 1, 2, \cdots \end{aligned} \qquad (2.7.6)$$

In most practical applications it is required that the conical main beams of S_{xM} and S_{yN} intersect and their maxima be directed toward the same direction. If it is desired to have only one main beam that is directed along $\theta = \theta_0$ and $\phi = \phi_0$, the progressive phase shift between the elements in the x- and y- directions must be equal to

$$\begin{aligned} \beta_x &= -kd_x \sin\theta_0 \cos\phi_0 \\ \beta_y &= -kd_y \sin\theta_0 \sin\phi_0 \end{aligned} \qquad (2.7.7)$$

When solved simultaneously, (2.7.7) can be expressed as

$$\tan\phi_0 = \frac{\beta_y d_x}{\beta_x d_y}$$

$$\sin^2\theta_0 = \left(\frac{\beta_x}{k d_x}\right)^2 + \left(\frac{\beta_y}{k d_y}\right)^2 \tag{2.7.8}$$

The principle maximum ($m = n = 0$) and the grating lobes can be determined by

$$\sin\theta\cos\phi - \sin\theta_0\cos\phi_0 = \pm\frac{m\lambda}{d_x}, \qquad m = 0, 1, 2, \cdots \tag{2.7.9-a}$$

$$\sin\theta\sin\phi - \sin\theta_0\sin\phi_0 = \pm\frac{n\lambda}{d_x}, \qquad n = 0, 1, 2, \cdots \tag{2.7.9-b}$$

When solved simultaneously, they reduce to

$$\phi = \tan^{-1}\left[\frac{\sin\theta_0\sin\phi_0 \pm n\lambda/d_y}{\sin\theta_0\cos\phi_0 \pm m\lambda/d_x}\right] \tag{2.7.10}$$

and

$$\theta = \sin^{-1}\left[\frac{\sin\theta_0\cos\phi_0 \pm m\lambda/d_x}{\cos\phi}\right] = \sin^{-1}\left[\frac{\sin\theta_0\sin\phi_0 \pm n\lambda/d_y}{\sin\phi}\right] \tag{2.7.11}$$

Using a similar approach as that in Section 2.3, we may obtain the condition to avoid the grating lobes. In (2.7.9-a), let $m = 1$, we have

$$\sin\theta\cos\phi = \sin\theta_0\cos\phi_0 \pm \frac{\lambda}{d_x} \tag{2.7.12}$$

Because $|\sin\theta\cos\phi| \leqslant 1$, we should have

$$-1 \leqslant \sin\theta_0\cos\phi_0 \pm \frac{\lambda}{d_x} \leqslant 1$$

$$d_x < \frac{\lambda}{1 + |\sin\theta_0\cos\phi_0|} \tag{2.7.13}$$

The maximum value of $\sin\theta_0\cos\phi_0$ is $\sin\theta_{\max}$, where $\theta_{\max}$ is the maximum scan angle. Therefore the condition to avoid the grating lobes is

$$d_x < \frac{\lambda}{1 + |\sin\theta_{\max}|} \tag{2.7.14}$$

The difference between (2.3.3) and (2.7.14) is due to the fact that (2.3.3) corresponds to the case of the elements being distributed along z axis while (2.7.14), along x axis. Similarly, from (2.7.9-a)

$$-1 \leqslant \sin\theta_0\sin\phi_0 \pm \frac{\lambda}{d_y} \leqslant 1 \tag{2.7.15}$$

$$d_y < \frac{\lambda}{1 + |\sin\theta_0\sin\phi_0|}$$

The maxium value of $\sin\theta_0\sin\phi_0$ is $\sin\theta_{\max}$, thus we have

$$d_y < \frac{\lambda}{1 + |\sin\theta_{\max}|} \tag{2.7.16}$$

2.7.2   Taylor Patterns of Circular Aperture

Consider a planar aperture with a circular boundary of radius a, as shown in Figure 2.18. If the tangential field distribution on the aperture is $E_x(r', \phi')$ and zero elsewhere, the radiation field may be found from following formulas (1.3.51), (1.3.52) as

$$E_\theta = A \cos\phi (1 + \cos\theta) F_2$$
$$E_\phi = -A \sin\phi (1 + \cos\theta) F_2$$

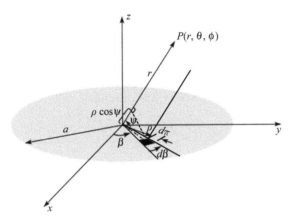

Figure 2.18   Cartesian/cylindrical/spherical coordinates.

From the above two formulas we have

$$\mathbf{E} = E_\theta \hat{\mathbf{a}}_\theta + E_\phi \hat{\mathbf{a}}_\phi$$

Therefore

$$E = A(1 + \cos\theta) F_2$$

where A and F_2 may be found from (1.3.49-b) and (1.3.50-b) as

$$A = \frac{je^{-jkr}}{2\lambda r}$$
$$F_2 = \iint\limits_S E_x^0(x', y') e^{jk\rho \cos\psi} ds'$$

where $\rho \cos\psi = x' \sin\theta \cos\phi + y' \sin\theta \sin\phi$. It is convenient to recast this integral in the polar coordinates illustrated in Figure 2.18 and defined by

$$x' = \rho \cos\beta, \quad y' = \rho \sin\beta$$

If $E_x^0(x', y')$ is designated as $E^0(\rho, \beta)$, the integral becomes

$$F_2(\theta, \phi) = \int_0^a \int_0^{2\pi} E^0(\rho, \beta) e^{jk\rho \sin\theta \cos(\phi - \beta)} \rho \, d\rho \, d\beta \qquad (2.7.17)$$

Usually the aperture distribution is circular symmetrical, that is $E^0(\rho, \beta) = E^0(\rho)$. Consequently, (2.7.17) yields

$$F_2(\theta) = 2\pi \int_0^a E^0(\rho) \, J_0(k\rho \sin\theta) \rho \, d\rho \qquad (2.7.18)$$

J_0 is the Bessel function of zero order.

Consider the commonly used parabolic type distribution with power N,

$$E^0(\rho) = \left[1 - \left(\frac{\rho}{a}\right)^2\right]^N \tag{2.7.19}$$

Substituting (2.7.19) into (2.7.18) and making use of the integration by parts and the following formula

$$\frac{d}{dx}[x^n \, J_n(x)] = x^n \, J_{n-1}(x)$$

We obtain finally

$$F_2(\theta) = \frac{(\pi a^2)N!2^{N+1} \, J_{N+1}(ka\sin\theta)}{(ka\sin\theta)^{N+1}} \tag{2.7.20}$$

For uniformly distributed aperture $N=0$, (2.7.20) becomes

$$F_2(\theta) = \frac{2\pi a^2 \, J_1(ka\sin\theta)}{ka\sin\theta} \tag{2.7.21}$$

The normalized form in terms of variable u is

$$F_2(u) = \frac{2 \, J_1(\pi u)}{\pi u} \tag{2.7.22}$$

$$u = \frac{2a}{\lambda}\sin\theta, \quad 0 \leqslant u \leqslant 2a/\lambda \tag{2.7.23}$$

The array factor $F_2(u)$ in (2.7.22) is with symmetrical side lobes whose field heights trail off as u^{-1}. Similar to what is done in Section 2.4.3, the modification of (2.7.22) is introduced as

$$F_T(u) = \frac{J_1(\pi u)\displaystyle\prod_{n=1}^{\bar{n}-1}\left(1 - u^2/u_n^2\right)}{\pi u\displaystyle\prod_{n=1}^{\bar{n}-1}\left(1 - u^2/\gamma_{1n}^2\right)} \tag{2.7.24}$$

where γ_{1n} are the roots of J_1 given by

$$J_1(\pi\gamma_{1n}) = 0, \quad n = 0, 1, 2, \cdots$$

One can see that (2.7.24) accomplishes the purpose of removing the first $\bar{n} - 1$ root pairs of (2.7.22) and replacing them by $\bar{n} - 1$ root pairs at the new positions $\pm u$. Other than the replacement $\sin u/u$ by $J_1(\pi u)/\pi u$ and $n\pi$ by γ_{1n} in (2.4.14), the other relationships remain the same, such as

$$\sigma = \frac{\gamma_{1\bar{n}}}{\sqrt{A^2 + (\bar{n} - 1/2)^2}} \tag{2.7.25}$$

$$\cosh^{-1} b = \pi A \tag{2.7.26}$$

The first five roots of $J_1(\pi\gamma_{1n})$ are listed in Table 2.6. Observing Table 2.6, it may be seen that $\gamma_{1\bar{n}} = \bar{n} + 1/4$, $\bar{n} = 1, 2, 3, 4, 5$. The approximate expression for (2.7.25) is

$$\sigma = \frac{\bar{n} + \dfrac{1}{4}}{\sqrt{A^2 + \left(\bar{n} - \dfrac{1}{2}\right)^2}}$$

Table 2.6   Roots of $J_1(\pi\gamma_{1n}.)$

n	0	1	2	3	4
$\pi\gamma_{1n}$	3.823	7.106	10.173	13.324	16.471
γ_{1n}	1.2197635	2.2619101	3.2381665	4.2411609	5.2428821

As $\bar{n} \to \infty$, $\sigma = 1$. From $\partial\sigma/\partial\bar{n} = 0$, it is seen that when $\bar{n} = (4A^2/3) + 1/2$, σ reaches the maximum. To keep the monotonic decrease of σ, $\bar{n}$ should be $\geqslant (4A^2/3) + 1/2$. This condition is slightly different from that given in Section 2.4.3.

To find the aperture distribution that will produce the Taylor pattern, $F_2(\theta)$ in (2.7.18) is rewritten as

$$F_2(u) = F_T(u) = \int_0^\pi p g_0(p)\, J_0(up)dp \tag{2.7.27}$$

where

$$p = \frac{\pi}{a}\rho, \quad g_0(p) = \frac{2a^2}{\pi} E^0(\rho)$$

It is helpful to express $g_0(p)$ as a series in the form

$$g_0(p) = \sum_{m=0}^\infty B_m\, J_0(\gamma_{1m}p), \quad p \leqslant \pi \tag{2.7.28}$$

Substituting (2.7.28) into (2.7.27) results in

$$F_T(u) = \sum_{m=0}^\infty B_m \int_0^\pi p\, J_0(\gamma_{1m}p)\, J_0(up)dp \tag{2.7.29}$$

Because of the orthogonality of the Bessel function, (2.7.29) may be written as

$$F_T(\gamma_{1k}) = B_k \int_0^\pi p\, J_0^2(\gamma_{1k}p)dp$$
$$= B_k \left\{ \frac{p^2}{2} \left[J_0^2(\gamma_{1k}p) + J_1^2(\gamma_{1k}p) \right] \right\}_0^\pi \tag{2.7.30}$$

from which

$$B_k = \frac{2}{\pi^2} \frac{F_T(\gamma_{1k})}{J_0^2(\gamma_{1k}\pi)} \tag{2.7.31}$$

In deriving B_k, $J_1(\gamma_{1k}\pi) = 0$ is used.

Because $F_T(\gamma_{1k}) \equiv 0$ for $k \geqslant \bar{n}$, the series in (2.7.28) is truncated, and the aperture distribution is given by

$$g_0(p) = \frac{2}{\pi^2} \sum_{m=0}^{\bar{n}-1} \frac{F_T(\gamma_{1m})}{J_0^2(\gamma_{1m}\pi)}\, J_0(\gamma_{1m}p) \tag{2.7.32}$$

where $F_T(\gamma_{1m})$ can be computed by using (2.7.24)[2].

Based on the continuous distributed circular aperture, it is possible to form the rectangular grid array of a circular boundary with a Taylor pattern.

Consider an array with radius $a = 5\lambda$. As shown in Figure 2.19, only a quarter is given due to the symmetry. The distance between the origin and mn element is

$$\gamma_{mn} = \left\{ \left[\frac{(2m-1)dx}{2} \right]^2 + \left[\frac{(2n-1)dy}{2} \right]^2 \right\}^{1/2} \tag{2.7.33}$$

If we choose $dx = dy = d = 0.5\lambda$, this circular array is actually formed by a square array by cutting the shadow part as shown in Figure 2.19.

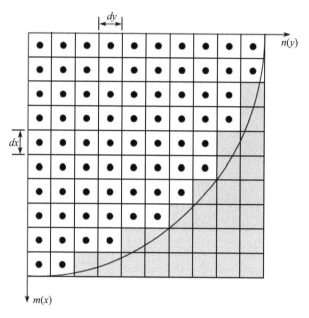

Figure 2.19   Rectangular grid array with circular boundary.

The excitation g_{mn} can be obtained by sampling of (2.7.28). Making use of the symmetry, the array factor can be written as

$$
\begin{aligned}
F_2(\theta, \varphi) &= 4 \sum_{m=1}^{M} \sum_{n=1}^{N} g_{mn} \cos\left[\frac{(2m-1)u_x}{2}\right] \cos\left[\frac{(2n-1)u_y}{2}\right] \\
&= 4 \sum_{m=1}^{10} \sum_{n=1}^{10} g(\gamma_{mn}) \cos\left[\frac{(2m-1)u_x}{2}\right] \cos\left[\frac{(2n-1)u_y}{2}\right]
\end{aligned}
$$

(2.7.34)

where

$$ u_x = kd_x \sin\theta \cos\phi, \quad u_y = kd_y \sin\theta \cos\phi \qquad (2.7.35) $$

2.8   Array Synthesis through Genetic Algorithm (GA)

2.8.1   Introduction to Genetic Algorithms

The evolution and natural selection of genetic algorithms may be used for the optimization techniques. Traditional optimization techniques search for the optimal solutions, using gradients and/or random guesses. Gradient methods quickly converge to a minimum, once an algorithm is close to that minimum. They have the disadvantages of getting stuck in local minima; requiring gradient calculations; working on only continuous parameters and being limited to optimizing a few parameters. Random-search methods do not require gradient calculations, but tend to be slow, and are susceptible to being stuck on local minima.

Electromagnetic-optimization problems often involve many parameters, and these parameters may be discrete. In addition, the number of possibilities is so large that an exhaustive search is impractical. Genetic algorithms can handle a large number of discrete parameters, and are easy to program and quickly give out gratifying results in a short period of time. Genes are the basic building blocks of genetic algorithms. A gene is a binary encoding of a parameter. A chromosome in a computer algorithm is an array of genes. Each chromosome has an associated cost function, assigning a relative merit to that chromosome. The algorithm begins with a large list of random chromosomes. Cost functions are evaluated for each chromosome. The chromosomes are ranked from the most-fit to the least-fit, according to their respective cost functions. Unacceptable chromosomes are discarded, leaving a superior species-subset of the original list. Genes that survive become parents, by swapping some of their genetic material to produce two new offspring. The parents reproduce enough to offset the discarded chromosomes. Thus, the total number of chromosomes remains constant after each iteration. Mutation causes small random changes in a chromosome. Cost functions are evaluated for the offspring and the mutated chromosome, and the process is repeated. The algorithm stops after a set number of iterations, or when an acceptable solution is obtained[18-21].

Figure 2.20 is a flow chart of a genetic algorithm. The algorithm begins by defining a chromosome as an array of parameter values to be optimized. If the chromosome has N_{par} parameters (an N-dimensional optimization problem), given by $p_1, p_2, \cdots, p_{N_{par}}$, then the chromosome is written as

$$chromosome = \left[\, p_1, p_2, p_3, \cdots, p_{N_{par}} \,\right] \tag{2.8.1}$$

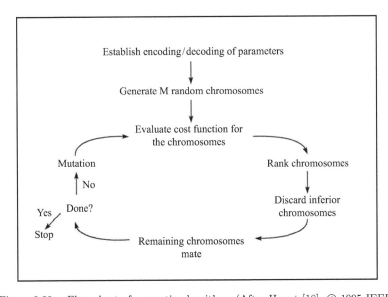

Figure 2.20   Flow chart of a genetic algorithm. (After Haupt [18], © 1995 IEEE)

Each chromosome has a cost function, found by evaluating a function f at $p_1, p_2, \cdots, p_{N_{par}}$. The cost function is represented by

$$cost = f(p_1, p_2, \cdots, p_{N_{par}}) \tag{2.8.2-a}$$

Sometimes the cost function is related to some type of error:

$$cost = a + b\,e \tag{2.8.2-a}$$

where e is the error and the constant a and b are chosen according to the end values of the cost function. Three types of errors can be established: absolute, relative, and mean square.

An alternative to the cost function is the fitness. By transforming the cost function, one obtains the fitness function normalized between 0 and 1:

$$fitness = \frac{a}{a + be} = \frac{1}{1 + pe} \qquad (2.8.3)$$

where $p = b/a$.

The parameters, p_n, can be discrete or continuous. If the parameters are continuous, either some limits need to be placed on the parameters, or they should be restricted to a few possible values. One way to limit the parameters is to encode them in a binary sequence, as follows

$$q_n = q_{\min} + \sum_{m=1}^{L_n} b_w[m]2^{1-m}Q \qquad (2.8.4)$$

where

L_n : number of quantization levels for q_n.

q_n : quantized version of p_n and does not have to be mathematically related to p_n.

$q_{\min}$: the minimum value of q_n.

b_w : array containing the L_n digital binary sequence representing q_n.

Q : half of the largest possible value of q_n.

$b_w[m]$: a number which is picked up from left to right of b_w.

For instance, if $L_n = 3$, p_n represents eight resistivity values, 100Ω, 200Ω, $\cdots$, 800Ω. Consequently, $q_{\min} = 100\Omega$ and $Q = 400\Omega$, then we have

b_w	$b_w[1]$	$b_w[2]$	$b_w[3]$	$q_n(\Omega)$
000	0	0	0	100
001	0	0	1	200
$\cdots$	$\cdots$	$\cdots$	$\cdots$	$\cdots$
111	1	1	1	800

The implementation of genetic algorithms described here only works with the binary encoding of the parameter, and not the parameters themselves. Whenever the cost function is evaluated, the chromosome must first be decoded. An example of a binary-encoded chromosome that has N_{par} parameters, each encoded with $N_{bit} = 10$ bits, is

$$chromosome = \left[\underbrace{1111001001}_{q_1} \underbrace{0011011111}_{q_2} \cdots \underbrace{0000101001}_{q_{N_{par}}} \right]$$

Substituting this binary representation into Eqn.(2.8.4) yields an array of quantized versions of the parameters. This chromosome has a total of $N_{gbit} = N_{pbit} \times N_{par}$ bits, where N_{pbit} is the number of bits for each parameter.

After devising a scheme to encode and decode the parameters, a list of random chromosomes is generated. Each chromosome has an associated cost, calculated from the cost function in Eqn. (2.8.2-a). An example of an arbitrary list of $N_{chro} = 8$ random chromosomes, with their associated costs, is shown in Table 2.7. The next step in the algorithm ranks the chromosomes from best to worst. Assuming a low cost is good, the ranking is shown in Table 2.8.

Table 2.7   A list of 8 random chromosomes and their associated costs.

Num.	Chromosome	Cost
1	011110010001	2.3
2	111001100101	1.7
3	111110001010	2.2
4	000101010111	12.6
5	001010100001	10.1
6	101010010101	1.5
7	111111110000	2.8
8	110011001100	3.0

Table 2.8   Chromosomes are ranked and selected.

Num.	Chromosome	Cost	Keep Half	Keep Costs<5
6	101010010101	1.5	keep	Keep
2	111001100101	1.7	keep	Keep
3	111110001010	2.2	keep	Keep
1	011110010001	2.3	keep	Keep
7	111111110000	2.8	discard	Keep
8	110011001100	3.0	discard	Keep
5	001010100001	10.1	discard	Discard
4	000101010111	12.6	discard	Discard

At this point, the unacceptable chromosomes are discarded. Unacceptable is user defined. Typically, the top x are kept (where x is even), and the bottom $n_{chro} - x$ are discarded. As an example, if 50% of the chromosomes are discarded, then chromosomes 1,2,3, and 6 are kept, while chromosomes 4,5,7 and 8 are discarded (column 4 in Table 2.8). Another possibility is to require the cost to meet a specified level. As an example, if the cost must be less than 5.0, then chromosomes 1,2,3,6,7, and 8 are kept, while chromosomes 4 and 5 are discarded (column 5 in Table 2.8). For this example, we will keep 50% of them.

The next step, after ranking and discarding the chromosomes, is to pair the remaining $N_{chro}/2$ chromosomes for mating. Any two chromosomes can mate. Some possible approaches are to pair the chromosomes from top to bottom of the list, pair them randomly, or pair them 1 with $N_{chro}/2$, 2 with $N_{chro}/2-1$, etc. Once paired, new offspring are formed from the pair-swapping genetic material. As an example, chromosome 6 is paired with 2, and 3 is paired with 1, from Table 2.8. If a random crossover point is between bits 5 and 6, the new chromosomes are formed from

$$\text{parent\#1(chromosome \#6) } \overbrace{10101}\,\underline{0010101}$$

$$\text{parent\#2(chromosome \#2) } \overline{11100}\,\underline{1100101}$$

$$\text{offspring\#1 } 10101\overbrace{\underbrace{1100101}}$$

$$\text{offspring\#2 } \overline{111000}\underline{010101}$$

After the surviving $N_{chro}/2$ chromosomes are paired and mated, the list of $N_{chro}/2$ parents and $N_{chro}/2$ offspring results in a total of N_{chro} chromosomes(the same number of chromosomes as at the start).

At this point, random mutations alter a small percentage of the bits in the list of chromosomes, by changing a "1" to a "0" or vice versa. A bit is randomly selected for mutation from the $N_{chro} \times N_{gbit}$ total number of bits in all the chromosomes. Increasing the number of mutations increases the algorithm's freedom to search outside the current region of parameter space. This freedom becomes more important as the algorithm begins to focus on a particular solution. Typically, about 1% of the bits mutate per iteration. Mutations do not occur on the final iteration.

After the mutations take place, the costs associated with the offspring and mutated chromosomes are calculated, and the process is repeated. The number of generations that evolve depends on whether an acceptable solution is reached, or a set number of iterations is exceeded. After a while, all of the chromosomes and associated costs become the same, except for those that are mutated. At this point, the algorithm should be stopped.

The MATLAB® code of the modified version in [18] that implements a very simple genetic algorithm is shown below:

```
                    ————————————————————————— Beginning of Code ——————————————————————
1
2    % This is a simple GA program for function optimization: a MATLAB implementation.
3    % Copyright (C) 2003 by Lingling Wang
4
5    N = 3;              % number of bits in a gene
6    M = 24;             % number of genes
7    Pop = 32;
8    MN = M*N;           % number of bits in a chromosome
9    last = 500;         % number of iteration
10   M2 = ceil(Pop/2);
11   % --- Creates an initial population
12   Gene = round(rand(Pop,M*N));
13   for ib = 1:last    % ---
14   Rank results and discards bottom 50%
15   % --------------------------------------------------
16   % Insert a cost function here, with a form:  %
17   %            Cost=function(Gene)             %
18   % where "Cost" is a Nx1 array                %
19   % --------------------------------------------------
20   [cost,ind] = sort(cost);    % sorts costs from best to worst
21   Gene = Gene(ind(1:M2),:);   % sorts Gene according to costs
22                               % and discards bottom half list
23   % --- Mate
24   cross = ceil((MN-1)*rand(M2,1));  % selects random cross over points
25   % --- Pairs genes and swaps binary digits to form offsprings
26   for ic = 1:2:ceil((Pop-1)/2)-1
27     crosst = cross(ic);
28     Gene(M2+ic,1:crosst) = Gene(ic,1:crosst);          % offspring #1
29     Gene(M2+ic,crosst+1:MN) = Gene(ic+1,crosst+1:MN);
30     Gene(M2+ic+1,1:crosst) = Gene(ic+1,1:crosst);      % offspring #2
31     Gene(M2+ic+1,crosst+1:MN) = Gene(ic,crosst+1:MN);
32   end
33   % --- Mutate (Here, we set the num. of mutated bits to be one per iteration)
34   ix = ceil(Pop*rand);        % random gene
35   iy = ceil(MN*rand);         % random bit in gene
36   Gene(ix,iy) = 1-Gene(ix,iy); % mutate bit iy in gene ix
37   end % for ib=1:last
38
                    ———————————————————————————— End of Code ————————————————————————————
```

After the first "for" statement, a function call must be inserted by the user, to calculate the cost of the chromosomes. Since the chromosomes are encoded in binary, the function must translate the binary chromosomes into continuous parameters before calculating the cost of the chromosomes. The number of chromosomes, number of bits per chromosome and number of iterations are set at the beginning of the program. From above, the use of a genetic algorithm requires determination of the following fundamental items: chromosome representation, selection function, reproduction function, creation of the initial population (i.e. M chromosomes mentioned before), termination criteria and the evaluation cost function.

For any GA, a chromosome representation scheme is needed. It determines how the problem is structured in the GA and how the genetic operators (crossover and mutation) are used. Each chromosome is made up of a sequence of genes from a certain alphabet, which could consist of binary digits (0 or 1), floating point numbers, integers, symbols (i.e., A, B, C, D), matrices, etc. Originally, the alphabet was limited to binary digits which we have discussed. Recently many investigations have shown that more natural representations are more efficient and produce better solutions[21].

One useful representation of an individual or chromosome for function optimization involves genes or variables from an alphabet of floating point numbers with values within the variables' lower and upper bounds. It initialized the parameters as

$$v_n = (v_{\max} - v_{\min}) \times rand + v_{\min}$$

where $rand$ is a random number between 0 and 1 created by the computer. As we can see, the floating pointing representation is more natural and is a real-valued representation. Accordingly, selection and reproduction functions are determined by the representation, which is discussed carefully in [20]. Michalewicz has done extensive experimentation comparing real-valued GA and binary GA and has shown that the real-valued GA is an order of magnitude more efficient in terms of CPU time, and that a real-valued representation moves the problem closer to the problem representation, which offers higher precision with more consistent results across replications[21].

So the following example using GA involves the floating point representation.

2.8.2 Optimized Design of Planar Array by Using the Combination of GA and Fast Fourier Transform (FFT)

A. The correspondence between GA characteristic elements and array parameters

To use GA in the synthesis of an antenna array, it is required to find the correspondence between GA characteristic elements and array parameters, such as

- Gene: Encoding of the current amplitude and phase values of the array element.

- Chromosome: Encoding of the current amplitude and phase values of the array.

- Population: A set of solutions (the number of the solutions corresponding to the scale of the population).

B. Array factor and fitness function

As is given in (2.7.1), the array factor of a $M \times N$ rectangular grid array may be written as

$$AF(\theta, \varphi) = \sum_{m=1}^{M} \sum_{n=1}^{N} \left\{ I_{mn} e^{j[(n-1)kd_x(\sin\theta\cos\varphi - \sin\theta_0\cos\varphi_0) + \Delta\phi_{mn}]} \right. $$

$$\left. \cdot e^{j[(m-1)kd_y(\sin\theta\sin\varphi - \sin\theta_0\sin\varphi_0)]} \right\} \qquad (2.8.5)$$

where, I_{mn} is the excitation amplitude of element (m,n); $\Delta\phi_{mn}$ is the initial phase relative to element $(1,1)$, (θ_0,φ_0) corresponds to beam direction.

Assume that there are Q sampling points of the radiation pattern, the fitness function is defined as

$$fitness = \frac{1}{1 + \sum_{i=1}^{Q} P_0\,|(S_{di} - S_{ci})|} \tag{2.8.6}$$

where P_0 is the penalty constant with the value within $(0,1)$, S_{di} is the expected value and S_{ci} is the computed value.

C. The application of FFT[22]

The computation of S_{ci} satisfying sampling theorem takes a lot of computer time. FFT is a good tool to reduce the computer time significantly. 2D FFT is defined as

$$X(p,q) = \sum_{m=1}^{M}\sum_{n=1}^{N} x(m,n)e^{-j2\pi(p-1)(\frac{n-1}{N})}e^{-j2\pi(q-1)(\frac{m-1}{M})} \tag{2.8.7}$$

Comparing (2.8.7) with (2.8.5), we have

$$S(p,q) = \sum_{m=1}^{M}\left[\sum_{n=1}^{N} I'_{mn}e^{j(n-1)\frac{2\pi}{N}p}\cdot e^{j(m-1)\frac{2\pi}{M}q}\right] \tag{2.8.8}$$

where

$$I'_{mn} = I_{mn}e^{j\Delta\phi_{mn}}e^{-j(n-1)kd_x\sin\theta_0\cos\varphi_0}e^{-j(m-1)kd_y\sin\theta_0\sin\varphi_0}$$

$$0 \leqslant p \leqslant N-1, \quad 0 \leqslant q \leqslant M-1$$

The correspondence between the (θ,φ) domain and the (p,q) domain is $p \leftrightarrow kd_x\sin\theta\cos\varphi$, $q \leftrightarrow kd_y\sin\theta\sin\varphi$. Thus, FFT may be carried out in the (p,q) domain and it can be transformed to the (θ,φ) domain through the mapping between these two domains.

Padding zeros to the series $x(m,n)$ we have the extended series $L, K(L > M, K > N)$. When L, K are the integer power of 2, the base 2 FFT is applied, otherwise, the mixed base FFT will be applied.

D. Example

Consider a rectangular planar array with Taylor distribution. The number of elements is $M \times N = 8 \times 8$, $dx = dy = 0.5\lambda$, the side lobe level is -20dB, $L = K = 64$, $N_n = 3$. The radiation pattern in the (p,q) domain, the corresponding current distribution and the 2D radiation patterns for $q = 0$, $p = 0$ are shown in Figure 2.21.

In the example, the floating coding is used. Thus, a chromosome representation has 128 columns including 64 for the amplitudes and 64 for the phases. The new population is chosen based on the fitness. The initial population is chosen to be 130, p_0 to be 0.01. So the population has 130 rows of chromosomes.

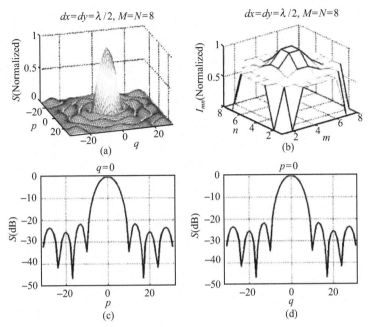

Figure 2.21   Rectangular planar array with Taylor distribution. (a) radiation pattern;
(b) current distribution; (c) $q = 0$ radiation pattern; (d) $p = 0$ radiation pattern.

(After Wang, Fang, and Sheng [22], © 2003 Wiley)

Up to 5000 generations, the radiation pattern, current distribution and the 2D radiation patterns for $q = 0$, $p = 0$ are shown in Figure 2.22. The maximum fitness versus number of

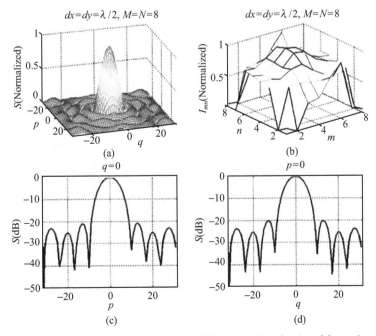

Figure 2.22   Results from GA. (a) radiation pattern; (b) current distribution; (c) $q = 0$ radiation pattern;
(d) $p = 0$ radiation pattern. (After Wang, Fang, and Sheng [22], © 2003 Wiley)

generations is shown in Figure 2.23. It is seen that when the number of generations reaches 5000, good results are obtained. In this example, the computation in the (θ, φ) domain and that in the (p, q) domain by FFT are compared and the ratio of computer time reduction is about $1/400$.

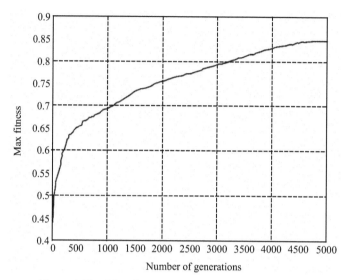

Figure 2.23   Max-fitness versus number of generations.

Bibliography

[1] R. E. Collin and F. J. Zucker, *Antenna Theory*, Part 1, McGraw-Hill Book Company, 1969.

[2] R. S. Elliott, *Antenna Theory and Design*, Prentice-Hall, Inc., 1981.

[3] C. A. Balanis, *Antenna Theory–Analysis and Design* (second edition), John Wiley & Sons, Inc., 1997.

[4] R. C. Hansen Edited, *Microwave Scanning Antennas*, vol. 1, Peninsula Publishing, 1964.

[5] S. M. Lin, *Mathematical Methods for the Design of Microwave Antennas* (in Chinese), North West University of Technology, 1983.

[6] M. G. Wang, S. W. Lu, and R. X. Liu, *Analysis and Synthesis of Array Antenna*, Xidian University Press, 1989.

[7] J. L. Butler, "Digital, matrix, and intermetiate frequency scanning," in R.C.Hansen, edited, *Microwave Scanning Arrays*, Academic Press, 1966.

[8] J. Litva and T. K. Y. Lo, *Digital Beamforming in Wireless Communications*, Artech House Publishers, 1996.

[9] W. X. Sheng, J. Zhou, D. G. Fang, and Y. Chris Gu, "Super-resolution DOA estimation in switch beam smart antenna," *Proceedings of 5th International symposium on Antennas, Propagation and EM Theory*, Beijing, 2000, pp. 603–606.

[10] J. Li, D. Zheng and P. Stotica, "Angle and waveform estimation via RELAX," *IEEE Trans On Aerospace and Electronic System*, vol. 33, no. 3, pp. 1077–1087, Jul., 1997.

[11] W. X. Sheng, D. G. Fang, D. J. Li, and P. K. Liang, "Angular superresolution for phased antenna array by phase weighting," *IEEE Trans. on Aerospace and Electronic Systems*, vol. 37. no. 4, pp. 32–40, Oct., 2001.

[12] D. G. Fang, W. X. Sheng, C. Zhang, and Z. Li, "Comparative study of two approaches in terms of improving cross range resolution," *IEEE Int. Symp. Antennas and Propagation*, 1997, pp. 2438–2442.

[13] W. X. Sheng, D. G. Fang, J. Y. Sun and S. H. Guo, "Angular super-resolution for scanning antenna by angle weighting method," *Proceedings of Cross Strait Tri-Regional Radio Science and Wireless Technology Conference*, Dec. 2000, pp. 79–82.

[14] B. D. Steinberg, *Principles of Aperture and Array System Design*, John Wiley & Sons, 1976.

[15] J. Treichler and B. Agee, "A new approach to multipath correction of constant modulus signals," *IEEE Trans. ASSP*, vol. 31, pp. 459-472, Apr., 1983.

[16] Y. Guo, D. G. Fang, and C. H. Liang, "Simple multiple interference direction-finding based on CMA under severe environment" *Chinese J. of Electronics*, vol. 10, no. 4. pp. 544–547, Oct., 2001.

[17] Y. Guo, D. G. Fang, N. C. Wang, and C. H. Liang, "A preprocessing LS-CMA in highly corruptive environment," *J. of Electronics*, vol. 19, no. 3, pp. 248–254, 2002.

[18] R. L. Haupt, "An introduction to genetic algorithms for electromagnetics," *IEEE Antennas Propagation Magazine*, vol. 37, no. 2, pp. 7–15, Apr., 1995.

[19] J. M. Johnson and Y. R. Samii, "Genetic Algorithm in engineering electromagnetics," *ibid*, vol. 39, no. 4, pp. 7–25, Aug., 1997.

[20] Y. R. Samii and E. Michielssen, *Electromagnetic Optimization by Genetic Algorithm*, John Wiley & Sons, Inc., 1999.

[21] Z. Michalewicz, *Genetic Algorithms + Data Structures = Evolution Programs*, AI Series, Springer-Verlag, New York, 1994.

[22] L. L. Wang, D. G. Fang and W. X. Sheng, "Combination of Genetic Algorithm (GA) and Fast Fourier Transform (FFT) for synthesis of arrays," *Microwave and Optical Technology Letters*, vol. 37, no. 1, pp. 56–59, Apr., 2003.

[23] T. K. Sarkar and N. Sangruzi, "An adaptive nulling system for a narrow-band signal with a look-direction constraint utilizing the CG method," *IEEE Trans. on Antennas Propagat.*, vol. 37, no. 7, pp. 940–944, 1989.

[24] G. Y. Zhang, *Phased Array Antenna Radar System*, Defense Industry Press, 1994.

[25] H. Wang, D. G. Fang, and Y. L. Chow, "Grating lobe reduction in a phased array of limited scanning," *IEEE Trans. on Antennas and Propagat.*, vol. 56, no. 6, pp. 1581–1586, Jun., 2008.

Problems

2.1 Prove the following two identities by using the concept of array factor. (E. A. Guillemin, Mathematics of Circuit Analysis, p. 437, 1949)

$$\sum_{n=1}^{N/2} 2\cos\frac{2n-1}{2}\psi = \frac{\sin(N\psi/2)}{\sin(\psi/2)}, \qquad \text{when } N \text{ is an even number;}$$

$$1 + \sum_{n=2}^{(N+1)/2} 2\cos(n-1)\psi = \frac{\sin(N\psi/2)}{\sin(\psi/2)}, \quad \text{when } N \text{ is an odd number.}$$

2.2 Prove that the radiation pattern $F(\theta)$ of the following two-element array is

$$F(\theta) = \left| \cos(45° + \theta) + e^{j(\alpha d \sin\theta + \beta)} \cos(45° - \theta) \right|$$

In this array, the radiation pattern of the element is a cosine function, and the element 2 has electric phase β ahead of element 1.

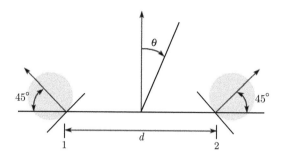

2.3 For a in-phase six-element linear array with spacing d and binomial current distribution 1:5:10:10:5:1, use the principle of pattern multiplication given in (1.3.34) to prove that the array factor $F(\theta)$ is

$$F(\theta) = \left[\cos\left(\frac{1}{2}\alpha d \cos\theta\right)\right]^5$$

2.4 Give the polar diagrams for the array factor of a two-element array for some values of d and β using the method given in Figure 2.2.

$$\beta = 0: \qquad d = 0, d = \lambda/4, d = \lambda/2, d = 3\lambda/4, d = \lambda$$
$$\beta = -\frac{\pi}{2}: \quad d = 0, d = \lambda/4, d = \lambda/2, d = 3\lambda/4, d = \lambda$$
$$\beta = -\pi: \quad d = 0, d = \lambda/4, d = \lambda/2, d = 3\lambda/4, d = \lambda$$

2.5 Prove (2.4.11-b) and (2.4.11-d). (Hint. Put the phase reference point at the center of the linear array.) For a four-element linear array, the side lobe level is specified as -15dB, using both the roots matching and coefficients matching methods to prove that the current distribution is 1:1.332:1.332:1. If spacing d is $\lambda/2$, give the radiation pattern.

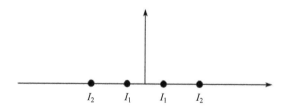

2.6 Find the current distribution of the Prob. 2.5 by using formula (2.4.11-c); do the same for the example of 5 element DT array given in Section 2.4.2 by using formula (2.4.11-a)

2.7 Write a general program for the design of DT array based on the root matching principle. Use this program to check the result of Prob. 2.5.

2.8 Design a -20dB Taylor distribution 13-element array with $\bar{n} = 5$ and $d = \lambda/2$ by using both the conventional sampling method based on (2.4.34) and root matching method. As is done in Section 2.4.2, give the pattern in terms of ψ.

2.9 Write a general program for the design of a Taylor array based on the root matching principle. Use this program to check the results of Prob. 2.8.

2.10 Check the values of I_n in Table 2.3.

2.11   1. If in Figure 2.13, $d = \lambda/2$, when the signal goes into port y_1, find X, that is, the phase distribution along a four-element array and prove that the radiation direction **r** makes an angle $\varphi = \cos^{-1} 3/4$ with the x axis;

2. If the direction of plane wave incident on the four-element array is along $-\mathbf{r}$ direction, prove that the output comes only from port y_1.

2.12   1. If $X = FY$, where F is an FFT matrix for four elements

$$F = \begin{pmatrix} 1 & 1 & 1 & 1 \\ 1 & e^{-j90} & e^{-j180} & e^{-j270} \\ 1 & e^{-j180} & e^{j0} & e^{-j180} \\ 1 & e^{-j270} & e^{-j180} & e^{-j90} \end{pmatrix}$$

and the number in the matrix is in degrees. Show that F may form four mutually orthogonal overlapped beams and find the directions of the beams.

2. If $F' = [\lambda]F$ is introduced, where

$$[\lambda] = \begin{pmatrix} e^{-j135^\circ} & 0 & 0 & 0 \\ 0 & e^{-j90^\circ} & 0 & 0 \\ 0 & 0 & e^{-j45^\circ} & 0 \\ 0 & 0 & 0 & e^{j0^\circ} \end{pmatrix}$$

prove that the same beam directions as those produced by Butler matrix can be obtained.

2.13   For a three-element linear array with spacing d, the directions of the incoming signal S and interference I make angle ϕ_s and ϕ_I with the array axis z respectively, where ϕ_I is unknown. Assume $V_i(i = 1, 2, 3)$ is the received voltage on each element, and W_1, W_2 are the weighting coefficients for the corresponding elements

1. Prove $a = V_1 - V_2\beta^{-1}$ and $b = V_2 - V_3\beta^{-1}$ have nothing to do with signal S, where $\beta = \exp(jkd\cos\phi_s)$ and element 1 is taken as the phase reference.

2. Assume $W_1 + \beta W_2 = C$, C is a constant related to the signal, prove if $aW_1 + bW_2 = 0$, we may have $W_1I + W_2I\exp(jkd\cos\phi_I) = 0$. It means that this weighting produces the null of the array pattern at the direction of interference.

3. In this case, prove $W_1V_1 + W_2V_2 = SC$, that is the sum of the received voltages from element 1 and 2 under the corresponding weights is related only with the signal. (For general proof, see [23].)

2.14   For a linear aperture $(-a/2, a/2)$, the required amplitude distribution is $\cos(\pi x/a)$. If eleven discrete elements are used to realize the density taper, find the location of each element.

2.15   For an N-element uniform linear array with spacing d, assume that $Nkd/2$ is large, to prove $D \approx 2Nd/\lambda$, where D is the directivity at $\theta = 90^\circ$ defined in (1.4.1).

Hint:

1. For linear array

$$D = \frac{4\pi f_{\max}^2(\theta, \phi)}{\int_0^{2\pi}\int_0^\pi f^2(\theta, \phi)\sin\theta d\theta d\phi} = \frac{2f_{\max}^2(\theta)}{\int_0^\pi f^2(\theta, \phi)\sin\theta d\theta}$$

2.

$$\int_0^\pi \left[\frac{\sin(Nkd\cos\theta/2)}{Nkd\cos\theta/2}\right]^2 \sin\theta d\theta = \frac{2}{Nkd}\int_{-Nkd/2}^{Nkd/2}\left[\frac{\sin(z)}{z}\right]^2 dz$$

and

$$\int_{-\infty}^\infty \left[\frac{\sin(z)}{z}\right]^2 dz = \pi$$

where $z = Nkd\cos\theta/2$, θ is defined in Figure 2.1.

2.16   For a multi-beam antenna array, the reciprocity for antenna patterns is discussed in Section 2.6.1. For a simple two-element linear array, prove the reciprocity for antenna patterns and discuss the conservation of energy in the following two cases:

1. The power divider is waveguide H-plane T-junction, the feeding port is H-arm (port3) and is matched, that is, the S parameter S_{33} equals to zero;

2. The power divider is waveguide magic T, the feeding port is H-arm, E-arm is terminated by a matched load. (Waveguide magic T corresponds to the Wilkinson microstrip power divider in Figure 6.10.)

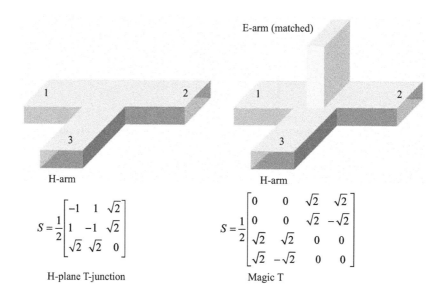

$$S = \frac{1}{2}\begin{bmatrix} -1 & 1 & \sqrt{2} \\ 1 & -1 & \sqrt{2} \\ \sqrt{2} & \sqrt{2} & 0 \end{bmatrix} \qquad S = \frac{1}{2}\begin{bmatrix} 0 & 0 & \sqrt{2} & \sqrt{2} \\ 0 & 0 & \sqrt{2} & -\sqrt{2} \\ \sqrt{2} & \sqrt{2} & 0 & 0 \\ \sqrt{2} & -\sqrt{2} & 0 & 0 \end{bmatrix}$$

H-plane T-junction   Magic T

2.17   In Figure 2.17. if $d_x = d_y = d$, consider two cases:

1. $M = 2, N = 1$, there are three elements on xoy plane located at $(0, 0)$, $(d, 2d)$, $(2d, 0)$ with phases 0, β, 2β respectively. Prove $\phi_0 = 0, \theta_0 = \sin^{-1}(-\beta/kd)$.
2. $M = 1, N = 2$, there are three elements on xoy plane located at $(0, 0)$, $(2d, d)$, $(0, 2d)$ with phases 0, β, 2β respectively. Prove $\phi_0 = \pi/2, \theta_0 = \sin^{-1}(-\beta/kd)$

2.18   In (2.3.4), two sub-arrays are formed by four left elements and four right elements. If two sub-arrays are formed by four odd-numbered elements and four even-numbered elements, prove that (2.3.4) becomes the following expression: $\left(\dfrac{\sin 4kd\cos\theta}{4\sin kd\cos\theta} \right) \cdot \left(\dfrac{\sin kd\cos\theta}{2\sin 0.5kd\cos\theta} \right)$

CHAPTER 3

Microstrip Patch Antennas

3.1   Introduction

The microstrip antennas and arrays have been widely used in recent years because of their good characteristics; they are electrically thin, lightweight, low cost, conformable and so on. However, the electrical performance of the basic microstrip antenna or array suffers from a number of serious drawbacks, including very narrow bandwidth, high feed network losses, high cross polarization, and low power handling capacity. With progress in both theory and technology, some of these drawbacks have been overcome, or at least alleviated to some extent. The rapidly developing markets, especially in personal communication systems (PCS), mobile satellite communications, direct broadcast (DBS), wireless local area networks(WLAN) and intelligent vehicle highway systems (IVHS), suggest that the demand for microstrip antennas and arrays will increase even further. In the meantime, the increasing demand calls for the further development of them.

A microstrip patch antenna element is very useful and also is the basic element of an array. In this chapter, we will focus on the description of the principle of the microstrip patch antenna. The approximate analysis introduced in this chapter is useful in two cases, one is to work out the design method which is good enough in some engineering application; the other is to derive the coarse models which are necessary in optimization through space mapping technique. The full wave analysis will be briefly introduced. All of them are useful for a good understanding of the recent development and innovative design of microstrip patch antennas.

3.2   Cavity Model and Transmission Line Model

The microstrip rectangular patch antenna may be considered to be a magnetic wall cavity. This is an approximate model, which in principle leads to a reactive input impedance, and it does not radiate any power. To account for radiation, a loss mechanism has to be introduced. Then we consider the perimeter of the patch as radiation slot. The radiation resistance caused by the radiation, and the loss resistance caused by the conduction-dielectric losses, result in the input impedance being complex. Although the separate consideration of the resistance and reactance implies some contradiction, this model has been verified by experiments to be an acceptable one, when the electric thickness is small.

3.2.1   Field Distribution from Cavity Model

The representation of the field configuration (mode) for a rectangular microstrip patch is similar to that of the waveguide cavity. The difference is that in the patch case the electric walls are only the top and bottom; the other walls are magnetic ones. The commonly used mode is TM_{010}^{z} as shown in Figure 3.1.

The excitation current density is $\mathbf{J}$

$$\mathbf{J} = \begin{cases} \hat{\mathbf{a}}_z, & c < x < d, \ y = 0 \\ 0, & \text{elsewhere} \end{cases} \tag{3.2.1}$$

which is similar to that of a waveguide cavity. According to the boundary condition, the fields in a magnetic wall cavity may be expressed as

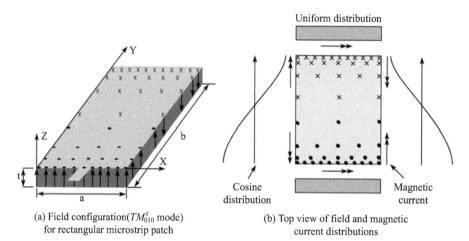

(a) Field configuration(TM_{010}^z mode)
for rectangular microstrip patch

(b) Top view of field and magnetic
current distributions

Figure 3.1 TM_{010}^z mode for a microstrip patch antenna.

$$E_{z,m} = A_m \cos \frac{m\pi x}{a} \cos \beta_m (y - b) \tag{3.2.2-a}$$

From Maxwell's equations

$$\mathbf{H} = \frac{j}{\omega \mu} \nabla \times \mathbf{E} = \frac{-j}{\omega \mu} \hat{\mathbf{a}}_z \times \nabla \mathbf{E}_z$$

we may have $H_{x,m}$ and $H_{y,m}$

$$H_{x,m} = -j A_m \frac{\beta_m}{\omega \mu_0} \cos \frac{m\pi x}{a} \sin \beta_m (y - b) \tag{3.2.2-b}$$

$$H_{y,m} = j A_m \frac{m\pi/a}{\omega \mu_0} \sin \frac{m\pi x}{a} \cos \beta_m (y - b) \tag{3.2.2-c}$$

where A_m is a constant which can be determined from the excitation, and β_m is the eigenvalue along y

$$\beta_m = \left[\omega^2 \epsilon \mu_0 - \left(\frac{m\pi}{a} \right)^2 \right]^{1/2} \tag{3.2.3}$$

Assuming a unit excitation at $x = x'$, from the boundary condition on $y = 0$

$$\delta (x - x') = \sum_{m=0}^{\infty} \frac{-j A_m \beta_m}{\omega \mu_0} \cos \frac{m\pi x}{a} \sin \beta_m b \tag{3.2.4}$$

where δ is the delta function. Multiplying both sides of (3.2.4) by $\cos (m\pi x/a)$, carrying out the integration between $x = 0$ and $x = a$ and applying the property of delta function and the orthogonality yields

$$\cos \frac{m\pi x'}{a} = -j A_m \frac{\beta_m}{\omega \mu_0} \sin \beta_m b \int_0^a \cos^2 \frac{m\pi x}{a} dx \tag{3.2.5}$$

$$A_m = \frac{\epsilon_{0m} \omega \mu_0 \cos \dfrac{m\pi x'}{a}}{-j \beta_m a \sin \beta_m b} \tag{3.2.6}$$

where $\epsilon_{0m} = 1$, when $m = 0$; $\epsilon_{0m} = 2$, when $m > 0$. If the excitation extends from $x' = c$ to $x' = d$, then

$$A_m = \frac{j\epsilon_{0m}\omega\mu_0}{a\beta_m \sin\beta_m b} \int_c^d \cos\frac{m\pi x'}{a} dx'$$

$$= \frac{j4\omega\mu_0}{m\pi\beta_m \sin\beta_m b} \sin\frac{m\pi(d-c)}{2a} \cos\frac{m\pi(d+c)}{2a}, m > 0 \qquad (3.2.7)$$

$$A_0 = \frac{j(d-c)}{a}\sqrt{\frac{\mu}{\epsilon}}\frac{1}{\sin\left(\omega\sqrt{\epsilon\mu_0}\,b\right)}, \qquad\qquad m = 0 \qquad (3.2.8)$$

Finally, we have[1]

$$E_z = \sum_{m=0}^{\infty} A_m \cos\frac{m\pi x}{a} \cos\beta_m(y-b) \qquad (3.2.9\text{-a})$$

$$H_x = \frac{1}{j\omega\mu}\sum_{m=0}^{\infty} A_m\beta_m \cos\frac{m\pi x}{a}\sin\beta_m(y-b) \qquad (3.2.9\text{-b})$$

$$H_y = -\frac{1}{j\omega\mu}\sum_{m=0}^{\infty} A_m \frac{m\pi}{a}\sin\frac{m\pi x}{a}\cos\beta_m(y-b) \qquad (3.2.9\text{-c})$$

Now we will focus on the examination of E_z. The series in (3.2.9-a) converges rapidly but possesses less physical insight. There is an alternative expression which is based on the following identity[2].

$$\frac{\cos(z-\pi)p}{\sin\pi p} = \frac{1}{\pi p} + \frac{2p}{\pi}\sum_{n=1}^{\infty}\frac{\cos nz}{p^2 - n^2} \qquad (3.2.10)$$

Let $z = \pi y/b$, $p = \beta_m b/\pi$, (3.2.10) becomes

$$\frac{\cos\beta_m(y-b)}{\sin\beta_m b} = \frac{1}{\beta_m b} + \frac{\beta_m b}{\pi^2}\sum_{n=1}^{\infty}\frac{2\cos(n\pi y)/b}{\left(\frac{\beta_m b}{\pi}\right)^2 - n^2} \qquad (3.2.11)$$

or

$$\frac{\cos\beta_m(y-b)}{\beta_m \sin\beta_m b} = \frac{1}{b}\left[\frac{1}{k^2 - k_{m0}^2} + \sum_{n=1}^{\infty}\frac{2\cos(n\pi y)/b}{k^2 - k_{mn}^2}\right] \qquad (3.2.12)$$

where $k^2 = \omega^2\epsilon\mu$, $k_{mn}^2 = (m\pi/a)^2 + (n\pi/b)^2$. When $m = 0$,

$$\frac{\cos k(y-b)}{\sin kb} = \frac{1}{b}\left[\frac{1}{k} + 2k\sum_{n=1}^{\infty}\frac{\cos(n\pi y/b)}{k^2 - k_{0n}^2}\right] \qquad (3.2.13)$$

The physical meaning of the mathematical identities in (3.2.12) and (3.2.13) is the resonance mode expansion in patch antenna. Actually, each term in the series represents a resonance mode. Rewrite (3.2.9-a) as

$$E_z = A_0 \cos k(y-b) + \sum_{m=1}^{\infty} A_m \cos\frac{m\pi x}{a}\cos\beta_m(y-b)$$

$$= \frac{j(d-c)}{a}\sqrt{\frac{\mu}{\epsilon}}\left[\frac{\cos k(y-b)}{\sin kb}\right] + \sum_{m=1}^{\infty}\frac{j4\omega\mu}{m\pi}\cos\frac{m\pi x}{a}\left[\frac{\cos\beta_m(y-b)}{\beta_m \sin\beta_m b}\right]$$

$$\cdot \sin\frac{m\pi(d-c)}{2a}\cos\frac{m\pi(d+c)}{2a}$$

$$(3.2.14)$$

Substituting (3.2.12) and (3.2.13) into (3.2.14) results in[1]

$$
E_z = j\omega\mu\left\{\frac{d-c}{k^2ab} + \sum_{n=1}^{\infty}\frac{2\,(d-c)}{ab\,(k^2-k_{0n}^2)}\cos\frac{n\pi y}{b}\right.
$$
$$
+ \sum_{m=1}^{\infty}\frac{[4\sin m\pi\,(d-c)/2a]\,[\cos m\pi\,(d+c)/2a]}{m\pi b\,(k^2-k_{m0}^2)}\cos\frac{m\pi x}{a}
$$
$$
\left.+ \sum_{m,n=1}^{\infty}\frac{[8\sin m\pi\,(d-c)/2a]\,[\cos m\pi\,(d+c)/2a]}{m\pi b\,(k^2-k_{mn}^2)}\cos\frac{m\pi x}{a}\cos\frac{n\pi y}{b}\right\}
$$

$$(3.2.15)$$

where

$$
k_{mn} = \sqrt{\left(\frac{m\pi}{a}\right)^2 + \left(\frac{n\pi}{b}\right)^2} \tag{3.2.16}
$$

Alternatively, (3.2.15) may be written as

$$
E_z = 2j\omega\mu\sum_{m,n=0}^{\infty}\frac{\epsilon_{0m}\epsilon_{0n}\sin\left[\frac{m\pi}{2a}(d-c)\right]\cos\left[\frac{m\pi}{2a}(d+c)\right]}{m\pi b\,(k^2-k_{mn}^2)}\cos\frac{m\pi x}{a}\cos\frac{n\pi y}{b}
$$
$$
= jk\eta\sum_{m,n=0}^{\infty}\frac{\omega\epsilon_{0m}\epsilon_{0n}\cos\dfrac{m\pi x_0}{a}}{ab\,(k^2-k_{mn}^2)}\mathrm{j}_0\left(\frac{m\pi w}{2a}\right)\cos\frac{m\pi x}{a}\cos\frac{n\pi y}{b} \tag{3.2.17}
$$

where $\mathrm{j}_0(x) = \sin(x)/x$. The source point is $(x_0,0)$, $x_0 = (d+c)/2$, $w = d-c$ is the width of feed segment. If in (3.2.1) $\mathbf{J}$ is defined as $(I/w)\hat{\mathbf{a}}_z$, (3.2.17) becomes

$$
E_z = jk\eta I\sum_{m,n=0}^{\infty}\frac{\epsilon_{0m}\epsilon_{0n}\cos\dfrac{m\pi x_0}{a}}{ab\,(k^2-k_{mn}^2)}\mathrm{j}_0\left(\frac{m\pi w}{2a}\right)\cos\frac{m\pi x}{a}\cos\frac{n\pi y}{b} \tag{3.2.18}
$$

where $\mathrm{j}_0(x) = \sin(x)/x$, $\epsilon_{0p} = 1$ for $p=0$ and 2 for $p \neq 0$, and I is a constant. When the microstrip patch antenna is fed at a point (x_0, y_0), interior to the patch by a coaxial cable, it is still convenient to think of the antenna as being fed by a uniform strip of vertically oriented electric current. In this case, E_z may be expressed as

$$
E_z = jk\eta I\sum_{m,n=0}^{\infty}\frac{\epsilon_{0m}\epsilon_{0n}\cos\dfrac{m\pi x_0}{a}\cos\dfrac{n\pi y_0}{b}}{ab\,(k^2-k_{mn}^2)}\mathrm{j}_0\left(\frac{m\pi w}{2a}\right)\cos\frac{m\pi x}{a}\cos\frac{n\pi y}{b}
$$

or[3][4]

$$
E_z = \frac{jk\eta I}{ab}\sum_{m,n=0}^{\infty}\frac{\epsilon_{0m}\epsilon_{0n}\psi_{mn}\,(x,y)\,\psi_{mn}\,(x_0,y_0)}{k^2-k_{mn}^2}\mathrm{j}_0\left(\frac{m\pi w}{2a}\right) \tag{3.2.19}
$$

where

$$
\psi_{mn}\,(x_0,y_0) = \cos\,(m\pi x_0/a)\cos\,(n\pi y_0/b)
$$
$$
\psi_{mn}\,(x,y) = \cos\,(m\pi x/a)\cos\,(n\pi y/b)
$$

where ω is the effective feed width and will be discussed later. For dominant mode TM_{010}^z, (3.2.19) reduces to

$$
E_z = \frac{2jk\eta I}{ab\left[k^2-(\pi/b)^2\right]}\cos\frac{\pi y}{b} \tag{3.2.20}
$$

The fields are cosine distributed along y and constant along x as shown in Figure 3.1(b).

3.2.2 Radiation Pattern

For the TM_{010}^z mode rectangular patch antenna shown in Figure 3.1, the field distribution is given by (3.2.20). It is seen that the radiation from the magnetic currents $\mathbf{E} \times \hat{\mathbf{n}}$ along y at $x = 0$ and $x = a$ is very small and those along x at $y = 0$ and $y = b$ are mainly responsible for the radiation. Consequently, the radiation from this antenna may be considered as that from two magnetic current slots both in the cavity model and in the transmission line model. In the cavity model, two slots are the walls of the cavity, which in the transmission line model are the loads of a transmission line with length b. For each slot, the length is a and the width is t. As shown in (1.3.51), (1.3.52) and Prob. 1.3, the representation of the radiation field is related to the coordinate system. In Prob. 1.3, only E_ϕ is nonzero. This representation is simpler and will be adopted here. Considering the effect of the ground plane, the magnitude of current should be doubled. From the results given in Prob. 1.3 and defining $E_x^0 = V/t$, where V is the peak voltage between the patch and the ground plane at the port, we have

$$E_\phi = -\frac{je^{-jkr}aV}{\lambda r}\frac{\sin(kt\sin\theta\cos\phi/2)}{kt\sin\theta\cos\phi/2}\frac{\sin(ka\cos\theta/2)}{ka\cos\theta/2}\sin\theta \qquad (3.2.21)$$

When $kt \ll 1$, (3.2.21) becomes[5]

$$E_\phi = -j\frac{Ve^{-jkr}}{\pi r}\frac{\sin(ka\cos\theta/2)}{\cos\theta}\sin\theta \qquad (3.2.22)$$

For the radiation from the two slots at $x = 0$ and $x = b$, the space factor, that is, the array factor here, should be multiplied. Similar to (1.3.38)

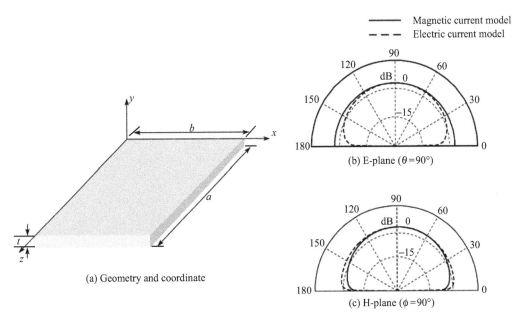

(a) Geometry and coordinate

(b) E-plane ($\theta = 90°$)

(c) H-plane ($\phi = 90°$)

—— Magnetic current model
– – – Electric current model

Figure 3.2   Radiation pattern of a microstrip patch antenna.

$$F_2' = \int_0^b [\delta(x' - 0) + \delta(x' - b)]e^{-jkx' \sin\theta \cos\phi} dx'$$

$$= 1 + e^{-jkb \sin\theta \cos\phi} = 2je^{-j(kb \sin\theta \cos\phi)/2} \cos\left(\frac{kb}{2} \sin\theta \cos\phi\right)$$

After removing the phase factor, the array factor of these two slot elements, F_2, will be

$$F_2 = 2\cos\left(\frac{kb}{2}\sin\theta\cos\phi\right) \tag{3.2.23}$$

The total radiation pattern may be obtained from $F_2 E_\phi$, where E_ϕ is given in (3.2.21) or (3.2.22)[6]. The E-plane pattern ($\theta = 90°$) and H-plane pattern ($\phi = 90°$) are shown in Figures 3.2(b) and (c). It is seen that the main difference between the result from an electric current model (IE3D^{TM}) and that from a magnetic current model (formulas (3.2.21) and (3.2.22)) occurs near the grazing angle. It is due to the fact that the image of electric current is in the opposite direction and the image of magnetic current is in the same direction. Actually the microstrip substrate is finite and the measured patterns are in between.

3.2.3   Radiation Conductance

To determine the total radiated power from one slot, we have to integrate the real part of the Poynting vector over a hemisphere with a large radius. Since the slot radiates only on one side, the integration is carried out over the hemisphere. With the field in (3.2.22) where $kt \ll 1$ is assumed, we get the radiation power P_r

$$P_r = \frac{1}{2}\int_0^\pi \int_0^\pi \frac{|E_\phi|^2}{\eta} r^2 \sin\theta d\theta d\phi$$

$$= \frac{1}{2}\sqrt{\frac{\epsilon}{\mu}} \frac{V^2}{\pi} \int_0^\pi \frac{\sin^2(ka\cos\theta/2)}{\cos^2\theta} \sin^3\theta d\theta \tag{3.2.24}$$

To obtain more precise results from (3.2.24), formula (3.2.22) may be replaced by (3.2.21). The integration will be carried out using numerical methods. Since V is the voltage across the center of the slot, a radiation conductance may be defined as the conductance placed across the center of the slot, which will dissipate the same power as that radiated by the slot. From (3.2.24)

$$G_s = \frac{2P_r}{V^2} = \frac{1}{\pi}\sqrt{\frac{\epsilon}{\mu}} \int_0^\pi \frac{\sin^2(ka\cos\theta/2)}{\cos^2\theta} \sin^3\theta d\theta$$

$$= \frac{S}{120\pi^2} \tag{3.2.25}$$

Making a change of variable $t = \cos\theta$, then the integral S in (3.2.25) may be carried out analytically, giving[4]

$$G_s = \frac{1}{120\pi^2}\left[xSi(x) + \cos x - 2 + \frac{\sin x}{x}\right] \tag{3.2.26}$$

where

$$x = ka, \quad Si(x) = \int_0^x \frac{\sin u}{u} du$$

Making use of the series expansion and neglecting the high order terms, the approximated results are as follows

$$
G_s = \begin{cases} \dfrac{1}{90}\left(\dfrac{a}{\lambda}\right)^2, & a < 0.35\lambda \\[2ex] \dfrac{1}{120}\dfrac{a}{\lambda} - \dfrac{1}{60\pi^2}, & 0.35\lambda \leqslant a < 2\lambda \\[2ex] \dfrac{1}{120}\dfrac{a}{\lambda}, & 2\lambda \leqslant a \end{cases} \tag{3.2.27}
$$

where $a = W$ is the radiation edge and λ is the wavelength in free space. In Figure 3.1, dimension $b = L$ is a non-radiation edge and is usually $b \approx \pi/\beta$, that is, $\beta b = \pi$ approximately. Therefore, the magnetic currents on both slots are with the same phase and magnitude. We may consider the conductance G_r of the patch antenna to be

$$
G_r = 2G_s \tag{3.2.28}
$$

3.2.4 Input Impedance from Cavity Model

The input impedance is defined as $Z_{in} = V/I$, where V and I are their values at the input port. In Figure 3.1, the feed segment is from c to d. Defining E_{z0} as E_z averaged over the feed segment, the corresponding voltage is V_o and the relationship between V_o and E_{z0} is $V_0 = -tE_{z0}$.

$$
Z_{in} = \frac{V_0}{I} = -\frac{tE_{z0}}{I} \tag{3.2.29}
$$

where E_{z0} is the average value of E_z given in (3.2.19) along the feed segment, and

$$
E_{z0} = \frac{1}{w}\int_{x_0-w/2}^{x_0+w/2} E_z dx \tag{3.2.30}
$$

where $w = d - c$, $x_o = (d+c)/2$, as shown in Figure 3.1. To account for all the losses, k in (3.2.19) should be replaced by[3]

$$
k_{eff} = k_0\sqrt{\epsilon_r\left(1 - j\tan\delta_{eff}\right)} \tag{3.2.31}
$$

where

$$
\tan\delta_{eff} = \frac{1}{Q} = \frac{P}{2\omega W_e} = \frac{P_r + P_c + P_d + P_{sw}}{2\omega W_e}
$$
$$
= \frac{1}{Q_r} + \frac{1}{Q_c} + \frac{1}{Q_d} + \frac{1}{Q_{sw}} \tag{3.2.32}
$$

P is the total loss power. P_r, P_c, P_d, P_{sw} are radiation, conductor, dielectric and surface wave power loss respectively. Q_r, Q_d, Q_c, Q_{sw} are the corresponding Q related to the losses and W_e is the average electric energy stored in the cavity.

According to the definition of Q_r[7],

$$
Q_r = \omega\frac{W_m + W_e}{P_r}
$$

where W_m is the average magnetic energy stored in the cavity. At resonance, $W_m = W_e$ which gives

$$
Q_r = \frac{2\omega W_e}{P_r} \tag{3.2.33}
$$

Also we have[7]

$$W_e = \frac{\epsilon_0 \epsilon_r}{4} \int_0^a \int_0^b \int_0^t |E_{z0}|^2 dx\,dy\,dz \tag{3.2.34}$$

When calculating W_e, we do not consider the excitation mechanism. Only the eigen modes at resonance are considered, that is $E_z(x, y, z)$ may be written as

$$
\begin{aligned}
E_z(x, y, z) &= E \cos \frac{m\pi x}{a} \cos \frac{n\pi y}{b} \\
&= \frac{V}{t} \cos \frac{m\pi x}{a} \cos \frac{n\pi y}{b}
\end{aligned} \tag{3.2.35}
$$

where V is the peak voltage between patch and ground plane, the same as that in (3.2.22). Substituting (3.2.35) into (3.2.34) yields

$$W_e = \frac{\epsilon_0 \epsilon_r ab V^2}{4t\epsilon_{0m}\epsilon_{0n}} \tag{3.2.36}$$

For the microstrip antenna, (3.2.35) should be

$$G_r = \frac{2P_r}{V^2} \tag{3.2.37}$$

From (3.2.33), (3.2.36) and (3.2.37), we have

$$Q_r = \frac{2\omega W_e}{P_r} = \frac{\omega \epsilon_0 \epsilon_r ab}{t\epsilon_{0m}\epsilon_{0n}G_r} = \frac{\epsilon_r ab}{60\lambda t\epsilon_{0m}\epsilon_{0n}G_r} \tag{3.2.38}$$

where $G_r \approx 2G_s$, and G_s is given in (3.2.27). Since W_e and P depend on δ_{eff} in a very complex manner, strictly speaking, the solution for δ_{eff} in this nonlinear equation is very complex. Fortunately, in the above derivation, an accurate value of Q can still be obtained even though W_e and P are computed by simply using k for k_{eff}[3].

The conductor loss P_c may be computed as follows

$$
\begin{aligned}
P_c &= 2 \int_0^a \int_0^b \frac{1}{2} |J_s|^2 R_s dx\,dy = R \int_0^a \int_0^b |H|^2 dx\,dy \\
&= \frac{R_s W_m}{\mu_0 t/4}
\end{aligned} \tag{3.2.39}
$$

At resonance, $W_m = W_e$, R_s is the surface resistance of the conductor, $R_s = 1/(\sigma_c \Delta_c)$, σ_c is the conductivity of the conductor and Δ_c is the skin depth. Then we have

$$P_c = 2\omega \Delta_c \frac{W_e}{t} \tag{3.2.40}$$

and

$$Q_c = \frac{t}{\Delta_c}, \quad \Delta_c = \sqrt{\frac{2}{\omega \mu_0 \sigma_c}} = \frac{1}{\pi}\sqrt{\frac{\lambda_0}{120\sigma_c}} \tag{3.2.41}$$

The dielectric loss P_d is

$$P_d = \frac{1}{2}\sigma_d \int_0^a \int_0^b \int_0^t |E_z|^2 dx\,dy\,dz = \frac{2\sigma_d W_e}{\epsilon_0 \epsilon_r} \tag{3.2.42}$$

$$Q_d = \frac{2\omega W_e}{P_d} = \frac{\omega \epsilon_0 \epsilon_r}{\sigma_d} = \frac{1}{\tan \delta} \tag{3.2.43}$$

where $\sigma_d = \omega\epsilon'' + \sigma = \omega\epsilon_0\epsilon_r\tan\delta$. σ and σ_d are the conductivity and the effective conductivity of the dielectric respectively. In a lossy dielectric, $\epsilon = \epsilon' - j\epsilon''$, $\epsilon' = \epsilon_0\epsilon_r$ and $\omega\epsilon''$ is the dielectric damping[7]. The surface wave power loss may be computed by the approximate closed form obtained using numerical fitting[8]

$$P_{sw} = P_r\frac{1-\eta_s}{\eta_s} \tag{3.2.44}$$

$$\eta_s = 1 - 3.4H_e, \qquad H_e = \frac{t}{\lambda}\sqrt{\epsilon_r - 1}, \qquad t/\lambda < 0.06 \tag{3.2.45}$$

$$\eta_s = 1 - 3.4H_e + \frac{1600}{\epsilon_r^3}(H_e^3 - 100H_e^{5.6}), \qquad 0.06 < t/\lambda < 0.16 \tag{3.2.46}$$

$$Q_{sw} = \frac{\eta_s}{1-\eta_s}Q_r \tag{3.2.47}$$

From all the results given above, the Q of the cavity may be computed. Specifically, the Q for the TM^z_{010} mode rectangular cavity is

$$Q = \left[\frac{120\lambda_0 tG_r}{\epsilon_r ab(1-3.4H_e)} + \frac{1}{\pi t}\sqrt{\frac{\lambda_0}{120\sigma_c}} + \tan\delta\right]^{-1} \tag{3.2.48}$$

The input impedance then can be computed from (3.2.19) and (3.2.29). In case of the coaxial cable excitation, when using (3.2.19), w is the effective feed width. It is pointed out in [3] that the impedance is unaffected by the orientation of the feed strip. Thus, in the case of a rectangle, the strip was always taken to be parallel to the x axis. An effective width of five times the diameter of the coaxial feed cable center conductor was used. The average voltage will be

$$V_0 = -tE_{z_0} = -\frac{t}{w}\int_{x_0-w/2}^{x_0+w/2} E_z(x, y_0)dx \tag{3.2.49}$$

and from (3.2.29)

$$Z_{in} = \frac{V_0}{I} = \frac{jk\eta t}{ab}\sum_{m,n=0}^{\infty}\frac{\epsilon_{0m}\epsilon_{0n}\psi_{mn}^2(x_0, y_0)}{k_{mn}^2 - k_{eff}^2}\, j_0^2\left(\frac{m\pi w}{2a}\right) \tag{3.2.50}$$

Let $k_{mn} = \omega_{mn}\sqrt{\mu_0\epsilon_0\epsilon_r}$, then (3.2.50) results in

$$Z_{in} = j\omega\frac{t}{\epsilon_0\epsilon_r ab}\sum_{m,n=0}^{\infty}\frac{\epsilon_{0m}\epsilon_{0n}\psi_{mn}^2(x_0, y_0)\, j_0^2(m\pi w/(2a))}{\omega_{mn}^2 - \omega^2(1 - j\tan\delta_{eff})}$$

$$= j\omega\sum_{m,n=0}^{\infty}\frac{\alpha_{mn}}{\omega_{mn}^2 - \omega^2(1 - j\tan\delta_{eff})}$$

$$= \sum_{m,n=0}^{\infty}\frac{1}{G_{mn} + j\left[\omega C_{mn} - 1/(\omega L_{mn})\right]} \tag{3.2.51}$$

where

$$\alpha_{mn} = \frac{\epsilon_{0m}\epsilon_{0n}t}{\epsilon_0\epsilon_r ab}\psi_{mn}^2(x_0, y_0)\, j_0^2\left(\frac{m\pi w}{2a}\right) \tag{3.2.52}$$

$$G_{mn} = \frac{\omega\tan\delta_{eff}}{\alpha_{mn}}, \qquad C_{mn} = \frac{1}{\alpha_{mn}}, \qquad L_{mn} = \frac{\alpha_{mn}}{\omega_{mn}^2} \tag{3.2.53}$$

The microstrip antenna is typically narrow-band. Thus over the bandwidth of the antenna operation, at any mode (m,n), $G_{mn}(\omega)$ can be approximated simply by $G_{mn}(\omega_{MN})$. Thus (3.2.51) represents a Foster expansion of a driving point impedance function and gives the equivalent circuit shown in Figure 3.3. The higher the model indices (m,n), the higher the ω_{mn}. Since $L_{mn} \to 0$ with increasing model indices (m,n), the infinite number of high order Foster sections can be simply replaced to form a single series inductance; for $(0,0)$ mode, $L_{mn} \to \infty$. Consequently, the inductance forms an open circuit and may be removed from Figure 3.3.

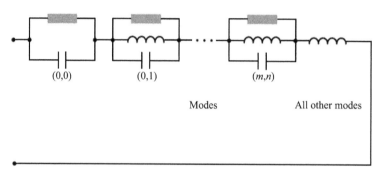

Figure 3.3   Network model for a microstrip antenna operating in a band around the (M,N) mode.

A simpler alternative representation, that would be adequate for frequencies near resonance of a mode (M,N) but sufficiently away from all other resonance, is to write Z_{in} in (3.2.51) as

$$Z_{in} \approx \frac{j\omega\alpha_{MN}}{\omega_{MN}^2 - \omega^2(1 - j\tan\delta_{eff})} + j\omega \sum_{(m,n)\neq(M,N)} \frac{\alpha_{mn}}{\omega_{mn}^2 - \omega_{MN}^2}$$

$$= \frac{1}{G_{MN} + j\left[\omega C_{MN} - 1/(\omega L_{MN})\right]} + j\omega L' \tag{3.2.54}$$

where

$$L' = \sum_{(m,n)\neq(M,N)} \frac{\alpha_{mn}}{\omega_{mn}^2 - \omega_{MN}^2} \tag{3.2.55}$$

The corresponding equivalent circuit is shown in Figure 3.4. The equivalent circuits shown in Figure 3.3 and Figure 3.4 work equally well for both a rectangular patch and a circular patch. For TM_{01} mode, due to large values of $\left|\omega_{mn}^2 - \omega_{MN}^2\right|$, L' is small. The equivalent is a simple RLC parallel resonance circuit. According to (3.2.32), (3.2.52) and (3.2.53)

$$R_{mn} = \frac{1}{G_{mn}} = \frac{\alpha_{mn}}{\omega_{mn}\tan\delta_{eff}} = \frac{\alpha_{mn}Q}{\omega_{mn}}$$

$$= \frac{\epsilon_{0m}\epsilon_{0n}tQ}{\omega_{mn}\epsilon_0\epsilon_r ab}\psi_{mn}^2(x_0, y_0)j_0^2\left(\frac{m\pi w}{2a}\right)$$

$$= \frac{60\epsilon_{0m}\epsilon_{0n}Qt\lambda_{mn}}{\epsilon_r ab}\psi_{mn}^2(x_0, y_0)j_0^2\left(\frac{m\pi w}{2a}\right) \tag{3.2.56-a}$$

$$R_{01} = \frac{120\lambda_{01}Qt}{\epsilon_r ab}\cos^2\left(\frac{\pi y_0}{b}\right) \tag{3.2.56-b}$$

$$R_{10} = \frac{120\lambda_{10}Qt}{\epsilon_r ab}\cos^2\left(\frac{\pi x_0}{a}\right)j_0^2\left(\frac{\pi w}{2a}\right) \tag{3.2.56-c}$$

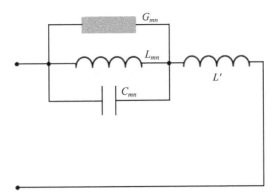

Figure 3.4   Simplified network mode, valid when ω_{MN} is well separated from all other resonant frequencies.

where $\lambda_{01}, \lambda_{10}$ are the resonant wavelengths of $(0,1)$ mode and $(1,0)$ mode.

$$\lambda_{01} = 2b\sqrt{\epsilon_e} \tag{3.2.57-a}$$

$$\lambda_{10} = 2a\sqrt{\epsilon_e} \tag{3.2.57-b}$$

In (3.2.56-a) and (3.2.57-a), ϵ_e is the effective dielectric constant related to the fringing field; $a = a' + 2\Delta l$ and $b = b' + 2\Delta l$ are the effective dimensions. To account for the fringing field, the end-effect extension Δl of an open end beyond its physical dimension a or b should be considered. It is assumed that all the edges expand equally out from the original patch. The correction of edge effect, that is, the solution of ϵ_e and Δl, will be fully discussed in Section 3.3.1.

If Q is approximated by Q_r, according to (3.2.38), formula (3.2.56-a) yields

$$R_{01} = \frac{1}{G_r} \cos^2\left(\frac{\pi y_0}{b}\right) \tag{3.2.58-a}$$

$$R_{10} = \frac{1}{G_r} \cos^2\left(\frac{\pi x_0}{a}\right) j_0^2\left(\frac{\pi w}{2a}\right) \tag{3.2.58-b}$$

$1/G_r$ is usually between $(100\text{-}300)\Omega$. The location of feed point (x_0, y_0) may be adjusted to obtain the required matching to the feed line.

3.2.5   Input Impedance from Transmission Line Model

The model described above is partially successful in predicting microstrip radiation performance and requires considerable calculations. The transmission line model of Derneryd[5] leads to results adequate for most engineering purposes and requires less computation. It provides a reasonable interpretation of the radiation mechanism while simultaneously giving simple expressions for the characteristics. The microstrip radiation element may be treated as a line resonator with no transverse field variations. The fields vary along the length, which is usually a half-wavelength, and radiation occurs mainly from the magnetic currents at the open-circuit ends. The equivalent circuit for the TM_{010} mode is shown in Figure 3.5, where b' is the physical dimension and Δl is the end-effect extension. The input admittance may be obtained through transmission line theory

$$Y_{in} = G_s + Y_c \frac{G_s + jY_c \tan \beta(b' + 2\Delta l)}{Y_c + jG_s \tan \beta(b' + 2\Delta l)} \tag{3.2.59}$$

As shown in (3.2.28), when $\beta b = \beta(b' + 2\Delta l) = \pi$, then

$$Y_{in} = 2G_s$$

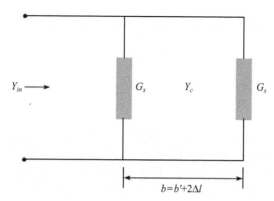

Figure 3.5   Equivalent circuit for a microstrip radiating element.

where $\beta = 2\pi/\lambda_e$, λ_e is the wavelength of the transmission line b and $\lambda_e = \lambda/\sqrt{\epsilon_e}$. The resonant frequency $f_r = f_{01}$ is given in (3.2.57-a).

Formula (3.2.59) may also be written as

$$Y_{in} = G_s + Y_c \frac{G_s + jY_c \tan(\pi f/f_r)}{Y_c + jG_s \tan(\pi f/f_r)} \tag{3.2.60}$$

3.2.6   Bandwidth of Input Impedance, Efficiency and Directivity

Based on the equivalent circuit in Figure 3.4 and the approximation of $L' \approx 0$, (3.2.54) reduces to

$$Z_{in} \approx \frac{1}{1/R_{mn} + j\left[\omega C_{MN} - 1/(\omega L_{MN})\right]} \tag{3.2.61}$$

where Z_{in} is the input impedance for TM_{mn} mode and $R_{mn} = 1/G_{mn}$. At resonance, Q is given by (3.2.33), and $\omega_{mn} = 1/\sqrt{L_{mn}C_{mn}}$

$$Q = 2\omega_{mn}\frac{W_e}{P} = 2\omega_{mn}\left(\frac{C_{mn}V^2/2}{V^2/R_{mn}}\right)$$

$$= \frac{R_{mn}}{\sqrt{L_{mn}/C_{mn}}} \tag{3.2.62}$$

Thus (3.2.61) may be rewritten as

$$Z_{in} = \frac{R_{mn}}{1 + jQ\left(f/f_{mn} - f_{mn}/f\right)} = \frac{R_{mn}}{1 + jQS} \tag{3.2.63}$$

Assuming that the characteristic impedance of the feed line equals to R_{mn}, then the norm of the reflection coefficient $|\Gamma|$ is

$$|\Gamma| = \left|\frac{Z_{in} - R_{mn}}{Z_{in} + R_{mn}}\right| \tag{3.2.64}$$

Substituting (3.2.63) into (3.2.64) results in

$$|\Gamma| = \left[1 + \frac{4}{Q^2 S^2}\right]^{-1/2} \tag{3.2.65}$$

Consequently

$$\frac{\rho+1}{\rho-1} = \frac{1}{|\Gamma|} = \left[1+\frac{4}{Q^2 S^2}\right]^{1/2} \tag{3.2.66}$$

and

$$S = \frac{f}{f_{mn}} - \frac{f_{mn}}{f} = \pm\frac{\rho-1}{Q\sqrt{\rho}} \tag{3.2.67}$$

where ρ is the Voltage Standing Wave Ratio (VSWR).

In (3.2.67), a minus sign is for the lower frequency f_L and a plus sign is for the higher one f_H. From (3.2.67), we may solve for both f_L and f_H and finally the relative bandwidth BW is

$$BW = \frac{f_H - f_L}{f_{mn}} \times 100\% = \frac{\rho-1}{Q\sqrt{\rho}} \times 100\% \tag{3.2.68}$$

The radiation efficiency is defined as

$$\eta_r = \frac{P_r}{P} = \frac{Q}{Q_r} \tag{3.2.69}$$

Substituting (3.2.32) into (3.2.69) results in

$$\eta_r = \frac{Q_c Q_d Q_{sw}}{Q_c Q_d Q_{sw} + Q_r Q_d Q_{sw} + Q_r Q_c Q_{sw} + Q_r Q_c Q_d} \tag{3.2.70}$$

The directivity D_0 may be obtained from (1.4.3), (3.2.22) and (3.2.33). Notice that from Prob. 1.3, the maximum radiation is at $\theta = \phi = 90°$ and

$$D_0 = \frac{|E_{\max}|^2 r^2}{60 P_r} = \frac{[2aV/(\lambda r)]^2 r^2}{60 \cdot 0.5 V^2 G_r} = \frac{2}{15 G_r}\left(\frac{a}{\lambda}\right)^2 = \frac{1}{15 G_s}\left(\frac{a}{\lambda}\right)^2 \tag{3.2.71}$$

G_s is given in (3.2.27), $a = w$ is the radiation edge and λ is the wavelength in free space.

3.2.7 Multiport Analysis

Once the antenna has been replaced by an ideal cavity with the appropriate loss tangent, it becomes a relatively simple matter to perform multiport analysis. Consider a rectangular microstrip with two feed points of coordinates (x_1, y_1) for port 1 and (x_2, y_2) for ports 2. The Z parameters for the two ports are simply

$$Z_{11} = \frac{V_1}{I_1}\bigg|_{port\ 2\ open} \tag{3.2.72}$$

$$Z_{22} = \frac{V_2}{I_2}\bigg|_{port\ 1\ open} \tag{3.2.73}$$

$$Z_{12} = \frac{V_1}{I_2}\bigg|_{port\ 1\ open} \tag{3.2.74}$$

$$Z_{21} = Z_{12} \tag{3.2.75}$$

Thus Z_{11} and Z_{22} are computed in precisely the same way as computing the input impedance for the one port given in (3.2.50). It is easy to see that the mutual impedance Z_{12} may be computed by

$$Z_{12} = \frac{jk\eta t}{ab} \sum_{m,n=0}^{\infty} \frac{\epsilon_{0m}\epsilon_{0n}\psi_{mn}(x_1, y_1)\psi_{mn}(x_2, y_2)}{k_{mn}^2 - k_{eff}^2} j_0^2\left(\frac{m\pi w}{2a}\right) \tag{3.2.76}$$

For some applications, it is of interest to find the input impedance at one port while the other is loaded. This would provide a means for tuning, matching, modifying patterns and so on. The corner cut may also be treated as loading and will be discussed in Section 3.3.2. In such case, from

$$\begin{pmatrix} V_1 \\ V_2 \end{pmatrix} = \begin{pmatrix} Z_{11} & Z_{12} \\ Z_{21} & Z_{22} \end{pmatrix} \begin{pmatrix} I_1 \\ I_2 \end{pmatrix} \tag{3.2.77}$$

and

$$\frac{V_2}{I_2} = Z_2 = Z_L \tag{3.2.78}$$

we have

$$Z_{in} = \frac{V_1}{I_1} = Z_{11} - \frac{Z_{12}^2}{Z_{22} - Z_L} \tag{3.2.79}$$

where Z_L is the loaded impedance at port 2.

3.3   Improvement and Extension of the Cavity Model

3.3.1   Correction of Edge Effect by DC Fringing Fields

The actual resonant frequency ω_r and the propagation constant β along y direction in Figure 3.1 are different from those of the ideal cavity due to the edge effect. The actual ones, often called the effective ones, may be obtained through the correction of the edge effect. A coaxial line has wave effect dependence only along the longitudinal direction. It has static (DC) effects in the transverse (cross-section) plane, e.g., the distributed capacitance and inductance, even at a high microwave (RF) frequency of, say, 30GHZ. This of course is a consequence of variable separation in the general solution of the partial-differential equation of wave. However, this consequence of variable separation may be applied here. For a patch resonator, the thinness of the substrate and its high dielectric constant ensure that wave effects only occur along the plane where the patch is located. Along the direction vertical to the patch plane, the static effects dominate. The static effects mean that the vertical distribution of an electric fringe field at the edges of a patch resonator at microwave frequency, can be approximated by the vertical distribution of the electric fringe field at the edges of the patch of grounded capacitor in DC. This is actually the quasi-static approximation. Therefore the problem is to find this static capacitance accounting for the edge effects.

The starting point of formulations is the root of area capacitance formula developed by Chow et al. in early 1982[9],

$$C = C_f \epsilon_0 \sqrt{4\pi S} \tag{3.3.1}$$

where S is the surface area and C_f is the "shape factor" dependent on the shape of the conductor. For a sphere, $C_f = 1$; for a square plate, $C_f = 0.9$; for a conducting cylinder with a length to diameter ratio of 7 to 1, $C_f = 1.1$. The plate with area A is a 3-D object, and A is the area of only one side of the plate. For both sides of the 3D plate, the total surface area is $S = 2A$. Considering a microstrip parallel plate capacitor of area A with infinite substrate, it is possible to set up two regular asymptotes for the thickness t of the substrate.

(a) The near asymptote C_a at $t \to 0$. It is clear from elementary electromagnetic that

$$C_a = \lim_{t \to 0} C = \frac{\epsilon_0 \epsilon_r A}{t} \tag{3.3.2}$$

(b) The far asymptote C_b, at $t \to \infty$. At infinite distance t, the two plates do not see each other. Each plate then appears to sit on the boundary of a half space of air and dielectric ϵ_r. This means that each plate is effectively sitting in a homogeneous medium with dielectric $(\epsilon_r+1)/2$. With the voltage $\pm V$ between two plates, the far asymptote of the capacitance is

$$C_b = \lim_{t\to\infty} C = C_f \epsilon_0 \frac{\epsilon_r+1}{2}\sqrt{8\pi A}\Big/2$$

$$= C_f \epsilon_0 \frac{\epsilon_r+1}{2}\sqrt{2\pi A} = 0.9\epsilon_0 \frac{\epsilon_r+1}{2}\sqrt{2\pi A} \qquad (3.3.3)$$

The synthetic asymptote is simply the sum of the two regular asymptotes. Therefore C_p may be expressed as

$$C_p = \left\{\left(\frac{\epsilon_0\epsilon_r A}{t}\right)^n + \left[0.9\epsilon_0 \frac{\epsilon_r+1}{2}\sqrt{2\pi A}\right]^n\right\}^{1/n} \qquad (3.3.4)$$

Obviously, there is error for intermediate values of t. If $n = 1$, around $t = \sqrt{A}$, the error is maximum and is about 10% when compared with the numerical solution by method of moments (MoM). If the exponent n is increased from 1 to 1.114, through matching with one numerical data point at an intermediate location of t, the maximum error is found to reduce to 2% over the whole range of t and ϵ_r.

For a microstrip patch on an infinite substrate with thickness t, the near asymptote is the same as that in (3.3.2). The far asymptote is just the capacitance of one plate sitting in a homogeneous medium with the dielectric constant $(\epsilon_r + 1)/2$, that is twice the second term in (3.3.4). The capacitance of a patch on infinite substrate is obtained as

$$C_p = \left\{\left(\frac{\epsilon_0\epsilon_r A}{t}\right)^n + \left[0.9\epsilon_0 \frac{\epsilon_r+1}{2}\sqrt{8\pi A}\right]^n\right\}^{1/n} \qquad (3.3.5)$$

The C_p value with a fringe effect included can be made to equal to that of a capacitor of the expanded plate area with only a parallel plate field and no fringe. The expanded area of the new plate is simply

$$A_e = \frac{C_p t}{\epsilon_0 \epsilon_r} \qquad (3.3.6)$$

For a rectangular patch, it is assumed that all the edges expand equally out from the original patch. If the physical dimensions of the plate are a' and b' and the effective ones are a and b then

$$A_e = ab = (a' + 2\Delta l)(b' + 2\Delta l) \qquad (3.3.7)$$

The solution of Δl from (3.3.7) is

$$\Delta l = -\frac{(a' + b')}{4} + \frac{1}{2}\sqrt{\left(\frac{a' + b'}{2}\right)^2 + (A_e - A)} \qquad (3.3.8)$$

where $A = a'b'$ is the physical surface area. When a', b', t and ϵ_r are specified, Δl may be obtained from (3.3.8).

A more accurate relation, which involves different extensions on the two perpendicular edges a and b, can be found in [10].

The patch may be considered as a resonating microstrip-line section along the y direction shown in Figure 3.1 for the TM_{01} mode. The geometry and the electric and magnetic flux lines are shown in Figure 3.6.

(a) Geometry   (b) Electric and magnetic flux

Figure 3.6   Microstrip transmission line.

If the dielectric were not present ($\epsilon_r = 1$), the line could be thought of as a two-wire line consisting of two flat strip conductors of width w, separated by a distance $2t$ (the ground plane is removed with image theory) with the propagation velocity v_ρ and propagation constant β will be simply $v_p = c$ and $\beta = k_0$.

The presence of the dielectric, and particularly the fact that the dielectric does not fill the air region above the strip as shown in Figure 3.6, complicates the behavior and analysis of the microstrip line, since the phase velocity of TEM fields in the dielectric region would be $c/\sqrt{\epsilon_r}$, but the phase velocity of TEM fields in the air region would be c. Thus, a phase match at the dielectric-air region would be impossible to attain for a TEM-type wave[7]. However, based on quasi-static approximation, the phase velocity can be expressed as

$$v_p = \frac{c}{\sqrt{\epsilon_e}} = \frac{1}{\sqrt{LC}} \qquad (3.3.9)$$

where L and C are the inductance and capacitance per unit length of the microstrip line when $\epsilon_r \neq 1$. Light velocity c may be expressed as

$$c = \frac{1}{\sqrt{LC_0}} \qquad (3.3.10)$$

where L is the same as that in (3.3.9); the reason is that L relates to the permeability not the permittivity. $C_0 = C\,|_{\epsilon_r=1}$. From (3.3.9), (3.3.10)

$$\epsilon_e = \frac{C}{C_0} \qquad (3.3.11)$$

Since some of the field lines are in the dielectric region and some are in the air, the effective dielectric constant ϵ_e satisfies the inequality

$$1 < \epsilon_e < \epsilon_r$$

and is dependent on the substrate thickness, t, and conductor width, w. For the usual microstrip transmission line, the width of the strip w is small; while for the patch antenna, it is large. Therefore, the patch effective dielectric constant $\epsilon_{e,p}$ will lie between ϵ_r and the line effective dielectric constant $\epsilon_{e,l}$. Through numerical experimenting, it is found that the best choice is

$$\epsilon_{e,p} = 0.7\epsilon_r + 0.3\epsilon_{e,l} \qquad (3.3.12)$$

The choice is not critical as $\epsilon_{e,p}$, $\epsilon_{e,l}$ and ϵ_r values are very close to each other. The closeness is due to the thinness of substrate, wide line width, and the high dielectric constant ϵ_r of the substrate.

In the following, the derivation of $\epsilon_{e,l}$ will be given. The subscript l will be dropped for simplicity and the capacitance is for unit length.

With very thin substrate of thickness t under a strip of width w, the parallel plate capacitance of the strip is given in (3.3.13) which is called the near asymptote of t.

$$C_a = \lim_{t \to 0} C = \frac{\epsilon_0 \epsilon_r w}{t} \tag{3.3.13}$$

With very thick substrate t, the strip of width w is equivalent to a cylindrical lead of radius $w/4$. The distributed capacitance is then that of a twin-lead sitting on the boundary of two half spaces of air and dielectric. This twin-lead sees an effective dielectric constant of $(\epsilon_r + 1)/2$ in a homogeneous medium. Thus, similar to (3.3.13), we have the far asymptote

$$C_b = \lim_{t \to \infty} C = 2\pi\epsilon_0 \frac{\epsilon_r + 1}{2} \frac{1}{\ln\left(\frac{2t}{w/4}\right)} \tag{3.3.14}$$

Let the synthetic asymptote be simply the sum of the near and far asymptotes. Such arrangement requires the condition that in the sum, the near asymptote should be dominant at the small parameter limit (of t), but negligible at the large parameter limit. The reverse is true for the far asymptote. This condition means that a synthetic asymptote can be constructed for C, that is[12]:

$$C = \left\{ \left(\frac{\epsilon_0 \epsilon_r w}{t}\right)^n + \left[2\pi\epsilon_0 \frac{\epsilon_r + 1}{2} \left(\frac{1}{\ln\left[(8t/w) + 1\right]} - \frac{1}{8t/w} \right) \right]^n \right\}^{1/n} \tag{3.3.15}$$

With $n = 1$, it is seen that the second term (far asymptote of t) requires two modifications for the above condition. The first is the $+1$ in the log function of the denominator, and the next is a third term to cancel the singularity of the modified second term at t. No modification is needed for the first (near asymptote) term, as it goes to zero already at $t \to 0$. With such modifications, the synthetic asymptote converges smoothly at the two limits of $t \to 0$ and $t \to \infty$, but still has a maximum error of 10% at the intermediate values of t. A match with one data point with numerical computation at an arbitrary t, enables us to set the exponent $n = 1.08$. This small effort reduces the maximum error to 2%.

With the formula of the capacitance known, the effective dielectric may be obtained from (3.3.11) by computing (3.3.15) twice with $\epsilon_r \neq 1$ and $\epsilon_r = 1$. The characteristic impedance Z_0 is obtained from (3.3.9) and (3.3.11) as

$$Z_0 = \frac{1}{v_p C} = \frac{\sqrt{\epsilon_e}}{c\,C} = \sqrt{\frac{\epsilon_0 \mu_0}{C C_0}}$$

3.3.2 Irregularly Shaped Patch as Perturbation of Regularly Shaped Patch

An irregular shaped patch (e.g., resonator or microstrip antenna) is usually not completely arbitrary but is a regular (rectangular or circular) patch with modification. For example, the modification may be a corner-cut as mentioned in Section 3.2.7. This means that the irregular shape is usually a perturbation of a regular shape. As an example of this approach, consider a corner-cut microstrip patch antenna fed by the microstrip line as shown in Figure 3.7(a). The formula for the frequency response of this patch antenna may be summarized in the following points:

(a) Figure 3.7(a) and (b) give the small outward expansion of patch edges, based on the technique introduced in Section 3.3.1. The expansion Δl is obtained from the rectangular patch.

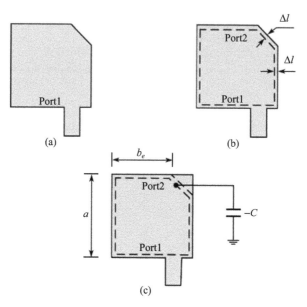

Figure 3.7   (a) Microstrip patch antenna with a 45° corner-cut, (b) The expanded patch with
magnetic wall in place of the original patch with a fringe field, (c) The cut-patch simulated
by an uncut-patch and a load of $-C$ at the unexpanded cut-line.

(b) Figure 3.7(b) and (c) create a virtual corner-cut, with the second port at the cut-line
(unmoved by the expansion) loaded by a $(-C)$ of parallel plate capacitor. C is a
simple parallel plate capacitor of the corner-cut with no fringe field, that is

$$C = \frac{\epsilon_0 \epsilon_r A_c}{h} \tag{3.3.16}$$

where the corner-cut area is $A_c = (a_e - b_e)^2/2$.

The current from port 2 into this $(-C)$ then must cancel the current from port 2
into C of the corner. As a result, there is no net current out of port 2 and the zero
current condition at the cut is now satisfied. If the corner becomes large, the charge
distribution on the corner-cut region decreases from the corner in a cosine manner. In
this case, a correction factor should be introduced[11].

(c) With the expanded patch of Figure 3.7(c), the applying of (3.2.79) results in

$$Z_{in} = \frac{V_1}{I_1} = Z_{11} - \frac{Z_{12}Z_{21}}{Z_{22} + 1/(j\omega C)} \tag{3.3.17}$$

In (3.3.17), the relationship $V_2/I_2 = 1/j\omega(-C)$ is used. Z_{11}, Z_{12}, Z_{21}, and Z_{22} may be
computed by using (3.2.72)–(3.2.75) based on the rectangular patch without a corner-cut.

3.4   Design Procedure of a Single Rectangular Microstrip Patch Antenna

In Figure 3.1(a), the mode is $(0,1)$ but in some cases, $(1,0)$ mode is preferable. In that
case, the feed point is located at the edge in which the field is with nonuniform distribution.
The design procedure for these two cases is largely the same.

Usually, the overall goal of a design is to achieve specific performance characteristics at a
specified operating frequency.

The first step is to choose the microstrip substrate. All the parameters of the substrate ϵ_r,
t, $\tan \delta$, σ_c are determined. The second step is to find the parameters of the microstrip patch

including the width w, the length L and the feed point based on the requirement of good matching with the given characteristic impedance of the microstrip feed line. Finally, the performance of the antenna should be examined for the efficiency, bandwidth, and radiation pattern.

3.4.1 Choice of the Microstrip Substrate

As is seen in Section 3.3, larger ϵ_r may result in smaller size. However, the bandwidth will be narrower. If the thickness increases, the bandwidth will increase accordingly, and may increase the weight and the surface wave excitation. All these factors are quite involved and there would be some trade-off.

Table 3.1 may provide a reference on choosing the parameters of a microstrip line for the given bandwidth[13]. The efficiency in terms of the parameters may also be computed (See Prob. 3.4). The commercial materials available should be taken into account.

Table 3.1   Bandwidth(BW) with VSWR(ρ) $\leqslant$ 2 of rectangular microstrip patch antenna.

h/λ_0	BW(%)			
	$\epsilon_r = 1.00$	$\epsilon_r = 2.55$	$\epsilon_r = 4.70$	$\epsilon_r = 10.2$
0.005	1.16	0.85	0.76	0.50
0.020	3.55	1.96	1.38	0.79
0.040	7.14	3.84	2.62	1.47
0.060	10.85	5.86	3.93	2.27
0.080	14.61	7.96	5.45	3.18
0.100	18.42	10.15	7.02	4.17

3.4.2 Coarse Determination of the Dimensions for Initial Patch Design

Two dimensions w and L should be determined; w is always related to the radiation edge and is equal to a or b and L is always related to the non-radiation edge and is equal to a or b as well. For example in Figure 3.1, $w = a$ and $L = b$.

For a dielectric substrate of thickness t, an antenna operating at frequency of f_r, and for an efficient radiator, the actual width is:

$$b' = w = \frac{c}{2f_r} \left(\frac{\epsilon_r + 1}{2} \right)^{-1/2} \tag{3.4.1}$$

where c is the velocity of light. For widths smaller than those selected according to (3.4.1), radiator efficiency is greater, but higher order modes may result, causing field distortions.

Next is to compute ϵ_e, Δl and L by formulas:

$$\epsilon_e = \frac{\epsilon_r + 1}{2} + \frac{\epsilon_r - 1}{2} \left(1 + \frac{10t}{w} \right)^{-1/2} \tag{3.4.2}$$

$$\Delta l = 0.412t \frac{(\epsilon_e + 0.3)(w/t + 0.264)}{(\epsilon_e - 0.258)(w/t + 0.8)} \tag{3.4.3}$$

$$a' = L = \frac{c}{2f_r\sqrt{\epsilon_e}} - 2\Delta l \tag{3.4.4}$$

3.4.3 Feeding Methods

Formula (3.2.51) can be used for both for the microstrip line feeding and for the coaxial-line feeding. In the design, there is no significant difference for these two feeding methods. To illustrate the design procedure, the microstrip line feeding method is given here. Assuming that the feed point is always along x axis, two lower modes $(0,1)$ and $(1,0)$ can be excited. If it is desired to assign the $(1,0)$ mode as the dominant mode, the $(0,1)$ mode will be considered as the cross-polarization and should be suppressed through proper choice of the dimensions. For $(1,0)$ mode, $w = b$, $L = a$. Figure 3.8 gives the cross-polarization against b'/a'. b' and a' are the physical dimensions of the patch; $a' = a - 2\Delta l$ and $b' = b - 2\Delta l$. It is seen from Figure 3.8 that when $b'/a' \approx 1.5$, the cross-polarization reaches the minimum. In design, this criterion should be satisfied as much as possible.

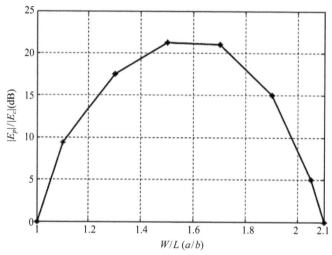

Figure 3.8   Plot of copolarization component E_p to cross-polarization component E_c versus aspect ratio a/b.

3.4.4   Matching Between the Patch and the Feed

For a given characteristic impedance of the microstrip line, the width w of the strip should be determined. The closed form for t/w was derived in [14]. From (3.3.13) and (3.3.16), the near inverse asymptote may be obtained as

$$\lim_{Z_0 \to 0} \left(\frac{t}{w} \right) = \lim_{\sqrt{C_a C_{a0}} \to 0} \left(\frac{t}{w} \right) = \frac{\epsilon_0 \sqrt{\epsilon_r}}{\sqrt{C_a C_{a0}}} = Z_0 \sqrt{\frac{\epsilon_0 \epsilon_r}{\mu_0}} \tag{3.4.5}$$

where C_{a0} is the distributed capacitance of the microstrip when $\epsilon_r = 1$ in (3.3.13). The far inverse asymptote can be obtained from (3.3.14) and (3.3.16) as

$$\lim_{Z_0 \to \infty} \left(\frac{t}{w} \right) = \frac{1}{8} e^{2\pi Z_0 \sqrt{\epsilon_0 (\epsilon_r + 1)/(2\mu_0)}} \tag{3.4.6}$$

Similar to (3.3.15), the synthetic inverse asymptote is

$$\frac{t}{w} = \left\{ \left(Z_0 \sqrt{\frac{\epsilon_0 \epsilon_r}{\mu_0}} \right)^n + \left[\frac{1}{8} \left(e^{2\pi Z_0 \sqrt{\frac{\epsilon_0 (\epsilon_r + 1)}{2\mu_0}}} - 1 - 2\pi Z_0 \sqrt{\frac{\epsilon_0 (\epsilon_r + 1)}{2\mu_0}} \right) \right]^n \right\}^{1/n} \tag{3.4.7}$$

The last two terms in (3.4.7) are added to ensure that the far asymptote drops to zero at $Z_0 \to 0$. These two terms are based on the approximate series $e^x \approx 1 + x$, when $x \to 0$.

Unlike the single value of n in (3.3.15), two match points with different ϵ_r are needed and from numerical computation results in a linear equation, $n = 0.911 - 0.00629\epsilon_r$.

A small change in Z_0 in the exponent in (3.4.7) gives a large swing in t/w. One can tolerate a larger error of t/w in (3.4.7), say a maximum of 10% instead of 4%. Such a large tolerance in t/w means easier fabrication.

Once Z_0 is specified, from (3.4.7), the width w of the strip is given because t is already fixed. The alternative formulas for determining t/w are

$$\frac{w}{t} = \frac{2}{\pi}\left\{R - 1 - \ln(2R - 1) + \frac{\epsilon_r - 1}{2\epsilon_r}\left[\ln(R-1) + 0.293 - \frac{0.517}{\epsilon_r}\right]\right\} \tag{3.4.8-a}$$

with $R = 377\pi/(2Z_0\sqrt{\epsilon_r})$, $Z_0 < (44 - 2\epsilon_r)\Omega$

$$\frac{w}{t} = \frac{8e^H}{e^{2H} - 2} \tag{3.4.8-b}$$

with

$$H = Z_0\sqrt{2(\epsilon_r + 1)}/120 + (0.2258 + 0.1208/\epsilon_r)(\epsilon_r - 1)/(\epsilon_r + 1), \quad Z_0 \geqslant (44 - 2\epsilon_r)\Omega$$

In using the above formulas, the value of Z_0 should be identified first.

The Wheeler's formula may serve as reference to compare the accuracy of (3.4.7) and (3.4.8-a)

$$Z_0 = \frac{377}{\sqrt{\epsilon_r}}\left\{\frac{w}{t} + 0.833 + 0.165\frac{\epsilon_r - 1}{\epsilon_r^2} + \frac{\epsilon_r + 1}{\pi\epsilon_r}\left[\ln\left(\frac{w}{t} + 1.88\right) + 0.758\right]\right\}^{-1}$$

$$w/t > 1 \quad (3.4.9\text{-a})$$

$$Z_0 = \frac{120}{\sqrt{2(\epsilon_r + 1)}}\left[\ln\left(\frac{8t}{w} + \frac{1}{32}\left(\frac{w}{t}\right)^2\right) - \frac{\epsilon_r - 1}{\epsilon_r + 1}\left(0.2258 + \frac{0.1208}{\epsilon_r}\right)\right]$$

$$w/t \leqslant 1 \quad (3.4.9\text{-b})$$

To match the feed and the patch, the input impedance of the patch should be computed. In the computation, more accurate results can be obtained by using (3.2.51) that takes into account the higher order modes.

3.4.5   Design Example

The specification is f_r=16.0GHz, BW($\rho \leqslant 2$) $\geqslant$3.0 %, and TM_{10} operating mode. The parameters for the microstrip substrate are ϵ_r= 2.65, $\tan\delta = 5 \times 10^{-4}$, $t = 0.5$mm, and $\sigma = 8020$S/mm. The characteristic impedance Z_0 of the feeding microstrip line is chosen as 70 Ω. Table 3.2 gives all the results.

Based on Table 3.2, the frequency response of the VSWR can be computed from (3.2.51), (3.2.64) and (3.2.66). The result is shown in Figure 3.9 (Line 1). To make further improvement, the optimization procedure may be used to adjust a', b' and x_0'. The objective function is chosen as

$$U = \max_i\left[\mathrm{abs}\left(A_i - B_i\right)\right], \quad i = 1, 2, 3 \tag{3.4.10}$$

where $i = 1, 2, 3$ are related to the frequency points $f_1 = 15.9$GHz, $f_2 = 16$GHz, $f_3 = 16.1$GHz. B_i are the VSWR computed at the corresponding frequency points and A_i are set to be $A_1 = 1.5$, $A_2 = 1.2$, $A_3 = 1.5$. The optimization procedure may minimize U to get the new a', b', x_0'. After it is done, we have $a' = 5.41$mm, $b' = 8.11$mm, $x_0' = 0.95$mm. The frequency response of the VSWR for the new set of parameters is shown in Figure 3.9 (Line 2). The results from IE3D^{TM} are also given in Figure 3.10 for comparison, Line 1 is for the

Table 3.2 A design example.

Parameters	$w' = b'$(mm)	ϵ_e	Δl(mm)	$L' = a'$(mm)	$L = a$(mm)	
Formulas	(3.4.1)	(3.4.2)	(3.4.3)	(3.4.4)	$a = a' + 2\Delta l$	
Data	6.94	2.45	0.25	5.49	5.98	
$w = b$(mm)*	Q_r	Q_c	Q_d	Q_{sw}	Q	$\eta\%$
$b = 1.5a' + 2\Delta l$	(3.2.38)	(3.2.41)	(3.2.43)	(3.2.47)	(3.2.32)	(3.2.69)
8.73	28.09	355.87	2000	213.07	22.94	81.6
BW($\rho \leqslant 2$)	D_0(dB)	G(dB)	w(mm)	x_0(mm)	x_0'(mm)	
(3.2.68)	(3.2.71)	$G = \eta D_0$	(3.4.7)	(3.2.56-c)	$x_0' = a' - (a - x_0)$	
3.08	8.19	7.31	0.78	1.73	1.23	

* To reduce the cross polarization, according to Figure 3.8, b' should be chosen as $1.5a'$.

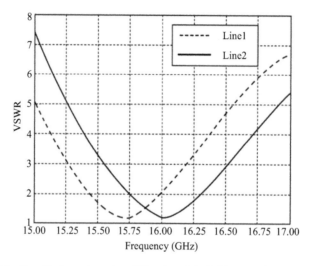

Figure 3.9   Results from formulas. (Line 1: original; Line 2: after optimization.)

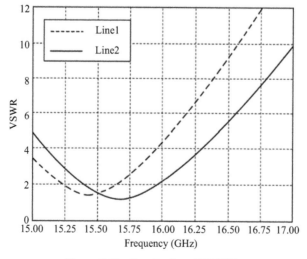

Figure 3.10   Results from IE3D^{TM}.

(Line 1: set of original parameters, Line 2: set of optimized parameters.)

original set of parameters and Line 2 for the new one. The optimization procedure indeed improves the design.

3.5 Example of LTCC Microstrip Patch Antenna

A conventional patch antenna is usually of narrow bandwidth. There are many types of broadband patch antennas with a slot on the patch. Among them the U-slot antenna with a U-slot on the patch and the H-slot antenna with an H-slot on the patch have attracted much attention because of their outstanding performance. In 1995, Huynh and Lee[15] presented an experimental study of the U-slot antennas with very good impedance bandwidth performance. The new antenna is a probe-fed rectangular microstrip patch antenna on a unity permittivity substrate with an internal U-shaped slot. But the cross polarization level in the H-plane of this type of antenna is a little higher than desired, i.e., above -10dB. Early in 1981, Dobust presented a H-slot antenna, which has very good polarization performance[16]. But its bandwidth performance is not as good as that of the U-slot antennas. In many applications, both broad bandwidth and low cross-polarization are needed simultaneously. Neither of the two types meets these requirements. Comparing the U-slot antenna with the H-slot antenna, we can find that geometry of the radiation patch for the H-slot antenna is symmetrical both in the E-plane and the H-plane, while the U-slot antenna is symmetrical only in the E-plane. A difference exists in their current distribution. The current distribution on the patch determines the radiation pattern, including the polarization performance. If the antenna is vertically polarized, then the vertical part of the current vectors determines the co-polarization intensity in the far field and the horizontal part determines the cross-polarization intensity. Looking at the current distribution in the U-slot antenna, we can find the horizontal part of the current distributed symmetrically in the E-plane (vertical plane) but asymmetrically in the H-plane (horizontal plane). As a result, in the E-plane, the cross polarized field components will cancel each other. Consequently, the cross-polarization level is low. But in the H-plane, the cross polarized field components will not cancel each other, so a high cross-polarization level will result. Applying a similar analysis to the H-slot antenna, their low cross-polarization levels both in E and H-plane come from the symmetrical current distribution both in the E and H-plane. Antennas and filters may be viewed as resonators from the point of view of frequency dependence, so they must have some features in common. It is predictable that the antenna performance may be improved by increasing the number of stages of the slots, just like the filter performance does. Based on the above ideas, it is seen that the bandwidth of the antenna may be improved with the number of slot pairs increased, and at the same time, a good polarization performance may be kept both in the E-plane and the H-plane for symmetrical current distribution. Consequently, the tooth-like-slot antenna was proposed[17]. Recently, the demand for integrated low-cost circuits and/or antennas has increased with the rapid progress of wireless communication systems. As a result, the LTCC (low temperature co-fired ceramics) technology is becoming more and more popular for its flexibility in the integration of arbitrary numbers of layers. However, the typical high dielectric constant of LTCC materials is not suitable for the bandwidth improvement of a microstrip antenna. Thus a broadband antenna based on LTCC technology needs careful design. The layout of this antenna is shown in Figure 3.11 and the photo of the prototype is shown in Figure 3.12. The substrate is a sandwich structure with the feed line in the middle. The total thickness is h. The thickness of the upper substrate is $h1$, the lower is $h2$. The tooth-like-slot on the patch is composed of several paralleled vertical rectangular slots and a horizontal rectangular slot; it is a symmetrical structure both in the vertical and horizontal planes.

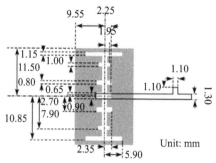

Figure 3.11   The layout of the tooth-like-slot
patch antenna.

Figure 3.12   A photo of the prototype of the
tooth-like-slot patch antenna.

The impedance performance of the tooth-like-slot antenna may be investigated through numerous numerical simulations by using the commercial software such as IE3D^{TM}. There are some key points in the adjustments.

1. The width of the patch and the length of the main slot pair must be adjusted first, as they are the most important variables affecting the central frequency.

2. The sizes of slot pairs and the location of the end of the feed line should be adjusted carefully, for they have important effects on the input impedance.

In this example, the central frequency is 5.76GHz. The LTCC multilayer structure consists of LTCC A6-S ceramic material ($\epsilon_r = 5.9$) of equal layer thickness (4 mils). And the total thickness of substrate is 2 millimeters and the other dimensions are shown in Figure 3.11. The measured return loss is shown in Figure 3.13. The measured radiation patterns for the E-plane and H-plane are shown in Figures 3.14(a) and 3.14(b) respectively. From the results it is seen that the bandwidth is about 20%, the cross-polarization level in the E-plane is about −35dB, and in the H-plane is nearly −22dB. All are sufficient to meet the requirements in most applications.

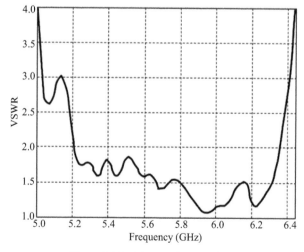

Figure 3.13   Measured return loss.

Wideband microstrip antenna has been a hot topic recently. Other methods for increasing the bandwidth include patches with thick substrate and low dielectric constant, planar gap-coupled and directly coupled multi-resonators, stacked electromagnetically coupled or aperture-coupled patches, impedance-matching techniques, log-periodic configurations and so on. There are many publications available for this topic[18–20].

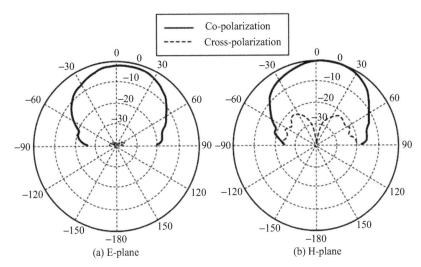

(a) E-plane   (b) H-plane

Figure 3.14   Radiation patterns.

Bibliography

[1] Y. T. Lo et al, "Theory and experiments on microstrip antennas," *IEEE Trans. on Antennas and Propagation*, vol. 27, no. 3, pp. 137–145, Mar., 1979.

[2] R. E. Collin, *Field Theory of Guided Waves*, Appendix, McGraw-Hill Book Co., 1960.

[3] W. F. Richards et al., "An improved theory for microstrip antennas and applications," *IEEE Trans. on Antennas and Propagation*, vol. 29, no. 1, pp. 38–46, Jan., 1981.

[4] S. S. Zhong, *Microstrip Antenna Theory and Applications* (in Chinese), Xidian University Press, 1991.

[5] A. G. Derneryd, "Linearly polarized microstrip antennas," *IEEE Trans. on Antennas and Propagation*, vol. 24, no. 11, pp. 846–851, Nov., 1976.

[6] I. J. Bahl and P. Bhartia, *Microstrip Antennas*, Artech House, 1980.

[7] D. M. Pozar, *Microwave Engineering* (second edition), John Wiley & Sons, Inc., 1998.

[8] B. Nauwelaers and A. Van de Capelle, "Surface wave losses of rectangular microstrip antennas," *Electronic Letters*, vol. 25, no. 11, pp. 696–697, 1989.

[9] Y. L. Chow and M. M. Yovanovich, "The shape factor of the capacitance of a conductor," *J. of Appl. Physics*, vol. 53, pp. 8470–8475, 1982.

[10] Y. X. Sun, Y. L. Chow and D. G. Fang,"Impedance formulation of RF patch resonators and antennas of cavity model using fringe extension of patch from DC capacitors," *Microwave and Optical Technology Letters*, vol. 35, no. 4, pp. 293–297, Nov., 2002.

[11] Y. L. Chow, Y. X. Sun and D. G. Fang, "Irregularly shaped patch as perturbation of regularly sharped patch," *Microwave and Optical Technology Letters*, vol. 28, pp. 70–74, Jan., 2002.

[12] W. C. Tang and Y. L. Chow, "CAD formulas, their inverses and interrelations for microstrip, CPW lines without and with backing ground plane, by successive synthetic asymptotes," *J. of EM. Waves and Appl.*, vol. 16, pp. 1–20, Jan., 2002.

[13] T. A. Milligan, *Modern Antenna Design*, McGraw-Hall, 1985.

[14] W. C. Tang and Y. L. Chow, "Formula and its inversion by synthetic asymptote — a simple example of microstrip structure," *2000 China-Japan Joint Meeting Conference Proceedings*, pp. 74–77, 2000.

[15] T. Huynh, K. F. Lee, "Single-layer Single Patch Wideband Microstrip Antenna," *Electron, Letters*, vol. 31, no. 1, pp. 1310–1312, 1995.

[16] G. Dubost, *Flat Radiating Dipoles and Their Applications to Arrays*, Research Studies Press, New York, 1981.

[17] G. B. Han, D. G. Fang and H. Wang, "Design of the LTCC tooth-like-slot antenna," 2004 *International Conference on Microwave and Millimeter Wave Technology Proceedings*, pp. 23–26, 2004.

[18] K. Girish, K. P. Ray, *Broadband Microstrip Antenna*, Artech House, 2003.

[19] K. L. Wong, *Compact and Broadband Microstrip Antennas*, Wiley, 2002.

[20] Y. X. Guo, *Design and Analysis of Microstrip Antennas and Dielectric Resonator Antennas for Wireless Communications*, Ph.D. dissertation, City University of Hong Kong, 2001.

Problems

3.1 Prove (3.2.68).

3.2 For the example given in Section 3.4.5, verify f_r, Δl, ϵ_e by using the method given in Section 3.3.

3.3 For TM_{10} mode, and $b/a=1.5$, compute the efficiency of the antenna when t/λ varies from $0 \rightarrow 0.10$ in the cases of $\epsilon_r = 1.00, 2.55, 4.70, 10.2$ (In the computation, the computed points may be gradually densified).

3.4 Check the data in Table 3.2.

3.5 In the example of Table 3.2, compute $E_{10\,max}/E_{01\,max}$, for $b=8.73$mm, 7.25mm and 10.25mm. E is the far field intensity.

3.6 In the example of Table 3.2, give the radiation pattern for both the E plane and the H plane.

3.7 In the example of Table 3.2, compute G_s by using (3.2.26) and the approximate formula In the range of $Z_0 = (50–75)\Omega$, $\epsilon_r = 2.20–2.70$, improve the accuracy of (3.4.7).

3.8 In the range of $Z_0 = (50–75)\Omega$, $\epsilon_r = 2.20–2.70$, improve the accuracy of (3.4.7).

3.9 Based on the design example, give out the frequency response within the 3% frequency bandwidth.

3.10 Based on the frequency response in Prob. 3.9, find ϵ_r and $\tan \delta$, assuming they are unknowns. If the frequency response is obtained from measurement, it may be considered as the measurement of the dielectric parameters.

CHAPTER 4

Spectral Domain Approach and Its Application to Microstrip Antennas

4.1 Introduction

The cavity model and other simple methods discussed in Chapter 3 provide a simple intuitive understanding of the performance of microstrip antennas. Although the accuracy of these approximate models is limited, they are nevertheless useful in providing the preliminary design and predicting the trends of these characteristics with the variation of the design parameters.

In contrast, the full-wave models are able to analyze arbitrarily-shaped geometries and can take into account the effect of various types of feeds. Over the last decade, a host of electromagnetic solvers have been developed for commercial applications. One common problem encountered in using these CAD software packages for the design of microstrip antennas is that most of these techniques are suitable for analyzing a given geometry but not directly for design purposes. In other words, given a structural configuration with all the relevant-dimensions, one can use an appropriate CAD software tool to simulate the response of the structure over a desired range of frequencies. However, the complementary synthesis problem of predicting the structural dimensions for a specified response is usually a much more daunting task.

Ideally, a CAD software package should require little preprocessing and be able to predict, efficiently and accurately, all the characteristics of a given practical antenna configuration. Also, it should have built-in optimization routines that enable it to choose among several geometrical shapes and estimate the effect of varying a design parameter. For multi-feed antennas or multiple patch array configurations, it is important for the software to model rigorously the mutual coupling effects as well.

Currently available software can rarely meet all of these requirements. Moreover, most of the commercial software packages are canned, thus they are difficult to communicate with. The full-wave modeling tool developed on one's own may easily incorporate the most recent progress in global optimization routines such as the genetic algorithm, evolution strategy[1] and space mapping technique[2]. The full-wave analysis is also the basis for using and developing the commercial software.

It should be noticed that both the CAD software and the full-wave analysis are not a substitute for antenna design experience or a thorough understanding of the principles of operation of microstrip antennas and arrays. A microstrip antenna design is based on sound scientific principle, but it also retains a significant component of intuitive understanding and a creative problem-solving approach that can only come from experience.

"Full-wave" may be used to describe finite difference time domain (FDTD) or finite element solutions, but most of the full-wave analyses of microstrip antennas have been the moment method solutions using the exact Green's function for the dielectric substrate. This technique enforces the boundary conditions at an air-dielectric interface and treats the contributions of space waves, surface waves, dielectric loss, and coupling to external structures in an accurate manner. It is also possible to apply the method to a wide variety of patch and substrate geometries, including arrays, mutual coupling effects, multiple layers, stacked elements, and various feeding methods, and it can be easily extended to infinite arrays of microstrip antennas. It is probably the most popular analysis technique for microstrip antennas and arrays. Therefore, in this chapter, the full-wave analysis of microstrip antennas using the moment method the will be systematically introduced, based on the spectral domain approach.

4.2   Basic Concept of Spectral Domain Approach[3, 4]

The spectral analysis approach has many applications and is also widely used in microstrip antennas and feed networks. This approach may be traced to the appearance of Fourier analysis more than one hundred years ago. For a periodic time variation function, it may be expanded into Fourier series containing different frequencies. For a non-periodic time variation function, the Fourier series is extended to the Fourier integral, and the corresponding frequency spectrum becomes a continuous one. In this case, two domains of the transform are time and frequency. This approach may be applied in spatial domain as well. For a given spatial distributed function, it may be superimposed by plane waves with both different frequencies and different incident angles, including real and complex components. Each plane wave is a member of the plane wave spectrum (PWS) family. This superposition corresponds to using the Fourier transform. This superposition may also be done by using cylindrical waves, which corresponds to Hankel transform. The plane wave, cylindrical wave and spherical wave may be expressed in terms of other waves, for example, a plane wave may be expressed by either cylindrical waves or spherical waves. This kind of wave transformation brings a lot of flexibility and simplicity in solving boundary value problems.

For infinite planar multilayered structures, it is difficult to find the spatial Green's function satisfying all the boundary conditions. If we expand the spherical wave produced by a point source into plane wave spectra or cylindrical wave spectra, this problem may easily be dealt with. For each plane wave or cylindrical wave spectrum, the planar multilayered structure problem may be considered as a transmission line problem and the spectral Green's functions are easy to find by spectral domain immittance (SDI) approach as will be seen later.

After the boundary condition matching is done in the spectral domain, the spatial domain Green's functions may be obtained through the inverse transform, such as inverse Fourier transform or inverse Hankel transform. Either of them is called the well-known Sommerfeld integral, which is the spectral representation of a spherical wave and will be shown in the following example.

We express any scalar field component as $f(\mathbf{r}) \equiv f(\boldsymbol{\rho}, z)$, where $\boldsymbol{\rho} = \hat{\mathbf{x}}x + \hat{\mathbf{y}}y$ is the projection of $\mathbf{r}$ on the xoy plane, and introduce the Fourier transform pair

$$Ff(\mathbf{r}) \equiv \tilde{f}(\mathbf{k}_\rho; z) = \int_{-\infty}^{\infty} \int_{-\infty}^{\infty} f(\mathbf{r}) e^{j\mathbf{k}_\rho \cdot \boldsymbol{\rho}} dx dy$$

$$= \int_{-\infty}^{\infty} \int_{-\infty}^{\infty} f(\mathbf{r}) e^{jk_x x} e^{jk_y y} dx dy \tag{4.2.1}$$

$$F^{-1} \tilde{f}(\mathbf{k}_\rho; z) \equiv f(\mathbf{r}) = \frac{1}{(2\pi)^2} \int_{-\infty}^{\infty} \int_{-\infty}^{\infty} \tilde{f}(\mathbf{k}_\rho; z) e^{-j\mathbf{k}_\rho \cdot \boldsymbol{\rho}} dk_x dk_y$$

$$= \frac{1}{(2\pi)^2} \int_{-\infty}^{\infty} \int_{-\infty}^{\infty} \tilde{f}(\mathbf{k}_\rho; z) e^{-jk_x x} e^{-jk_y y} dk_x dk_y \tag{4.2.2}$$

where $\mathbf{k}_\rho = \hat{\mathbf{x}}k_x + \hat{\mathbf{y}}k_y$, $k_\rho = \sqrt{k_x^2 + k_y^2}$. Formula (4.2.2) is known as the Sommerfeld integral.

For a spherical wave function e^{-jkr}/r, we may find its Fourier transform which results in the following Sommerfeld identity[5]

$$\frac{e^{-jkr}}{r} = \frac{1}{2\pi j} \int_{-\infty}^{\infty} \int_{-\infty}^{\infty} \frac{e^{-jk_z|z|}}{k_z} e^{-jk_x x} e^{-jk_y y} dk_x dk_y \tag{4.2.3}$$

where $r = \sqrt{x^2 + y^2 + z^2}$, $k_z = \sqrt{k^2 - k_x^2 - k_y^2}$, k is the propagation constant along $\hat{\mathbf{k}}$ direction and k_x, k_y, k_z are the propagation constants along x y z directions. They may be expressed in spherical coordinates as $k_x = k\sin\beta\cos\alpha$, $k_y = k\sin\beta\sin\alpha$ and $k_z = k\cos\beta$, with β being the angle between $\hat{\mathbf{k}}$ and $\hat{\mathbf{z}}$, α being the angle between $\hat{\mathbf{x}}$ and the direction of the projection vector of $\hat{\mathbf{k}}$ onto the xoy plane. Parameter k_z may also be expressed as

$$k_z = \begin{cases} \sqrt{k^2 - (k_x^2 + k_y^2)}, & k_x^2 + k_y^2 \leqslant k^2 \\ -j\sqrt{(k_x^2 + k_y^2) - k^2}, & k_x^2 + k_y^2 > k^2 \end{cases} \tag{4.2.4}$$

Waves for which $k_x^2 + k_y^2 \leqslant k^2$, corresponding to real angles α, β, contribute to the propagating wave. This region is called the visible region. Whereas those for which $k_x^2 + k_y^2 > k^2$ correspond to real α and complex angle $\beta = \pm(\pi/2) + j\beta_i$. In that case, $k_x = \pm k\cos\alpha\cosh\beta_i$, $k_y = \pm k\sin\alpha\cosh\beta_i$ are real numbers with $|k_x| \geqslant k, |k_y| \geqslant k; k_z = \pm jk\sinh\beta_i$. Due to the pure imaginary k_z, waves in this region are evanescent. Therefore, this region is called the invisible region. Formula (4.2.3) shows that the spherical wave function may be superimposed by the plane wave spectra $e^{-jk_z z}/k_z$. The evanescent waves are mainly responsible for the near field including the singularity at the source point. Formula (4.2.3) contains all the waves needed, therefore it is the full-wave representation.

It will be seen that in (4.2.2), only one radial variable k_ρ will suffice in place of two independent variables x and y. The appropriate formulation of such problems is in terms of the Hankel transform, a one-dimensional transform with a Bessel function kernel. For this purpose let $k_x = k_\rho\cos\alpha$, $k_y = k_\rho\sin\alpha$, $x = \rho\cos\varphi$, $y = \rho\sin\varphi$ which transform the Cartesian coordinates into cylindrical ones. According to the Jacobian determinant[6] $dk_x dk_y = k_\rho d\alpha dk_\rho$ and we have

$$f(\mathbf{r}) = \frac{1}{(2\pi)^2} \int_{-\infty}^{\infty} \int_{-\infty}^{\infty} \tilde{f}(k_\rho; z) e^{-j(k_x x + k_y y)} dk_x dk_y$$

$$= \frac{1}{(2\pi)^2} \int_{0}^{\infty} \int_{0}^{2\pi} \tilde{f}(k_\rho; z) e^{-jk_\rho\rho\cos(\alpha-\varphi)} k_\rho d\alpha dk_\rho \tag{4.2.5}$$

Letting $\alpha - \varphi = \xi$, and making use of the following identity

$$J_n(\eta) = \frac{(\mp j)^n}{2\pi} \int_{0}^{2\pi} e^{\pm j(\eta\cos\xi - n\xi)} d\xi \tag{4.2.6}$$

formula (4.2.5) may be written as

$$f(\mathbf{r}) = \frac{1}{2\pi} \int_{0}^{\infty} \tilde{f}(k_\rho; z) \left(\frac{1}{2\pi} \int_{-\varphi}^{2\pi-\varphi} e^{-jk_\rho\rho\cos\xi} d\xi \right) k_\rho dk_\rho$$

$$= \frac{1}{2\pi} \int_{0}^{\infty} \tilde{f}(k_\rho; z) \left(\frac{1}{2\pi} \int_{0}^{2\pi} e^{-jk_\rho\rho\cos\xi} d\xi \right) k_\rho dk_\rho$$

$$= \frac{1}{2\pi} \int_{0}^{\infty} \tilde{f}(k_\rho; z) J_0(k_\rho\rho) k_\rho dk_\rho \tag{4.2.7}$$

where J_0 is the zero order Bessel function which corresponds to the cylindrical standing wave.

Notice the following identities

$$\mathrm{J}_0(u) = \frac{1}{2}\left[\mathrm{H}_0^{(1)}(u) + \mathrm{H}_0^{(2)}(u)\right] \tag{4.2.8}$$

$$\mathrm{H}_0^{(1)}(u) = -\mathrm{H}_0^{(2)}(e^{-\pi j}u) \tag{4.2.9}$$

where $\mathrm{H}_0^{(1)}$ and $\mathrm{H}_0^{(2)}$ are zero order first kind and second kind Hankel functions which correspond to the inward and outward cylindrical travelling waves respectively. Considering (4.2.8), (4.2.9) and $\widetilde{f}(k_\rho; z) = \widetilde{f}(-k_\rho; z)$, formula (4.2.7) may be expressed in terms of $\mathrm{H}_0^{(2)}$

$$f(\mathbf{r}) = \frac{1}{4\pi}\int_{-\infty}^{\infty} \widetilde{f}(k_\rho; z)\,\mathrm{H}_0^{(2)}(k_\rho\rho)k_\rho dk_\rho \tag{4.2.10}$$

Both (4.2.7) and (4.2.10) are known as the Hankel transform, and also as the Sommerfeld integrals.

Formula (4.2.3) thus may be written in terms of Hankel transform

$$\frac{e^{-jkr}}{r} = \frac{1}{j}\int_0^{\infty} \frac{e^{-jk_z|z|}}{k_z}\,\mathrm{J}_0(k_\rho\rho)k_\rho dk_\rho \tag{4.2.11}$$

$$= \frac{1}{2j}\int_{-\infty}^{\infty} \frac{e^{-jk_z|z|}}{k_z}\,\mathrm{H}_0^{(1)}(k_\rho\rho)k_\rho dk_\rho$$

$$= \frac{1}{2j}\int_{-\infty}^{\infty} \frac{e^{-jk_z|z|}}{k_z}\,\mathrm{H}_0^{(2)}(k_\rho\rho)k_\rho dk_\rho \tag{4.2.12}$$

which is another kind of Sommerfeld identity showing that the spherical wave function may also be superimposed by cylindrical wave spectra.

4.3   Some Useful Transform Relations

Since the layered medium is invariant along the x and y coordinates and all the quantities depend on $x-x'$ and $y-y'$, where superscript prime denotes the source point, it is convenient to introduce the "shifted" Fourier transform pair

$$F\{f(x - x', y - y')\} = \widetilde{f}(k_x, k_y)$$

$$= \int_{-\infty}^{\infty}\int_{-\infty}^{\infty} f(x - x', y - y')e^{j[k_x(x-x')+k_y(y-y')]}dxdy \tag{4.3.1}$$

$$F^{-1}\{\widetilde{f}(k_x, k_y)\} = f(x - x', y - y')$$

$$= \frac{1}{(2\pi)^2}\int_{-\infty}^{\infty}\int_{-\infty}^{\infty} \widetilde{f}(k_x, k_y)e^{-j[k_x(x-x')+k_y(y-y')]}dk_x dk_y \tag{4.3.2}$$

By changing to the polar coordinates in both the transform and space domains according to

$$x - x' = \rho\cos\varphi, \quad y - y' = \rho\sin\varphi \tag{4.3.3}$$

$$k_x = k_\rho\cos\alpha, \quad k_y = k_\rho\sin\alpha \tag{4.3.4}$$

where

$$\rho = \sqrt{(x - x')^2 + (y - y')^2}, \quad \varphi = \tan^{-1}\left(\frac{y - y'}{x - x'}\right) \tag{4.3.5}$$

$$k_\rho = \sqrt{k_x^2 + k_y^2}, \quad \alpha = \tan^{-1}\left(\frac{k_y}{k_x}\right) \tag{4.3.6}$$

we can conveniently express various inverse Fourier integrals that arise in terms of the Sommerfeld-type integrals of the form[7]

$$S_n[\tilde{f}(k_\rho)] = \frac{1}{2\pi} \int_0^\infty \tilde{f}(k_\rho) \, J_n(k_\rho\rho) k_\rho^{n+1} dk_\rho \tag{4.3.7}$$

Using the expression in (4.3.7), (4.2.7) may be rewritten as

$$\mathrm{F}^{-1}\{\tilde{f}(k_\rho)\} = S_0[\tilde{f}(k_\rho)] \tag{4.3.8}$$

Some other useful transformations expressed in terms of S_n are as follows

$$\mathrm{F}^{-1}\left\{jk_x\tilde{f}(k_\rho)\right\} = \cos\varphi S_1\left[\tilde{f}(k_\rho)\right] \tag{4.3.9}$$

$$\mathrm{F}^{-1}\left\{jk_y\tilde{f}(k_\rho)\right\} = \sin\varphi S_1\left[\tilde{f}(k_\rho)\right] \tag{4.3.10}$$

$$\mathrm{F}^{-1}\left\{k_x^2\tilde{f}(k_\rho)\right\} = -\frac{1}{2}\left\{\cos2\varphi S_2\left[\tilde{f}(k_\rho)\right] - S_0\left[k_\rho^2\tilde{f}(k_\rho)\right]\right\} \tag{4.3.11}$$

$$\mathrm{F}^{-1}\left\{k_y^2\tilde{f}(k_\rho)\right\} = \frac{1}{2}\left\{\cos2\varphi S_2\left[\tilde{f}(k_\rho)\right] + S_0\left[k_\rho^2\tilde{f}(k_\rho)\right]\right\} \tag{4.3.12}$$

$$\mathrm{F}^{-1}\left\{k_xk_y\tilde{f}(k_\rho)\right\} = -\frac{1}{2}\sin2\varphi S_2[\tilde{f}(k_\rho)] \tag{4.3.13}$$

The derivation of (4.3.11) is given as an example

$$\mathrm{F}^{-1}\left\{k_x^2\tilde{f}(k_\rho)\right\} = \frac{1}{(2\pi)^2}\int_0^\infty\int_0^{2\pi} k_x^2\tilde{f}(k_\rho)e^{-jk_\rho\rho\cos(\alpha-\varphi)}k_\rho d\alpha dk_\rho$$

$$= \frac{1}{(2\pi)^2}\int_0^\infty\left[\int_0^{2\pi}\left(\frac{e^{j\alpha}+e^{-j\alpha}}{2}\right)^2\tilde{f}(k_\rho)\right.$$

$$\left. e^{-jk_\rho\rho\cos(\alpha-\varphi)}d\alpha\right]k_\rho^3 dk_\rho$$

$$= \frac{1}{(2\pi)^2}\int_0^\infty\left[\int_{-\varphi}^{-\varphi+2\pi}\frac{1}{4}e^{j2(\xi+\varphi)}e^{-jk_\rho\rho\cos\xi}\tilde{f}(k_\rho)d\xi\right.$$

$$+ \int_{-\varphi}^{-\varphi+2\pi}\frac{1}{4}e^{-j2(\xi+\varphi)}e^{-jk_\rho\rho\cos\xi}\tilde{f}(k_\rho)d\xi$$

$$\left.+ \int_{-\varphi}^{-\varphi+2\pi}\frac{1}{2}\tilde{f}(k_\rho)e^{-jk_\rho\rho\cos\xi}d\xi\right]k_\rho^3 dk_\rho$$

$$= -\frac{1}{2}\left\{\frac{1}{2}e^{j2\varphi}\frac{1}{2\pi}\int_0^\infty\tilde{f}(k_\rho)\,J_2(k_\rho\rho)k_\rho^3 dk_\rho\right.$$

$$+ \frac{1}{2}e^{-j2\varphi}\frac{1}{2\pi}\int_0^\infty\tilde{f}(k_\rho)\,J_2(k_\rho\rho)k_\rho^3 dk_\rho$$

$$\left.- \frac{1}{2\pi}\int_0^\infty k_\rho^2\tilde{f}(k_\rho)\,J_0(k_\rho\rho)k_\rho dk_\rho\right\}$$

$$= -\frac{1}{2}\left\{\cos2\varphi S_2\left[\tilde{f}(k_\rho)\right] - S_0\left[k_\rho^2\tilde{f}(k_\rho)\right]\right\}$$

In the above derivation, the periodicity

$$e^{\pm j2\xi}e^{-jk_\rho\rho\cos\xi} = e^{\pm j2(\xi+2\pi)}e^{-jk_\rho\rho\cos(\xi+2\pi)} \tag{4.3.14}$$

and the property of the Bessel function[8]

$$\mathrm{J}_n(-x) = \mathrm{J}_{-n}(x) = (-1)^n\,\mathrm{J}_n(x) \tag{4.3.15}$$

are used.

Following a similar derivation, we obtain another type of Sommerfeld integral[9]

$$
\begin{aligned}
\mathrm{F}^{-1}\{e^{jn\alpha}\widetilde{f}(k_\rho)\} &= \frac{1}{(2\pi)^2}\int_0^\infty\int_0^{2\pi}e^{jn\alpha}\widetilde{f}(k_\rho)e^{-jk_\rho\rho\cos(\alpha-\varphi)}k_\rho\,d\alpha\,dk_\rho \\
&= \frac{1}{2\pi}e^{jn\varphi}(-j)^n\int_0^\infty\widetilde{f}(k_\rho)\,\mathrm{J}_n(k_\rho\rho)k_\rho\,dk_\rho \\
&= e^{jn\varphi}(-j)^n S_n\{\widetilde{f}(k_\rho)\}
\end{aligned}
\tag{4.3.16}
$$

where

$$S_n\{\widetilde{f}(k_\rho)\} = \frac{1}{2\pi}\int_0^\infty\widetilde{f}(k_\rho)\,\mathrm{J}_n(k_\rho\rho)k_\rho\,dk_\rho \tag{4.3.17}$$

which is different from that in (4.3.7). Formula (4.3.16) results in

$$\mathrm{F}^{-1}\left\{\begin{matrix}\sin\\\cos\end{matrix}\ n\alpha\widetilde{f}(k_\rho)\right\} = (-j)^n\ \begin{matrix}\sin\\\cos\end{matrix}\ n\varphi S_n\{\widetilde{f}(k_\rho)\} \tag{4.3.18}$$

where $S_n\{\widetilde{f}(k_\rho)\}$ is defined in (4.3.17).

The Sommerfeld identity may also be extended into some modified forms. For example, if we take the derivatives $\partial^2 f/\partial z\partial\rho$ on both sides of the Sommerfeld identity (4.2.12)

$$\frac{\partial^2}{\partial z\partial\rho}\left(\frac{e^{-jkr}}{r}\right) = \frac{\partial^2}{\partial z\partial\rho}\left(\frac{1}{2j}\int_{-\infty}^\infty\frac{e^{-jk_z|z|}}{k_z}\mathrm{H}_0(k_\rho\rho)k_\rho\,dk_\rho\right) \tag{4.3.19}$$

We have the following useful modified Sommerfeld identity[10]

$$\frac{e^{-jkr}}{r^5}(3+j3kr-k^2r^2)\rho|z| = \frac{1}{2}\int_{-\infty}^\infty e^{-jk_z|z|}H_1^{(2)}(k_\rho\rho)k_\rho^2\,dk_\rho \tag{4.3.20}$$

4.4   Scalarization of Maxwell's Equations

Consider a uniaxially anisotropic, possibly lossy medium, which is transversely unbounded with respect to the z axis and is characterized, relative to free space, by z-dependence, in general complex-valued permeability and permittivity dyadic, $\underline{\underline{\mu}} = \underline{\underline{I}}_t\mu_t + \hat{z}\hat{z}\mu_z$ and $\underline{\underline{\epsilon}} = \underline{\underline{I}}_t\epsilon_t + \hat{z}\hat{z}\epsilon_z$, respectively, where $\underline{\underline{I}}_t$ is the transverse unit dyadic. We wish to compute the fields $(\mathbf{E}, \mathbf{H})$ at an arbitrary point $\mathbf{r}$ due to a specified current distribution $(\mathbf{J}, \mathbf{M})$[9]. For the isotropic medium, $\mu_t = \mu_z = \mu_r$, $\epsilon_t = \epsilon_z = \epsilon_r$.

These fields are governed by the Maxwell's equations

$$\nabla\times\mathbf{E} = -j\omega\mu_0\underline{\underline{\mu}}\cdot\mathbf{H} - \mathbf{M} \tag{4.4.1}$$

$$\nabla\times\mathbf{H} = j\omega\epsilon_0\underline{\underline{\epsilon}}\cdot\mathbf{E} + \mathbf{J} \tag{4.4.2}$$

Since the medium is homogeneous and of infinite extent in any transverse (to z) plane, the analysis is facilitated by the Fourier transformation of all fields with respect to the transverse coordinates. Upon applying (4.2.1) to (4.4.1), (4.4.2) and separating the transverse and longitudinal parts of the resulting equations[11]

$$\left(\widetilde{\nabla}_t + \hat{\mathbf{z}}\frac{d}{dz}\right) \times (\widetilde{\mathbf{E}}_t + \widetilde{\mathbf{E}}_z) = -j\omega\mu_0\underline{\boldsymbol{\mu}} \cdot (\widetilde{\mathbf{H}}_t + \widetilde{\mathbf{H}}_z) - \widetilde{\mathbf{M}} \qquad (4.4.3)$$

$$\left(\widetilde{\nabla}_t + \hat{\mathbf{z}}\frac{d}{dz}\right) \times (\widetilde{\mathbf{H}}_t + \widetilde{\mathbf{H}}_z) = j\omega\epsilon_0\underline{\boldsymbol{\epsilon}} \cdot (\widetilde{\mathbf{E}}_t + \widetilde{\mathbf{E}}_z) + \widetilde{\mathbf{J}} \qquad (4.4.4)$$

where $\widetilde{\nabla}_t = -j(k_x\hat{\mathbf{x}} + k_y\hat{\mathbf{y}})$. After several following steps of derivation

$$\widetilde{\nabla}_t \times \widetilde{\mathbf{E}}_t + \frac{d}{dz}\hat{\mathbf{z}} \times \widetilde{\mathbf{E}}_t + \widetilde{\nabla}_t \times \widetilde{\mathbf{E}}_z + \frac{d}{dz}\hat{\mathbf{z}} \times \widetilde{\mathbf{E}}_z$$
$$= -j\omega\mu_0\mu_t\widetilde{\mathbf{H}}_t - j\omega\mu_0\mu_z\widetilde{\mathbf{H}}_z - \widetilde{\mathbf{M}}_z - \widetilde{\mathbf{M}}_t \qquad (4.4.5)$$

$$\widetilde{\nabla}_t \times \widetilde{\mathbf{H}}_t + \frac{d}{dz}\hat{\mathbf{z}} \times \widetilde{\mathbf{H}}_t + \widetilde{\nabla}_t \times \widetilde{\mathbf{H}}_z + \frac{d}{dz}\hat{\mathbf{z}} \times \widetilde{\mathbf{H}}_z$$
$$= j\omega\epsilon_0\epsilon_t\widetilde{\mathbf{E}}_t + j\omega\epsilon_0\epsilon_z\widetilde{\mathbf{E}}_z + \widetilde{\mathbf{J}}_z + \widetilde{\mathbf{J}}_t \qquad (4.4.6)$$

$$\Downarrow$$

$$\widetilde{\nabla}_t \times \widetilde{\mathbf{E}}_z + \frac{d}{dz}\hat{\mathbf{z}} \times \widetilde{\mathbf{E}}_t = -j\omega\mu_0\mu_t\widetilde{\mathbf{H}}_t - \widetilde{\mathbf{M}}_t \qquad (4.4.7)$$

$$\widetilde{\nabla}_t \times \widetilde{\mathbf{H}}_z + \frac{d}{dz}\hat{\mathbf{z}} \times \widetilde{\mathbf{H}}_t = j\omega\epsilon_0\epsilon_t\widetilde{\mathbf{E}}_t + \widetilde{\mathbf{J}}_t \qquad (4.4.8)$$

$$\widetilde{\nabla}_t \times \widetilde{\mathbf{E}}_t = -j\omega\mu_0\mu_z\widetilde{\mathbf{H}}_z - \widetilde{\mathbf{M}}_z = j\mathbf{k}_\rho \cdot (\hat{\mathbf{z}} \times \widetilde{\mathbf{E}}_t)\hat{\mathbf{z}} \qquad (4.4.9)$$

$$\widetilde{\nabla}_t \times \widetilde{\mathbf{H}}_t = j\omega\epsilon_0\epsilon_z\widetilde{\mathbf{E}}_z + \widetilde{\mathbf{J}}_z = -j\mathbf{k}_\rho \cdot (\widetilde{\mathbf{H}}_t \times \hat{\mathbf{z}})\hat{\mathbf{z}} \qquad (4.4.10)$$

$$\Downarrow$$

$$\widetilde{\nabla}_t \times \widetilde{\mathbf{E}}_z \times \hat{\mathbf{z}} + \frac{d}{dz}\hat{\mathbf{z}} \times \widetilde{\mathbf{E}}_t \times \hat{\mathbf{z}} = -j\omega\mu_0\mu_t\widetilde{\mathbf{H}}_t \times \hat{\mathbf{z}} - \widetilde{\mathbf{M}}_t \times \hat{\mathbf{z}} \qquad (4.4.11)$$

$$\widetilde{\nabla}_t \times \widetilde{\mathbf{H}}_z \times \hat{\mathbf{z}} + \frac{d}{dz}\hat{\mathbf{z}} \times \widetilde{\mathbf{H}}_t \times \hat{\mathbf{z}} = j\omega\epsilon_0\epsilon_t\widetilde{\mathbf{E}}_t \times \hat{\mathbf{z}} + \widetilde{\mathbf{J}}_t \times \hat{\mathbf{z}} \qquad (4.4.12)$$

$$-j\omega\epsilon_0\epsilon_z\widetilde{E}_z = j\mathbf{k}_\rho \cdot (\widetilde{\mathbf{H}}_t \times \hat{\mathbf{z}}) + \widetilde{J}_z \qquad (4.4.13)$$

$$-j\omega\mu_0\mu_z\widetilde{H}_z = j\mathbf{k}_\rho \cdot (\hat{\mathbf{z}} \times \widetilde{\mathbf{E}}_t) + \widetilde{M}_z \qquad (4.4.14)$$

we arrive at

$$\frac{d\widetilde{\mathbf{E}}_t}{dz} = \frac{1}{j\omega\epsilon_0\epsilon_t}(k_t^2 - \nu^e\mathbf{k}_\rho\mathbf{k}_\rho\cdot)(\widetilde{\mathbf{H}}_t \times \hat{\mathbf{z}})$$

$$+ \mathbf{k}_\rho\frac{\widetilde{J}_z}{\omega\epsilon_0\epsilon_z} - \widetilde{\mathbf{M}}_t \times \hat{\mathbf{z}} \qquad (4.4.15)$$

$$\frac{d\widetilde{\mathbf{H}}_t}{dz} = \frac{1}{j\omega\mu_0\mu_t}(k_t^2 - \nu^h\mathbf{k}_\rho\mathbf{k}_\rho\cdot)(\hat{\mathbf{z}} \times \widetilde{\mathbf{E}}_t)$$

$$+ \mathbf{k}_\rho\frac{\widetilde{M}_z}{\omega\epsilon_0\epsilon_z} - \hat{\mathbf{z}} \times \widetilde{\mathbf{J}}_t \qquad (4.4.16)$$

$$-j\omega\epsilon_0\epsilon_z\widetilde{E}_z = j\mathbf{k}_\rho \cdot (\widetilde{\mathbf{H}}_t \times \hat{\mathbf{z}}) + \widetilde{J}_z \qquad (4.4.17)$$

$$-j\omega\mu_0\mu_z\widetilde{H}_z = j\mathbf{k}_\rho \cdot (\hat{\mathbf{z}} \times \widetilde{\mathbf{E}}_t) + \widetilde{M}_z \qquad (4.4.18)$$

where $k_t = k_0\sqrt{\mu_t\epsilon_t}$ and $k_0 = \omega\sqrt{\mu_0\epsilon_0}$ (being the free-space wavenumber), $\nu^e = \epsilon_t/\epsilon_z$ and $\nu^h = \mu_t/\mu_z$ are referred to as, respectively, the electric and magnetic anisotropy ratios. The derivation of (4.4.15) and (4.4.16) is not quite straightforward and thus needs some explanation. Take the derivation of (4.4.15) from (4.4.11) as an example

$$\frac{d\widetilde{\mathbf{E}}_t}{dz} = -\nabla_t \times \widetilde{\mathbf{E}}_z \times \hat{\mathbf{z}} - j\omega\mu_0\mu_t(\widetilde{\mathbf{H}}_t \times \hat{\mathbf{z}}) - \widetilde{\mathbf{M}}_t \times \hat{\mathbf{z}}$$

$$= -(jk_y\hat{\mathbf{y}} + jk_x\hat{\mathbf{x}})\widetilde{E}_z - j\omega\mu_0\mu_t(\widetilde{\mathbf{H}}_t \times \hat{\mathbf{z}}) - \widetilde{\mathbf{M}}_t \times \hat{\mathbf{z}}$$

$$= -j\mathbf{k}_\rho \frac{j\mathbf{k}_\rho \cdot (\widetilde{\mathbf{H}}_t \times \hat{\mathbf{z}}) + \widetilde{J}_z}{-j\omega\epsilon_0\epsilon_z} - j\omega\mu_0\mu_t(\widetilde{\mathbf{H}}_t \times \hat{\mathbf{z}}) - \widetilde{\mathbf{M}}_t \times \hat{\mathbf{z}}$$

$$= \frac{1}{j\omega\epsilon_0\epsilon_t}\left[\omega^2\mu_0\epsilon_0\epsilon_t\mu_t - \frac{\epsilon_t}{\epsilon_z}\mathbf{k}_\rho\mathbf{k}_\rho\cdot\right](\widetilde{\mathbf{H}}_t \times \hat{\mathbf{z}}) + \frac{\mathbf{k}_\rho\widetilde{J}_z}{\omega\epsilon_0\epsilon_z} - \widetilde{\mathbf{M}}_t \times \hat{\mathbf{z}}$$

The subsequent analysis is greatly simplified if one defines a rotated spectrum-domain coordinate system based on $(\mathbf{k}_\rho, \hat{\mathbf{z}} \times \mathbf{k}_\rho)$ (see Figure 4.1), with the unit vectors $(\hat{\mathbf{u}}, \hat{\mathbf{v}})$ given by

$$\hat{\mathbf{u}} = \frac{k_x}{k_\rho}\hat{\mathbf{x}} + \frac{k_y}{k_\rho}\hat{\mathbf{y}} \tag{4.4.19}$$

$$\hat{\mathbf{v}} = -\frac{k_y}{k_\rho}\hat{\mathbf{x}} + \frac{k_x}{k_\rho}\hat{\mathbf{y}} \tag{4.4.20}$$

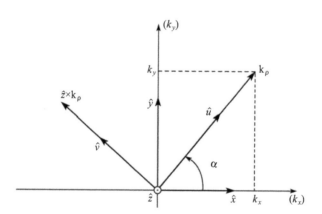

Figure 4.1   Rotated spectrum-domain coordinate system.

where $k_\rho = \sqrt{k_x^2 + k_y^2}$. If we now express the transverse electric and magnetic fields as

$$\widetilde{\mathbf{E}}_t = \hat{\mathbf{u}}V^e + \hat{\mathbf{v}}V^h \tag{4.4.21}$$

$$\widetilde{\mathbf{H}}_t \times \hat{\mathbf{z}} = \hat{\mathbf{u}}I^e + \hat{\mathbf{v}}I^h \tag{4.4.22}$$

and project (4.4.15) and (4.4.16) onto $\hat{\mathbf{u}}$ and $\hat{\mathbf{v}}$, we find that these equations are decoupled into two sets of transmission line equations of the form

$$\frac{dV^p}{dz} = -jk_z^p Z^p I^p + v^p \tag{4.4.23}$$

$$\frac{dI^p}{dz} = -jk_z^p Y^p V^p + i^p \tag{4.4.24}$$

where the superscript p may be e or h. Hence, the components of $\widetilde{\mathbf{E}}_t$ and $\widetilde{\mathbf{H}}_t$ in the (u, v) plane may be interpreted as voltages and currents on a transmission-line analog of the medium along the z axis, which was anticipated in the notation introduced in (4.4.21) and (4.4.22). The propagation wavenumbers and the characteristic impedances and admittances of this transmission line are given as

$$k_z^p = \sqrt{k_t^2 - \nu^p k_\rho^2} \tag{4.4.25}$$

$$Z^e = \frac{1}{Y^e} = \frac{k_z^e}{\omega\epsilon_0\epsilon_t} \tag{4.4.26}$$

$$Z^h = \frac{1}{Y^h} = \frac{\omega\mu_0\mu_t}{k_z^h} \tag{4.4.27}$$

where the square root branch in (4.4.25) is specified by the condition that $-\pi < \arg\{k_z^p\} \leqslant 0$. The voltage and current sources in (4.4.23) and (4.4.24) are given by

$$v^e = \frac{k_\rho}{\omega\epsilon_0\epsilon_z}\widetilde{J}_z - \widetilde{M}_v, \qquad i^e = -\widetilde{J}_u \tag{4.4.28}$$

$$i^h = -\frac{k_\rho}{\omega\mu_0\mu_z}\widetilde{M}_z - \widetilde{J}_v, \qquad v^h = \widetilde{M}_u \tag{4.4.29}$$

In view of (4.4.21), (4.4.22), (4.4.13) and (4.4.14), the spectral fields may now be expressed as

$$\widetilde{\mathbf{E}} = \hat{\mathbf{u}}V^e + \hat{\mathbf{v}}V^h - \hat{\mathbf{z}}\frac{1}{j\omega\epsilon_0\epsilon_z}(jk_\rho I^e + \widetilde{J}_z) \tag{4.4.30}$$

$$\widetilde{\mathbf{H}} = -\hat{\mathbf{u}}I^h + \hat{\mathbf{v}}I^e + \hat{\mathbf{z}}\frac{1}{j\omega\mu_0\mu_z}(jk_\rho V^h - \widetilde{M}_z) \tag{4.4.31}$$

which indicate that outside the source region (V^e, I^e) and (V^h, I^h) represent fields that are, respectively, TM and TE to z. The space-domain fields $(\mathbf{E}, \mathbf{H})$ are obtained from (4.4.30) and (4.4.31) via the inverse transform (4.2.2).

The original vector problem has thus been reduced to the scalar transmission line problem. Note that, since superscript p represents e or h, two transmission lines are involved and associated, respectively, with the TM and TE partial fields.

The decoupling may also be implemented through the diagnalization of the matrix which relates $[E_x, E_y]^T$ and $[H_x, H_y]^{T\,[3]}$. The simultaneous implementation of Fourier transformation and diagnalization yields the vector Fourier transform[12]. The vector transform may also be extended to Hankel transform which yields the vector Hankel transform[12]. For structures with circular symmetry, this kind of transform is convenient to use[13].

4.5   Dyadic Green's Function (DGF)

Consider the solutions of the transmission line equation (4.4.23), (4.4.24) for unit-strength impulsive sources. Hence, let $V_i^p(z|z')$ and $I_i^p(z|z')$ denote the voltage and current, respectively, at z due to a 1-A shunt current source at z', and let $V_v^p(z|z')$, and $I_v^p(z|z')$ denote the voltage and current, respectively, at z due to a 1-V series voltage source at z' (see Figure

4.2). Then, from (4.4.23), (4.4.24) it follows

$$\frac{dV_i^p}{dz} = -jk_z^p Z^p I_i^p \tag{4.5.1}$$

$$\frac{dI_i^p}{dz} = -jk_z^p Y^p V_i^p + \delta(z - z') \tag{4.5.2}$$

$$\frac{dV_v^p}{dz} = -jk_z^p Z^p I_v^p + \delta(z - z') \tag{4.5.3}$$

$$\frac{dI_v^p}{dz} = -jk_z^p Y^p V_v^p \tag{4.5.4}$$

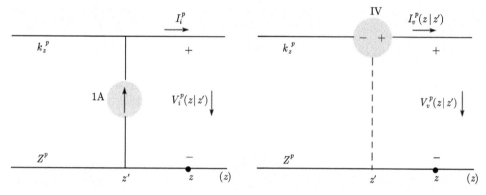

Figure 4.2　Network problems for the determination of the transmission-line Green's function.

where δ is the Dirac delta function, and they possess the reciprocity properties.

$$V_i^p(z|z') = V_i^p(z'|z) \tag{4.5.5}$$

$$I_v^p(z|z') = I_v^p(z'|z) \tag{4.5.6}$$

$$V_v^p(z|z') = -I_i^p(z'|z) \tag{4.5.7}$$

$$I_i^p(z|z') = -V_v^p(z'|z) \tag{4.5.8}$$

The above relationship can be easily seen from Figure 4.3. The linearity of the transmission line equations (4.4.23), (4.4.24) allows one to obtain (V^p, I^p) at any point z via the superposition integrals.

$$V^p = \langle V_i^p, i^p \rangle + \langle V_v^p, v^p \rangle \tag{4.5.9}$$

$$I^p = \langle I_i^p, i^p \rangle + \langle I_v^p, v^p \rangle \tag{4.5.10}$$

The notation $\langle , \rangle$ is used for integrals of products of two functions separated by the comma over their common spatial support. Upon substituting these equations into (4.4.30), (4.4.31) and using (4.4.28), (4.4.29), one obtains

$$\widetilde{\mathbf{E}} = \langle \widetilde{\underline{\mathbf{G}}}^{EJ}; \widetilde{\mathbf{J}} \rangle + \langle \widetilde{\underline{\mathbf{G}}}^{EM}; \widetilde{\mathbf{M}} \rangle \tag{4.5.11}$$

$$\widetilde{\mathbf{H}} = \langle \widetilde{\underline{\mathbf{G}}}^{HJ}; \widetilde{\mathbf{J}} \rangle + \langle \widetilde{\underline{\mathbf{G}}}^{HM}; \widetilde{\mathbf{M}} \rangle \tag{4.5.12}$$

where a dot over the comma in $\langle \ \rangle$ indicates a dot product. The derivation of $\widetilde{\underline{\mathbf{G}}}^{PQ}(\mathbf{k}_\rho; z|z')$ will be given in the following. From (4.4.30)

$$\widetilde{\mathbf{E}} = \hat{\mathbf{u}} V^e + \hat{\mathbf{v}} V^h - \hat{\mathbf{z}} \frac{1}{j\omega\epsilon_0\epsilon_z}(jk_\rho I^e + \widetilde{J}_z \delta(z - z'))$$

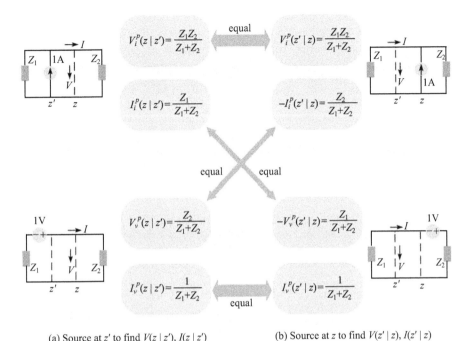

(a) Source at z' to find $V(z \mid z')$, $I(z \mid z')$ (b) Source at z to find $V(z' \mid z)$, $I(z' \mid z)$

Figure 4.3 Pictorial proof of the reciprocity properties in (4.5.5)–(4.5.8).

$$
\begin{aligned}
&= \hat{\mathbf{u}}(\langle V_i^e, i^e\rangle + \langle V_v^e, v^e\rangle) + \hat{\mathbf{v}}(\langle V_i^h, i^h\rangle + \langle V_v^h, v^h\rangle) \\
&\quad - \hat{\mathbf{z}}\frac{jk_\rho}{j\omega\epsilon_0\epsilon_z}(\langle I_i^e, i^e\rangle + \langle I_v^e, v^e\rangle) - \hat{\mathbf{z}}\frac{\widetilde{J}_z\delta(z-z')}{j\omega\epsilon_0\epsilon_z'}
\end{aligned}
\tag{4.5.13}
$$

We derive $\hat{\mathbf{u}}, \hat{\mathbf{v}}$ and $\hat{\mathbf{z}}$ components separately. For $\hat{\mathbf{u}}$ component

$$
\begin{aligned}
\langle V_i^e, i^e\rangle + \langle V_v^e, v^e\rangle &= \langle V_i^e, -\widetilde{J}_u\rangle + \langle V_v^e, \frac{k_\rho}{\omega\epsilon_0\epsilon_z'}\widetilde{J}_z - \widetilde{M}_v\rangle \\
&= -\langle V_i^e, \widetilde{J}_u\rangle + \langle V_v^e, \frac{k_\rho}{\omega\epsilon_0\epsilon_z'}\widetilde{J}_z\rangle - \langle V_v^e, \widetilde{M}_v\rangle
\end{aligned}
$$

where $\widetilde{J}_u$, $\widetilde{J}_z$ and $\widetilde{M}_v$ are the sources along $\hat{\mathbf{u}}$, $\hat{\mathbf{z}}$ and $\hat{\mathbf{v}}$ respectively. To find the Green's function, we put all of them to be 1. When the source point and field point are located in different media and to keep the continuity of the normal component of the electric flux density D on the boundary between two media, the dielectric constant related to the normal electric source should be evaluated at the source coordinate z' and is primed. Then we have the $\hat{\mathbf{u}}$ component of the DGF's $\widetilde{\underline{\mathbf{G}}}^{EJ}$ and $\widetilde{\underline{\mathbf{G}}}^{EM}$

$$
-\hat{\mathbf{u}}\hat{\mathbf{u}}V_i^e + \hat{\mathbf{u}}\hat{\mathbf{z}}\frac{k_\rho V_v^e}{\omega\epsilon_0\epsilon_z'} - \hat{\mathbf{u}}\hat{\mathbf{v}}V_v^e
$$

Through similar derivation, we obtain $\hat{\mathbf{v}}$ and $\hat{\mathbf{z}}$ components of $\widetilde{\underline{\mathbf{G}}}^{EJ}$ and $\widetilde{\underline{\mathbf{G}}}^{EM}$. From (4.4.31)

$$
\begin{aligned}
\widetilde{\mathbf{H}} &= -\hat{\mathbf{u}}I^h + \hat{\mathbf{v}}I^e + \hat{\mathbf{z}}\frac{1}{j\omega\mu_0\mu_z}(jk_\rho V^h - \widetilde{M}_z\delta(z-z')) \\
&= -\hat{\mathbf{u}}(\langle I_i^h, i^h\rangle + \langle I_v^h, v^h\rangle) + \hat{\mathbf{v}}(\langle I_i^e, i^e\rangle + \langle I_v^e, v^e\rangle) \\
&\quad + \hat{\mathbf{z}}\frac{jk_\rho}{j\omega\mu_0\mu_z}(\langle V_i^h, i^h\rangle + \langle V_v^h, v^h\rangle) - \hat{\mathbf{z}}\frac{\widetilde{M}_z\delta(z-z')}{j\omega\mu_0\mu_z'}
\end{aligned}
\tag{4.5.14}
$$

In (4.5.14), when the source point and field point are located in different media and to keep the continuity of the normal component of the magnetic flux density on the boundary between two media, the permeability related to the normal magnetic source should also be evaluated at source coordinates z' and is primed. From (4.5.14), we obtain $\widetilde{\underline{\mathbf{G}}}^{HJ}$ and $\widetilde{\underline{\mathbf{G}}}^{HM}$. Alternatively, $\widetilde{\underline{\mathbf{G}}}^{HJ}$ and $\widetilde{\underline{\mathbf{G}}}^{HM}$ may also be obtained from $\widetilde{\underline{\mathbf{G}}}^{EJ}$ and $\widetilde{\underline{\mathbf{G}}}^{EM}$ by duality. The complete results are as follows

$$\widetilde{\underline{\mathbf{G}}}^{EJ} = -\hat{\mathbf{u}}\hat{\mathbf{u}}V_i^e - \hat{\mathbf{v}}\hat{\mathbf{v}}V_i^h + \hat{\mathbf{z}}\hat{\mathbf{u}}\frac{k_\rho}{\omega\epsilon_0\epsilon_z}I_i^e + \hat{\mathbf{u}}\hat{\mathbf{z}}\frac{k_\rho}{\omega\epsilon_0\epsilon_z'}V_v^e$$

$$+\hat{\mathbf{z}}\hat{\mathbf{z}}\frac{1}{j\omega\epsilon_0\epsilon_z}\left[\frac{k_\rho^2}{j\omega\epsilon_0\epsilon_z'}I_v^e - \delta(z - z')\right] \tag{4.5.15}$$

$$\widetilde{\underline{\mathbf{G}}}^{HJ} = \hat{\mathbf{u}}\hat{\mathbf{v}}I_i^h - \hat{\mathbf{v}}\hat{\mathbf{u}}I_i^e - \hat{\mathbf{z}}\hat{\mathbf{v}}\frac{k_\rho}{\omega\mu_0\mu_z}V_i^h + \hat{\mathbf{v}}\hat{\mathbf{z}}\frac{k_\rho}{\omega\epsilon_0\epsilon_z'}I_v^e \tag{4.5.16}$$

$$\widetilde{\underline{\mathbf{G}}}^{EM} = -\hat{\mathbf{u}}\hat{\mathbf{v}}V_v^e + \hat{\mathbf{v}}\hat{\mathbf{u}}V_v^h + \hat{\mathbf{z}}\hat{\mathbf{v}}\frac{k_\rho}{\omega\epsilon_0\epsilon_z}I_v^e - \hat{\mathbf{v}}\hat{\mathbf{z}}\frac{k_\rho}{\omega\mu_0\mu_z'}V_i^h \tag{4.5.17}$$

$$\widetilde{\underline{\mathbf{G}}}^{HM} = -\hat{\mathbf{u}}\hat{\mathbf{u}}I_v^h - \hat{\mathbf{v}}\hat{\mathbf{v}}I_v^e + \hat{\mathbf{z}}\hat{\mathbf{u}}\frac{k_\rho}{\omega\mu_0\mu_z}V_v^h + \hat{\mathbf{u}}\hat{\mathbf{z}}\frac{k_\rho}{\omega\mu_0\mu_z'}I_i^h$$

$$+\hat{\mathbf{z}}\hat{\mathbf{z}}\frac{1}{j\omega\mu_0\mu_z'}\left[\frac{k_\rho^2}{j\omega\mu_0\mu_z}V_i^h - \delta(z - z')\right] \tag{4.5.18}$$

(4.5.15)–(4.5.18) may be easily transformed to a (x, y, z) coordinate system through (4.4.19) and (4.4.20). For example,

$$\widetilde{G}_{xx}^{EJ} = -V_i^e\frac{k_x^2}{k_\rho^2} - V_i^h\frac{k_y^2}{k_\rho^2} \tag{4.5.19}$$

$$\widetilde{G}_{yy}^{EJ} = -V_i^e\frac{k_y^2}{k_\rho^2} - V_i^h\frac{k_x^2}{k_\rho^2} \tag{4.5.20}$$

4.6 Mixed Potential Representations

The hypersingular behavior of some integral equation kernels causes difficulties in the solution procedure[14], which may be avoided if the fields are expressed in terms of vector and scalar potentials with weakly singular kernels. This led to the development of mixed-potential integral equations (MPIEs) for arbitrarily shaped scatterers in free space[15]. In layered media, an important advantage of the MPIEs is that the spectral Sommerfeld-type integrals (or series, in the case of laterally shielded environments) appearing in the potential kernels converge more rapidly and are easier to accelerate than those associated with the field forms that are obtained by differentiation of the potentials. For mixed potential representations, consider first the case where only electric current is present. It is then permissible to express the fields in terms of vector and scalar potentials through the equations

$$\mu_0\underline{\mu} \cdot \mathbf{H} = \nabla \times \mathbf{A} \tag{4.6.1}$$

$$\mathbf{E} = -j\omega\mathbf{A} - \nabla\phi \tag{4.6.2}$$

According to the definition of vector potential DGF $\underline{\mathbf{G}}^A(\mathbf{r}|\mathbf{r}')$

$$\mathbf{A} = \mu_0\langle\underline{\mathbf{G}}^A; \mathbf{J}\rangle \tag{4.6.3}$$

The fields due to arbitrary current distribution $\mathbf{J}$ may be expressed as

$$\mathbf{H} = \langle \underline{\underline{\mathbf{G}}}^{HJ}; \mathbf{J} \rangle \tag{4.6.4}$$

Upon substituting (4.6.3) (4.6.4) into (4.6.1), one obtains

$$\langle \underline{\underline{\boldsymbol{\mu}}} \cdot \underline{\underline{\mathbf{G}}}^{HJ}; \mathbf{J} \rangle = \langle \nabla \times \underline{\underline{\mathbf{G}}}^{A}; \mathbf{J} \rangle$$

It follows that

$$\underline{\underline{\boldsymbol{\mu}}} \cdot \underline{\underline{\mathbf{G}}}^{HJ} = \nabla \times \underline{\underline{\mathbf{G}}}^{A} \tag{4.6.5}$$

Since $\underline{\underline{\mathbf{G}}}^{HJ}$ has already been determined, we will use this relationship to obtain $\underline{\underline{\mathbf{G}}}^{A}$. The derivations are simplified in the spectrum domain, where the operator ∇ becomes $\widetilde{\nabla} = -jk_\rho\hat{\mathbf{u}} + \hat{\mathbf{z}}d/(dz)$. The relationship in (4.6.5) does not uniquely specify $\underline{\underline{\mathbf{G}}}^{A}$, making different formulations possible. We postulate the following form which possesses clear advantage over others[7]

$$\widetilde{\underline{\underline{\mathbf{G}}}}^{A} = \hat{\mathbf{u}}\hat{\mathbf{u}}\widetilde{G}_{vv}^{A} + \hat{\mathbf{v}}\hat{\mathbf{v}}\widetilde{G}_{vv}^{A} + \hat{\mathbf{z}}\hat{\mathbf{u}}\widetilde{G}_{zu}^{A} + \hat{\mathbf{z}}\hat{\mathbf{z}}\widetilde{G}_{zz}^{A} \tag{4.6.6}$$

when (4.6.6) is projected on the Cartesian-coordinate system via (4.4.19), (4.4.20)

$$\widetilde{\underline{\underline{\mathbf{G}}}}^{A} = \hat{\mathbf{x}}\hat{\mathbf{x}}\widetilde{G}_{vv}^{A} + \hat{\mathbf{y}}\hat{\mathbf{y}}\widetilde{G}_{vv}^{A} + \hat{\mathbf{z}}\hat{\mathbf{x}}\frac{k_x}{k_\rho}\widetilde{G}_{zu}^{A} + \hat{\mathbf{z}}\hat{\mathbf{y}}\frac{k_y}{k_\rho}\widetilde{G}_{zu}^{A} + \hat{\mathbf{z}}\hat{\mathbf{z}}\widetilde{G}_{zz}^{A} \tag{4.6.7}$$

As is known[16], for horizontal dipoles, z components are required to satisfy the boundary conditions at the interfaces. To calculate $\nabla \times \widetilde{\underline{\underline{\mathbf{G}}}}^{A}$, we rewrite $\widetilde{\underline{\underline{\mathbf{G}}}}^{A}$ as

$$\begin{aligned}
\widetilde{\underline{\underline{\mathbf{G}}}}^{A} &= (\widetilde{G}_{vv}^{A}\hat{\mathbf{u}} + \widetilde{G}_{zu}^{A}\hat{\mathbf{z}})\hat{\mathbf{u}} + (\widetilde{G}_{vv}^{A}\hat{\mathbf{v}})\hat{\mathbf{v}} + (\widetilde{G}_{zz}^{A}\hat{\mathbf{z}})\hat{\mathbf{z}} \\
&= \mathbf{C}_1\hat{\mathbf{u}} + \mathbf{C}_2\hat{\mathbf{v}} + \mathbf{C}_3\hat{\mathbf{z}}
\end{aligned}$$

According to the definition of $\nabla \times \widetilde{\underline{\underline{\mathbf{G}}}}^{A}$[17] and $\widetilde{\nabla} = -jk_\rho\hat{\mathbf{u}} + \hat{\mathbf{z}}d/(dz)$, we have

$$\begin{aligned}
\nabla \times \widetilde{\underline{\underline{\mathbf{G}}}}^{A} &= (\nabla \times \mathbf{C}_1)\,\hat{\mathbf{u}} + (\nabla \times \mathbf{C}_2)\,\hat{\mathbf{v}} + (\nabla \times \mathbf{C}_3)\,\hat{\mathbf{z}} \\
&= \left[\left(-jk_\rho\hat{\mathbf{u}} + \hat{\mathbf{z}}\frac{d}{dz}\right) \times \mathbf{C}_1\right]\hat{\mathbf{u}} + \left[\left(-jk_\rho\hat{\mathbf{u}} + \hat{\mathbf{z}}\frac{d}{dz}\right) \times \mathbf{C}_2\right]\hat{\mathbf{v}} \\
&\quad + \left[\left(-jk_\rho\hat{\mathbf{u}} + \hat{\mathbf{z}}\frac{d}{dz}\right) \times \mathbf{C}_3\right]\hat{\mathbf{z}} \\
&= \left(jk_\rho\widetilde{G}_{zu}^{A} + \frac{d}{dz}\widetilde{G}_{vv}^{A}\right)\hat{\mathbf{v}}\hat{\mathbf{u}} - \frac{d}{dz}\widetilde{G}_{vv}^{A}\hat{\mathbf{u}}\hat{\mathbf{v}} - jk_\rho\widetilde{G}_{vv}^{A}\hat{\mathbf{z}}\hat{\mathbf{v}} \\
&\quad + jk_\rho\widetilde{G}_{zz}^{A}\hat{\mathbf{v}}\hat{\mathbf{z}}
\end{aligned} \tag{4.6.8}$$

where $\partial/\partial v = 0$ is used. $\underline{\underline{\boldsymbol{\mu}}} \cdot \widetilde{\underline{\underline{\mathbf{G}}}}^{HJ}$ of the left side of (4.6.5) may be calculated via (4.5.16)

$$\begin{aligned}
\underline{\underline{\boldsymbol{\mu}}} \cdot \widetilde{\underline{\underline{\mathbf{G}}}}^{HJ} &= (\mu_t\hat{\mathbf{u}}\hat{\mathbf{u}} + \mu_t\hat{\mathbf{v}}\hat{\mathbf{v}} + \mu_z\hat{\mathbf{z}}\hat{\mathbf{z}}) \cdot \widetilde{\underline{\underline{\mathbf{G}}}}^{HJ} \\
&= -\mu_t I_i^e\hat{\mathbf{v}}\hat{\mathbf{u}} + \mu_t I_i^h\hat{\mathbf{u}}\hat{\mathbf{v}} - \frac{k_\rho}{\omega\mu_0}V_i^h\hat{\mathbf{z}}\hat{\mathbf{v}} + \mu_t\frac{k_\rho}{\omega\epsilon_0\epsilon_z'}I_v^e\hat{\mathbf{v}}\hat{\mathbf{z}}
\end{aligned} \tag{4.6.9}$$

The comparison between (4.6.8) and (4.6.9) yields

$$\frac{d}{dz}\widetilde{G}_{vv}^{A} + jk_\rho\widetilde{G}_{zu}^{A} = -\mu_t I_i^e \tag{4.6.10}$$

$$\frac{d}{dz}\widetilde{G}_{vv}^{A} = -\mu_t I_i^h \tag{4.6.11}$$

$$j\omega\mu_0\widetilde{G}_{vv}^{A} = V_i^h \tag{4.6.12}$$

$$j\omega\mu_0\widetilde{G}_{zz}^{A} = \eta_0^2\frac{\mu_t}{\epsilon_z'}I_v^e \tag{4.6.13}$$

where $\eta_0 = \sqrt{\mu_0/\epsilon_0}$ is the intrinsic impedance of free space. Substituting (4.6.11) into (4.6.10) results in

$$j\omega\mu_0\widetilde{G}_{zu}^A = \frac{\omega\mu_0\mu_t}{k_\rho}(I_i^h - I_i^e) \tag{4.6.14}$$

the solution of $\widetilde{\mathbf{G}}^A$ in (4.6.6), (4.6.7) are found from (4.6.12)–(4.6.14) to be

$$\widetilde{\mathbf{G}}^A = \hat{\mathbf{x}}\hat{\mathbf{x}}\frac{1}{j\omega\mu_0}V_i^h + \hat{\mathbf{y}}\hat{\mathbf{y}}\frac{1}{j\omega\mu_0}V_i^h + \hat{\mathbf{z}}\hat{\mathbf{x}}\frac{\mu_t k_x}{jk_\rho^2}(I_i^h - I_i^e) + \hat{\mathbf{z}}\hat{\mathbf{y}}\frac{\mu_t k_y}{jk_\rho^2}(I_i^h - I_i^e) + \hat{\mathbf{z}}\hat{\mathbf{z}}\frac{\mu_t}{j\omega\epsilon_0\epsilon_z'}I_v^e \tag{4.6.15}$$

The scalar potential may be found from the auxiliary condition

$$\nabla \cdot (\mu_t^{-1}\mu_z^{-1}\underline{\boldsymbol{\mu}} \cdot \mathbf{A}) = -j\omega\mu_0\epsilon_0\epsilon_t\phi \tag{4.6.16}$$

which can be shown to be consistent with the vector potential obtained above. To arrive at the mixed-potential form of $\mathbf{E}$, we postulate the decomposition

$$\epsilon_t^{-1}\nabla \cdot (\mu_t^{-1}\mu_z^{-1}\underline{\boldsymbol{\mu}} \cdot \underline{\mathbf{G}}^A) = -\nabla'G^\phi + C^\phi\hat{\mathbf{z}} \tag{4.6.17}$$

where G^ϕ is the scalar potential kernel and C^ϕ is the correction factor, which arises generally when both horizontal and vertical current components are present[7]. To find G^ϕ and C^ϕ, we substitute (4.6.6) into spectrum domain counterpart of (4.6.16) and use $\widetilde{\nabla}' = \hat{\mathbf{u}}jk_\rho + \hat{\mathbf{z}}d/dz'$, and we obtain

$$\widetilde{G}^\phi = \frac{1}{j\omega\mu_0\mu_z\epsilon_t}V_i^h + \frac{1}{k_\rho^2\epsilon_t}\frac{d}{dz}(I_i^h - I_i^e) \tag{4.6.18}$$

$$\widetilde{C}^\phi = \frac{1}{j\omega\epsilon_0\epsilon_z'\epsilon_t}\frac{d}{dz}I_v^e + \frac{d}{dz'}\widetilde{G}^\phi \tag{4.6.19}$$

To eliminate d/dz and d/dz' in (4.6.18) and (4.6.19), we use (4.5.2), (4.4.25)–(4.4.27) and (4.6.18). Then (4.6.18) becomes

$$\begin{aligned}
\widetilde{G}^\phi &= \frac{V_i^h}{j\omega\mu_0\mu_z\epsilon_t} + \frac{1}{k_\rho^2\epsilon_t}\left(jk_z^eY^eV_i^e - jk_z^hY^hV_i^h\right) \\
&= \frac{j\omega\epsilon_0 V_i^e}{k_\rho^2} - \frac{j\omega\epsilon_0 V_i^h}{k_\rho^2}\left(\frac{k_\rho^2}{\omega^2\epsilon_0\mu_0\epsilon_t\mu_z} + \frac{k_t^2 - v^h k_\rho^2}{\omega^2\epsilon_0\mu_0\epsilon_t\mu_t}\right) \\
&= \frac{j\omega\epsilon_0}{k_\rho^2}\left(V_i^e - V_i^h\right)
\end{aligned} \tag{4.6.20}$$

We use (4.6.20), (4.4.25)–(4.4.27), (4.5.1), (4.5.4) and (4.5.7); then (4.6.19) becomes

$$\begin{aligned}
\widetilde{C}^\phi &= \frac{1}{j\omega_0\epsilon_0\epsilon_z'\epsilon_t}\left(-jk_z^eY^eV_v^e\right) + \frac{j\omega\epsilon_0}{k_\rho^2}\left(\frac{dV_i^e}{dz'} - \frac{dV_i^h}{dz'}\right) \\
&= -\frac{1}{\epsilon_z'}V_v^e + \frac{j\omega\epsilon_0}{k_\rho^2}\left[-j\frac{(k_z^{e'})^2}{\omega\epsilon_0\epsilon_t}I_i^{e'} + j\omega\mu_0\mu_t'I_i^{h'}\right] \\
&= -\frac{1}{\epsilon_z'}V_v^e + \frac{j\omega\epsilon_0}{k_\rho^2}\left[j\frac{(k_z^{e'})^2}{\omega\epsilon_0\epsilon_t}V_v^e - j\omega\mu_0\mu_t'V_v^h\right] \\
&= \frac{\omega^2\epsilon_0\mu_0\mu_t'}{k_\rho^2}V_v^h - \left(\frac{1}{\epsilon_z'} + \frac{k_0^2\epsilon_t'\mu_t'\epsilon_z' - \epsilon_t'k_\rho^2}{k_\rho^2\epsilon_t'\epsilon_z'}\right)V_v^e \\
&= \frac{\omega^2\epsilon_0\mu_0\mu_t'}{k_\rho^2}\left(V_v^h - V_v^e\right)
\end{aligned} \tag{4.6.21}$$

where as usual prime refers to the source point.

We next substitute (4.6.15) and (4.6.16) into (4.6.2) and obtain

$$
\mathbf{E} = -j\omega\mathbf{A} - \nabla\phi
$$

$$
= -j\omega\mu_0\langle \underline{\underline{\mathbf{G}}}^A; \mathbf{J}\rangle + \nabla\frac{\nabla\cdot(\underline{\underline{\mu}}\cdot\mathbf{A})}{j\omega\epsilon_0\mu_0\epsilon_t\mu_t\mu_z}
$$

$$
= -j\omega\mu_0\langle \underline{\underline{\mathbf{G}}}^A; \mathbf{J}\rangle + \frac{1}{j\omega\epsilon_0}\nabla(-\langle\nabla' G^\phi, \mathbf{J}\rangle + \langle C^\phi\hat{\mathbf{z}}; \mathbf{J}\rangle) \tag{4.6.22}
$$

Using the Gauss' theorem, we have

$$
\int_s \nabla'\cdot(G^\phi\mathbf{J})ds' = \int_s \nabla' G^\phi\cdot\mathbf{J}ds + \int_s G^\phi\nabla'\cdot\mathbf{J}ds' = \int_L G^\phi\mathbf{J}\cdot\hat{n}dl = 0
$$

namely,

$$
\langle\nabla' G^\phi, \mathbf{J}\rangle = -\langle G^\phi, \nabla'\cdot\mathbf{J}\rangle
$$

then (4.6.22) may be expressed as

$$
\mathbf{E} = -j\omega\mu_0\langle\underline{\underline{\mathbf{G}}}^A; \mathbf{J}\rangle + \frac{1}{j\omega\epsilon_0}\nabla(\langle G^\phi, \nabla'\cdot\mathbf{J}\rangle + \langle C^\phi\hat{\mathbf{z}}; \mathbf{J}\rangle) \tag{4.6.23}
$$

which is the desired mixed-potential representation of $\mathbf{E}$. Note that kernels given above are expressed in terms of Sommerfeld integrals of spectral functions, for which explicit expressions in terms of the TLGFs have been derived.

When only magnetic currents are present, the analysis is similar to that given above. The mixed-potential representation for $\mathbf{H}$ may be obtained from the above formulas by the following replacement of symbols: $\mathbf{E}\to\mathbf{H}$, $\mathbf{J}\to\mathbf{M}$, $\mathbf{A}\to\mathbf{F}$, $\phi\to\psi$, $\epsilon\to\mu$, $\mu\to\epsilon$, $V\to I$, $I\to V$, $v\to i$, $i\to v$, $e\to h$, and $h\to e$. In the general case, where both electric and magnetic currents are present, we may use superposition to get the results.

The correction term in (4.6.23) may be grouped with the vector potential term, resulting in an alternative mixed-potential representation as[23]

$$
\mathbf{E} = -j\omega\mu_0\langle\underline{\underline{\mathbf{K}}}^A; \mathbf{J}\rangle + \frac{1}{j\omega\epsilon_0}\nabla\langle G^\phi, \nabla'\cdot\mathbf{J}\rangle \tag{4.6.24}
$$

where the new vector potential Green's function $\underline{\underline{\mathbf{K}}}^A$ can be expressed as

$$
\underline{\underline{\mathbf{K}}}^A = \begin{bmatrix} G_{xx}^A & 0 & G_{xz}^A \\ 0 & G_{xx}^A & G_{yz}^A \\ G_{zx}^A & G_{zy}^A & G_{zz}^A \end{bmatrix} \tag{4.6.25}
$$

The spectral domain counterparts of G_{xx}^A, G_{zx}^A and G_{zy}^A have the same forms as those in (4.6.15), respectively. The spectral domain counterparts of G_{xz}^A, G_{yz}^A, and G_{zz}^A can be written as

$$
\tilde{G}_{xz}^A = \frac{\mu_t' k_x}{j k_\rho^2}(V_v^h - V_v^e) \tag{4.6.26}
$$

$$
\tilde{G}_{yz}^A = \frac{\mu_t' k_y}{j k_\rho^2}(V_v^h - V_v^e) \tag{4.6.27}
$$

$$
\tilde{G}_{zz}^A = \frac{1}{j\omega\epsilon_0}\left[\left(\frac{\mu_t}{\epsilon_z'} - \frac{\mu_t'(k_z^e)^2}{\epsilon_t k_\rho^2}\right)I_v^e + \frac{k_0^2\mu_t\mu_t'}{k_\rho^2}I_v^h\right] \tag{4.6.28}
$$

Components $\widetilde{G}_{xz}^{A}$ and $\widetilde{G}_{yz}^{A}$ are related to $\widetilde{G}_{zx}^{A}$ and $\widetilde{G}_{zy}^{A}$ by

$$\widetilde{G}_{xz}^{A}(\boldsymbol{\rho}, z | z') = -\frac{\mu_t'}{\mu_t} \widetilde{G}_{zx}^{A}(\boldsymbol{\rho}, z' | z) \tag{4.6.29}$$

$$\widetilde{G}_{yz}^{A}(\boldsymbol{\rho}, z | z') = -\frac{\mu_t'}{\mu_t} \widetilde{G}_{zy}^{A}(\boldsymbol{\rho}, z' | z) \tag{4.6.30}$$

Therefore, the Green's functions required are $\widetilde{G}_{xx}^{A}, \widetilde{G}_{zz}^{A}, \widetilde{G}_{zx}^{A}, \widetilde{G}_{zy}^{A}$, and $\widetilde{G}^{\phi}$. Also, $\widetilde{G}_{zx}^{A}$ and $\widetilde{G}_{zy}^{A}$ have the same kernels so that a total of four Green's functions are required to evaluate.

4.7   Transmission-Line Green's Functions

In the above, we have shown that the problem of finding the spectrum Green's function for multilayered dielectric structures may reduce to that of finding the Green's function for the transmission line. The formulation developed so far is for an unspecified stratification, since no assumption has been made regarding the z dependance of the media parameters. We now specialize it to the case of a multilayered medium with piecewise-constant parameters. The parameters pertaining to layer n with boundaries at z_n and z_{n+1} are distinguished by subscript n. The transmission line analog of the layered medium consists of a cascade connection of uniform transmission line sections, where section n with terminals at z_n and z_{n+1} has propagation constant γ_n^p and characteristic admittance Y_n^p. To find the transmission-line Green's functions (TLGFs), we excite the transmission line network by unit-strength voltage and current sources at z' in section m and compute the voltage and current at z in section n. Hence, the primed media parameters are assumed to be the values pertaining to layer m, while the unprimed ones are those of layer n. In the original problems, the real sources will always be the electric current sources. In cases when the equivalence principle is used, the magnetic currents will be introduced[18]. The magnetic current density is actually the tangential electric field. It should be noticed that although in dealing with the slot or aperture problems, the electric fields in the equation may be removed from the left side to the right side, that is, the source side by the inverse operation of the matrix, however, the electric field should never be considered as the magnetic current density. The only case to introduce the magnetic current is the application of the equivalence principle. Figure 4.4(a) shows the fields

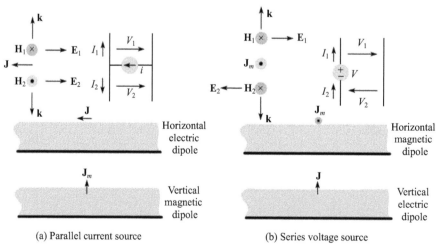

(a) Parallel current source   (b) Series voltage source

Figure 4.4   Two kinds of sources.

produced by the horizontal electric current or vertical magnetic current. The tangential electric fields are continuous but the tangential magnetic fields are not. The source, therefore, behaves like a parallel current source. Similarly in the case of Figure 4.4(b) according to the field analysis, the source behaves like a series voltage source.

4.7.1 Parallel Current Source

In the analysis of a transmission line circuit, only one formula is enough, that is

$$T = Inc. + Ref. \tag{4.7.1}$$

where T denotes the total wave, $Inc.$ and $Ref.$ denote the incident wave and reflected wave respectively. If we take the incident wave to be V_{inc} and take the source layer as the reference of the phase, then from (4.7.1) and Figure 4.5, we have

$$V_{i,00} = V_{inc} \left(1 + \Gamma e^{-2\gamma_1 h_1}\right)$$
$$= V_{inc} \left(1 + \frac{Y_1 - Y_{u1}}{Y_1 + Y_{u1}} e^{-2\gamma_1 h_1}\right) \tag{4.7.2}$$
$$V_{i,10} = V_{inc} e^{-\gamma_1 h_1} \left(1 + \Gamma\right)$$
$$= V_{inc} e^{-\gamma_1 h_1} \frac{2Y_1}{Y_1 + Y_{u1}} \tag{4.7.3}$$

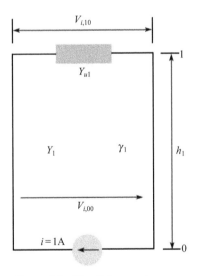

Figure 4.5 Parallel current source.

where $V_{i,00}$ is the voltage at layer 0 due to the current source at layer 0, $V_{i,10}$ is the voltage at layer 1 due to the current source at layer 0, Γ is the voltage reflection coefficient at reference plane 1, Y_{u1} is the load admittance on the upper side, Y_1 and γ_1 are the characteristic admittance and propagation constant respectively. The superscript p (p represent e or h) is omitted here and hereafter. After a simple derivation, we have

$$\frac{V_{i,10}}{V_{i,00}} = \frac{Y_1 / \sinh \gamma_1 h_1}{Y_{u1} + Y_1 \coth \gamma_1 h_1} \tag{4.7.4}$$

This formula gives the relationship of the voltages between the adjacent layers. This is called the voltage translation formula which translates the voltage from one layer to the other. The load admittance Y_{u1} could be the input admittance, of the next cascade section, for example the section with parameters Y_2, γ_2, h_2 and load admittance Y_{u2}. The formula of input admittance Y_{u1} can be derived by using (4.7.1) or found from the textbook [19] as

$$Y_{u1} = Y_2 \frac{Y_2 + Y_{u2} \coth \gamma_2 h_2}{Y_{u2} + Y_2 \coth \gamma_2 h_2} \tag{4.7.5}$$

For the special case when $Y_{u2} = \infty$

$$Y_{u1} = Y_2 \coth \gamma_2 h_2 \tag{4.7.6}$$

Similar to (4.7.4), we have

$$\frac{V_{i,21}}{V_{i,11}} = \frac{Y_2 / \sinh \gamma_2 h_2}{Y_{u2} + Y_2 \coth \gamma_2 h_2} \tag{4.7.7}$$

The meaning of the notations is similar to that in (4.7.4).

If the source I is located at 0 layer, the cascade of these two sections corresponds to the multiplication of both sides of (4.7.4) and (4.7.7). Assigning $V_{i,10} = V_{i,11}$, $V_{i,21} = V_{i,20}$, this results in

$$\frac{V_{i,20}}{V_{i,00}} = \frac{Y_1 / \sinh \gamma_1 h_1}{Y_{u1} + Y_1 \coth \gamma_1 h_1} \cdot \frac{Y_2 / \sinh \gamma_2 h_2}{Y_{u2} + Y_2 \coth \gamma_2 h_2} \tag{4.7.8}$$

(4.7.8) is also a voltage translation formula. For unit current source, $i = 1$A, therefore

$$V_{i,00} = \frac{1}{Y_{u0} + Y_{d0}} \tag{4.7.9}$$

and (4.7.8) becomes

$$V_{i,20} = \frac{1}{Y_{u0} + Y_{d0}} \cdot \frac{Y_1 / \sinh \gamma_1 h_1}{Y_{u1} + Y_1 \coth \gamma_1 h_1} \cdot \frac{Y_2 / \sinh \gamma_2 h_2}{Y_{u2} + Y_2 \coth \gamma_2 h_2} \tag{4.7.10}$$

where Y_{u0} and Y_{d0} are input admittances at 0 layer looking up (field layer side) and looking down respectively.

In the general case, the source coordinate is located at mth layer, field coordinate is located at nth layer $(n \geqslant m)$ and there are $N + 1$ sections between these two layers, the voltage at n due to the source at m is $V_{i,nm}$ and we have

$$V_{i,nm} = \frac{1}{Y_{um} + Y_{dm}} \prod_{j=1}^{N+1} \frac{Y_j / \sinh \gamma_j h_j}{Y_{uj} + Y_j \coth \gamma_j h_j} \tag{4.7.11}$$

where the subscript u and d have the same meaning as those in (4.7.10). From (4.7.11) we can find the voltage anywhere due to the parallel unit current source. If the source and field coordinate are located at the same layer m, then in (4.7.11) we just keep one term with the index m in the product and set h_m to be zero, we have

$$V_{i,mm} = \frac{1}{Y_{um} + Y_{dm}} \tag{4.7.12}$$

To find the current $I_{i,nm}$ at n due to the source I at m, we may follow the same procedure as above. Alternatively, we can obtain the formula directly from (4.7.11) by noticing that

$$I_{i,nm} = Y_{un} V_{i,nm} \tag{4.7.13}$$

Substituting (4.7.11) into (4.7.13) yields

$$I_{i,nm} = \frac{Y_{un}}{Y_{um} + Y_{dm}} \prod_{j=1}^{N+1} \frac{Y_j / \sinh \gamma_j h_j}{Y_{uj} + Y_j \coth \gamma_j h_j} \tag{4.7.14}$$

From (4.7.14) we may find the current anywhere, due to the parallel unit current source. If the source and field coordinates are located at the same layer m, then in (4.7.14) we just keep one term with the index m in the product and set h_m to be zero, and we have

$$I_{i,mm} = \frac{Y_{um}}{Y_{um} + Y_{dm}} \tag{4.7.15}$$

4.7.2   Series Voltage Source

For the series voltage source shown in Figure 4.6, we have

$$I_{v,10} = \frac{Y_{u0} Y_{d0}}{Y_{u0} + Y_{d0}} \tag{4.7.16}$$

$$V_{v,00} = \frac{I_{v,10}}{Y_{u0}} = \frac{Y_{d0}}{Y_{u0} + Y_{d0}} \tag{4.7.17}$$

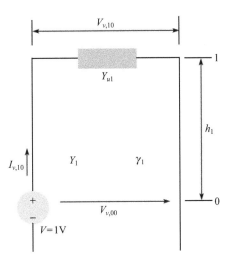

Figure 4.6   Series voltage source.

where the subscript u and d have the same meaning as those in (4.7.10). Comparing (4.7.17) with (4.7.9), and from (4.7.11), we have

$$V_{v,nm} = \frac{Y_{dm}}{Y_{um} + Y_{dm}} \prod_{j=1}^{N+1} \frac{Y_j / \sinh \gamma_j h_j}{Y_{uj} + Y_j \coth \gamma_j h_j} \tag{4.7.18}$$

If the source and field coordinates are located at the same layer m, similar to (4.7.12), we have

$$V_{v,mm} = \frac{Y_{dm}}{Y_{um} + Y_{dm}} \tag{4.7.19}$$

Noticing that

$$I_{v,nm} = Y_{un} V_{v,nm} \tag{4.7.20}$$

we have

$$I_{v,nm} = \frac{Y_{dm}Y_{un}}{Y_{um}+Y_{dm}} \prod_{j=1}^{N+1} \frac{Y_j/\sinh\gamma_j h_j}{Y_{uj}+Y_j\coth\gamma_j h_j} \qquad (4.7.21)$$

If the source and field coordinates are located at the same layer m, similar to (4.7.15), we have

$$I_{v,mm} = \frac{Y_{dm}Y_{um}}{Y_{um}+Y_{dm}} \qquad (4.7.22)$$

Formulas (4.7.11), (4.7.14), (4.7.18) and (4.7.21) are the TLGFs that are enough to deal with any transmission line problems. Especially, it is very convenient to apply them to finding the spectral Green's functions associated with problems with multi-layer media.

4.7.3   Example

Consider the microstrip substrate with thickness d, parameters $\epsilon = \epsilon_r\epsilon_0$ and μ_0. The unit electric dipole is located at $z' = 0$ and directed along x (see Figure 4.7). To find the spectral mixed potentials for solving planar problem, assume $\gamma = jk_z$. In our case, we may neglect the superscript p on k_z in (4.4.25), because $\nu^e = \nu^h = 1$.

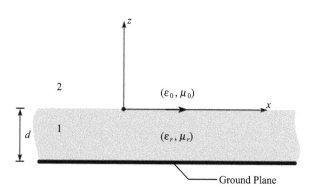

Figure 4.7   Microstrip substrate.

Solution: In this case, the source is the horizontal electric current. Consequently, $V_v = 0$, $I_v = 0$ and $\widetilde{C}^\phi = 0$. In addition, for the planar problem, $\widetilde{G}^A_{zx}$ is not involved. Therefore, only $\widetilde{G}^A_{xx} = V_i^h/j\omega\mu_0$ and $\widetilde{G}^\phi = j\omega\epsilon_0(V_i^e - V_i^h)/k_\rho^2$ are to be found.

According to (4.7.6) and (4.7.12), let $\gamma_1 = jk_{z1}, \gamma_2 = jk_{z2} = jk_{z0}$, we have

$$V_i^h = \frac{1}{Y_u^h + Y_d^h}$$

$$= \frac{1}{\dfrac{\gamma_2}{j\omega\mu_0} + \dfrac{\gamma_1}{j\omega\mu_0}\coth\gamma_1 d}$$

$$= \frac{j\omega\mu_0}{D_e} \qquad (4.7.23)$$

where $D_e = k_{z1}\cot k_{z1}d + jk_{z0}$, $k_{z0}/\omega\mu_0$ or $k_{z1}/\omega\mu_0$ is the characteristic admittance for $h(TE)$ wave and

$$V_i^e = \frac{1}{\dfrac{j\omega\epsilon_0}{\gamma_2} + \dfrac{j\omega\epsilon_r\epsilon_0}{\gamma_1}\coth\gamma_1 d}$$

$$= -\frac{k_{z1}k_{z0}\tan k_{z1}d}{\omega\epsilon_0 D_m} \tag{4.7.24}$$

where $D_m = -k_{z1}\tan k_{z1}d + j\epsilon_r k_{z0}$, $\omega\epsilon_0/k_{z0}$ or $\omega\epsilon_r\epsilon_0/k_{z1}$ is the characteristic admittance for $e(TM)$ wave. Finally, we have

$$\tilde{G}_{xx}^A = \frac{V_i^h}{j\omega\mu_0} = \frac{1}{D_e} \tag{4.7.25}$$

$$\tilde{G}^\phi = \frac{j\omega\epsilon_0}{k_\rho^2}(V_i^e - V_i^h)$$

$$= \frac{D_m k_0^2 - jD_e k_{z1}k_{z0}\tan k_{z1}d}{D_m D_e k_\rho^2} \tag{4.7.26}$$

V_i^h and V_i^e may also be expressed in terms of voltage reflection coefficient R or voltage transmission coefficient $T = 1 + R$. In this example, according to the transmission line theory $Y_d = Y_0(1 - R)/(1 + R)$, so

$$V_i = \frac{1}{Y_0 + Y_0(1 - R)/(1 + R)}$$

$$= \frac{1 + R}{2Y_0} = \frac{T}{2Y_0} \tag{4.7.27}$$

where R is the voltage reflection coefficient at $z = 0^+$ looking down, and Y_0 is the characteristic admittance in free space. For h wave, $Y_0^h = k_{z0}/\omega\mu_0$,

$$V_i^h = \frac{1 + R^h}{2Y_0^h} = \frac{\omega\mu_0(1 + R^h)}{2k_{z0}} = \frac{\omega\mu_0 T^h}{2k_{z0}} \tag{4.7.28}$$

For e wave, $Y_0^e = \omega\epsilon_0/k_{z0}$, and

$$V_i^e = \frac{1 + R^e}{2Y_0^e} = \frac{k_{z0}(1 + R^e)}{2\omega\epsilon_0} = \frac{k_{z0}T^e}{2\omega\epsilon_0} \tag{4.7.29}$$

More generally, if the source and field coordinates are at $z' \neq 0$, and $z > z'$, (4.7.28) and (4.7.29) may be modified as

$$V_i^h = \frac{\omega\mu_0}{2k_{z0}}\left[e^{-jk_{z0}(z-z')} + R^h e^{-jk_{z0}(z+z')}\right] \tag{4.7.30}$$

$$V_i^e = \frac{k_{z0}}{2\omega\epsilon_0}\left[e^{-jk_{z0}(z-z')} + R^e e^{-jk_{z0}(z+z')}\right] \tag{4.7.31}$$

Instead of using immittance (impedance or admittance), R^h, R^e may also be obtained by using wave matrix technique[20] or many others, such as that introduced in [9]. All the methods are equivalent to each other. From the author's teaching experience, the method introduced in this book is more accessible. The results for R^h and R^e from any of the methods are

$$R^p = \frac{r^p - e^{-j2k_{z1}d}}{1 - r^p e^{-j2k_{z1}d}} \tag{4.7.32}$$

where $\gamma_1 = jk_{z1}$, p stands for e or h and

$$r^h = \frac{k_{z0} - k_{z1}}{k_{z0} + k_{z1}} \tag{4.7.33}$$

$$r^e = \frac{k_{z1} - \epsilon_r k_{z0}}{k_{z1} + \epsilon_r k_{z0}} \tag{4.7.34}$$

It is seen from (4.7.29), (4.7.31) that T^e is a function of $\exp(-jk_{z0}z)$, thus we may write

$$\frac{\partial T^e}{\partial z} = -jk_{z0}T^e \tag{4.7.35}$$

Consequently (4.7.29) becomes

$$V_i^e = \frac{j}{2\omega\epsilon_0}\frac{\partial T^e}{\partial z} \tag{4.7.36}$$

Using the formulas (4.7.28) and (4.7.36), $\widetilde{G}_{xx}^A$ and $\widetilde{G}^\phi$ in (4.7.25) and (4.7.26) may have other forms which also appear in some literatures

$$\widetilde{G}_{xx}^A = \frac{1}{2jk_{z0}}T^h \tag{4.7.37}$$

$$\widetilde{G}^\phi = \frac{j\omega\epsilon_0}{k_\rho^2}\left(\frac{j}{2\omega\epsilon_0}\frac{\partial T^e}{\partial z} - \frac{\omega\mu_0}{2k_{z0}}T^h\right)$$

$$= -\frac{1}{2k_\rho^2}\frac{\partial T^e}{\partial z} - \frac{jk_0^2}{2k_\rho^2 k_{z0}}T^h$$

$$= -\frac{1}{2k_\rho^2}\frac{\partial T^e}{\partial z} - j\frac{k_\rho^2 + k_{z0}^2}{2k_\rho^2 k_{z0}}T^h$$

$$= \frac{-j}{2k_{z0}}\left[T^h + \frac{k_{z0}^2}{k_\rho^2}\left(T^h + \frac{1}{jk_{z0}}\frac{\partial T^e}{\partial z}\right)\right] \tag{4.7.38}$$

This example is also useful in deriving the higher order impedance boundary condition (HOIBC)[47, 48] used in solving electromagnetic scattering or radiation problems of coated objects. For the same structure as is shown in Figure 4.7, considering a two-dimensional scattering problem and assuming $k_y = 0$ or $k_x^2 = k_\rho^2$, we may find the Green's function in spectrum domain, that is the relationship between tangential electric field and magnetic field. For example, we derive the relationship between $\widetilde{E}_x$ and $\widetilde{H}_y$. Actually, from the equivalence principle, $\widetilde{H}_y$ is the source $\widetilde{J}_x$, so we have

$$Z_{xy} = \frac{\widetilde{E}_x}{\widetilde{H}_y} = \frac{\widetilde{E}_x}{\widetilde{J}_x} = \widetilde{G}_{xx}^{EJ} \tag{4.7.39}$$

From (4.5.19), we have

$$Z_{xy} = -V_i^e\frac{k_x^2}{k_\rho^2} = -V_i^e \tag{4.7.40}$$

Notice that in this case the sources are the plane wave spectra propagating along one direction, not like those in the example which is to solve the spectrum Green's function for current source propagating along both directions. Therefore in using (4.7.23), Y_u^h should be set to zero. Finally we have

$$Z_{xy} = -\frac{\gamma_1}{j\omega\epsilon_r\epsilon_0}\tanh\gamma_1 d = -j\frac{k_{z1}}{\omega\epsilon_r\epsilon_0}\tan k_{z1}d \tag{4.7.41}$$

This impedance is approximated as ratios of polynomials in the transform variable k_x as follows

$$Z_{xy}(k_x) \approx -\frac{c_3 + c_5 k_x^2}{1 + c_2 k_x^2} \tag{4.7.42}$$

The coefficients are determined by matching the impedances exactly at $k_x = 0$, $k_x = k_0/2$, $k_x = k_0$. Using an elementary property of the Fourier transform

$$\frac{\partial^n f(y)}{\partial y^n} \rightarrow (-j)^n k_y^n \widetilde{f}(k_y) \tag{4.7.43}$$

The approximate spatial domain boundary condition for $\partial/\partial y = 0$ is finally obtained as

$$\left(1 - c_1 \frac{\partial^2}{\partial x^2}\right) E_x(x, y) = \left(-c_3 + c_5 \frac{\partial^2}{\partial x^2}\right) H_y(x, y) \tag{4.7.44}$$

Following a similar procedure, we may establish the relationship between E_y and H_x.

4.8 Introduction to Complex Integration Techniques

After the spectral dyadic Green's functions are found, in order to return back to the spatial domain, we have to go through the integration of Sommerfeld integrals. The computation of these integrals is a difficult task because of the oscillatory and divergent behavior of the integrands and the occurrence of singularities in the complex k_ρ plane. There are two kinds of singularities, the branch points and the poles. To properly deal with the singularities is the key to the success of computing the Sommerfeld integrals.

4.8.1 Branch Points and Branch Cuts

From 4.7 it is seen that the integrand in Sommerfeld integral contains the term $e^{-jk_z z}$. In general, consider the integral

$$\int \widetilde{f}(k_\rho) e^{-jk_z z} dk_\rho \tag{4.8.1}$$

where $k_z = \sqrt{k^2 - k_\rho^2}$ and $k^2 = \omega^2 \epsilon \mu$. This raises the question of whether we should take $k_z = \sqrt{k^2 - k_\rho^2}$ or $-\sqrt{k^2 - k_\rho^2}$. To have a clear picture of the branch point, first we consider a real function[21]

$$y^2 = x \tag{4.8.2}$$

For a given value of x, there are two values of y : $y_1 = +\sqrt{x}$ and $y_2 = -\sqrt{x}$ as shown in Figure 4.8. We may call y_1 the first branch, and y_2 the second branch. The point where these two branches meet is called the branch point. By using these two branches y_1 and y_2, we can keep the different branches apart so that we know exactly which branch we are dealing with. Secondly, we consider a more general complex function W of complex variable h for an example

$$W = \sqrt{h} \tag{4.8.3}$$

where we use polar coordinates, $h = re^{j\theta}$ and thus

$$W_1 = r^{1/2} e^{j(\theta/2)}$$

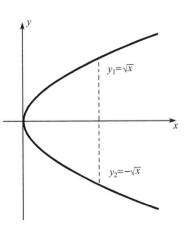

Figure 4.8   Two branches.

But this is not the only value of W. In the h plane, we can go around the origin and thus add 2π to θ without changing the value of h: $h = re^{j\theta} = re^{j(\theta + 2\pi)}$. But this second form of h gives a different value of W

$$W_2 = r^{1/2}e^{j(\theta/2+\pi)} = -W_1$$

If another 2π is added to θ, W will go back to W_1. Therefore, for a given h, there are two W's : W_1 and W_2. We note that to transfer from W_1 to W_2, we must go around the origin once in the h plane. The origin in this example is the branch point.

To describe this situation more clearly and to tell which of these two values W_1 and W_2, we are dealing with, we introduce the idea of the branch cut. In the h plane, we imagine a cut extending from the branch point $h = 0$ to infinity. We say that as long as we do not cross the branch cut, we are on the first branch (the first or top sheet of Riemann surface) and the value of the function is W_1. To get W_2, we must cross the branch cut. The branch cut can be drawn in any direction from the origin, and it needs not be a straight line. For convenience, we normally choose a cut along the negative real axis as shown in Figure 4.9(a). According to this choice of the branch cut, we define

$$\sqrt{h} = \begin{cases} W_1, & -\pi < \theta < \pi \\ W_2, & \pi < \theta < 3\pi \\ W_1, & 3\pi < \theta < 5\pi \end{cases}$$

This situation is pictured in the two Riemann sheets in Figure 4.9. The top sheet represents

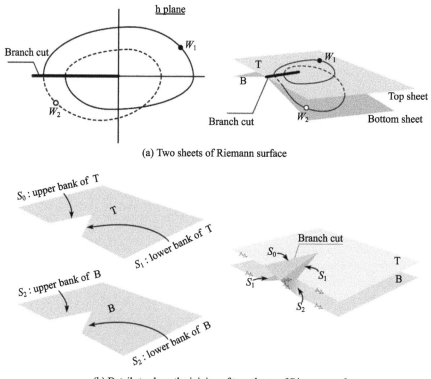

(a) Two sheets of Riemann surface

(b) Details to show the joining of two sheets of Riemann surface

Figure 4.9   Illustration of Riemann surface and branch cut.

W_1 and the bottom sheet W_2. These two sheets are put together at the branch cut, indicating that when we cross the branch cut, W_2 must be used. As the branch cut is crossed the second time, we go back to the top sheet of Riemann surface and obtain W_1. The two-sheeted Riemann surface cannot be constructed from physical sheets of paper without allowing intersections. The two sheets are only joined in the sense indicated by the double arrows as shown in Figure 4.9(b)[22]. The lower bank of the top sheet S_1 and the upper bank of the bottom sheet S_1 are stuck to each other. The lower bank of the bottom sheet S_2 and the upper bank of the top sheet S_0 are also stuck to each other. Of course it is impossible to do that in three-dimensional space. It is mathematically reasonable to imagine that the two points on different banks with the same coordinates as one point. We may also see that the cut through from S_0 to S_2 does not change the sign of W. For other multivalued functions, such as $W = \ln h$, there will be infinite sheets of Riemann surface.

Now consider $k_z = \sqrt{k^2 - k_\rho^2}$. We note that there are two branch points $k_\rho = +k$ and $-k$. To assure a unique specification of integrands in the complex k_ρ plane, it is necessary to discuss in detail the analytic properties of the square-root function $k_z(k_\rho)$[24]. When k_ρ is real and $|k_\rho| < k$ and k being assumed real for the moment, the guided wave along z is propagating and hence the propagation constant k_z is real and positive, consistent with a positive modal characteristic impendence [see Eqs. (4.4.26), (4.4.27)]. Thus we require a definition of $k_z(k_\rho)$ such that

$$\sqrt{k^2 - k_\rho^2} > 0, \quad -k < k_\rho < k \tag{4.8.4}$$

To ensure that integrands remain bounded as $|k_z(k_\rho)||z-z'| \to \infty$, it is necessary to impose restrictions on the imaginary part of k_z. For the time dependence $\exp(j\omega t)$, the required restriction for real k_ρ is $k_z = -j|k_z|$ (i.e., $\operatorname{Im} k_z < 0$ when $|k_\rho| > k$). This requirement, and also (4.8.4), follows from the radiation condition, which demands that the energy radiated by the source to distant observation points is bounded and outgoing. If k_ρ is allowed to be complex, the condition $\operatorname{Im} k_z < 0$ will be imposed for all permitted complex values of k_z. The analytic continuation of k_ρ from real to complex values is required for subsequent deformation of the integration contours.

To make the definition of the double-valued function $k_z(k_\rho)$ unique, a two-sheeted complex k_ρ plane is necessary, with branch cuts providing a means of passing from one Riemann sheet to the other. The choice of branch cuts is arbitrary but determines the disposition of those regions of the complex k_ρ plane in which $\operatorname{Im} k_z < 0$, or $\operatorname{Im} k_z > 0$. In the following, we are going to discuss the most important choice. Let us define

$$k - k_\rho = |k - k_\rho|e^{j\alpha}, \quad k + k_\rho = |k + k_\rho|e^{j\beta}, \quad \alpha, \ \beta \ are \ real \tag{4.8.5}$$

with the angles α and β selected as in Figure 4.10 so as to make $\alpha=0$ and $\beta=0$ when k_ρ is real and $|k_\rho| < k$ on the top sheet. Hence,

$$\sqrt{k^2 - k_\rho^2} = \left|\sqrt{k^2 - k_\rho^2}\right| e^{j(\alpha+\beta)/2} \tag{4.8.6}$$

where we have chosen the positive sign of the square root. To satisfy condition (4.8.4), it is required that

$$\alpha + \beta = 0, \quad -k < k_\rho < k \tag{4.8.7}$$

With the angles α and β defined as in Figure 4.10, it is evident that condition (4.8.7) is met for $\alpha = \beta = 0$ when $-k < k_\rho < k$.

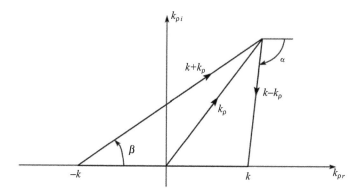

Figure 4.10   Complex k_ρ plane (without loss).

In the above discussion, k was assumed to be real. Since all physical media have some loss, it is also relevant to treat the case of a lossy dielectric with an assumed time dependence $\exp(j\omega t)$ and $\epsilon = \epsilon_r - j\sigma/\omega$, where σ is the conductivity of medium. Correspondingly $k = \omega\sqrt{\mu\epsilon}$ has a negative imaginary part, that is, $k = k_r + jk_i$ and $k_i < 0$. Now we determine the branch cut according to the condition $\operatorname{Re} k_z = 0$ or $\operatorname{Im} k_z = 0$. From Figure 4.11, we have

$$\tan\alpha = \frac{k_i - k_{\rho i}}{k_r - k_{\rho r}}, \quad \tan\beta = \frac{k_i + k_{\rho i}}{k_r + k_{\rho r}}, \quad k_r > 0, \; k_i < 0 \tag{4.8.8}$$

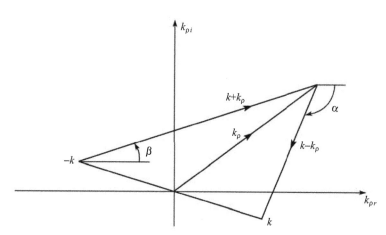

Figure 4.11   Complex k_ρ plane (with loss).

We also have

$$k_z = \left| \sqrt{k^2 - k_\rho^2} \right| \left[\cos\left(\frac{\alpha + \beta}{2}\right) + j\sin\left(\frac{\alpha + \beta}{2}\right) \right] \tag{4.8.9}$$

We want to find the contour on which $\operatorname{Re} k_z$ or $\operatorname{Im} k_z$ vanishes. From (4.8.9), it is seen that these contours are determined by enforcing $\alpha + \beta = 0, \pm 2\pi, \cdots$ for $\operatorname{Im} k_z = 0$ and $\alpha + \beta = \pm\pi, \pm 3\pi, \cdots$, for $\operatorname{Re} k_z = 0$. Since

$$\tan(\alpha + \beta) = \frac{\tan\alpha + \tan\beta}{1 - \tan\alpha\tan\beta} = 0 \tag{4.8.10}$$

One finds from (4.8.8), (4.8.10) that both $\operatorname{Re} k_z = 0$ and $\operatorname{Im} k_z = 0$ along the hyperbolas

$$k_{\rho r}k_{\rho i} = k_r k_i \tag{4.8.11}$$

where $k_{\rho r}$, k_r and $k_{\rho i}$, k_i denote the real and imaginary parts, respectively, of k_ρ and k. It is easy to verify that $\operatorname{Re} k_z = 0$ on those portions of the curves for which $|k_{\rho r}| > k_r$, and for which $\operatorname{Im} k_z = 0$ for $|k_{\rho r}| < k_r$. The corresponding behavior of k_z in the complex k_ρ plane is shown in Figure 4.12(a), (b) for $k_i < 0$.

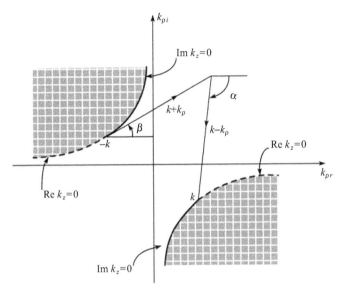

(a) Branch cut $\operatorname{Im} (k_z) = 0$; $\operatorname{Im} (k_z) < 0$ on entire top sheet

    $\operatorname{Re} (k_z) > 0$ in unshaded region; $\operatorname{Re} (k_z) < 0$ in shaded region;

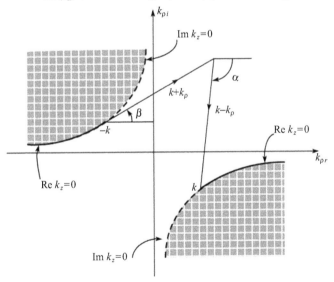

(b) Branch cut $\operatorname{Re} (k_z) = 0$; $\operatorname{Re} (k_z) > 0$ on entire top sheet

    $\operatorname{Im} (k_z) < 0$ in unshaded region; $\operatorname{Im} (k_z) > 0$ in shaded region;

Figure 4.12   Analytic properties of $\sqrt{k^2 - k_\rho^2}$.

A simple rule may be stated for the assignment of the algebra signs of $\operatorname{Re} k_z$ and $\operatorname{Im} k_z$ in various portions of the complex k_ρ plane. It follows from the recognition that sign changes in $\operatorname{Re} k_z$ or $\operatorname{Im} k_z$ can occur only when k_ρ crosses the contours $\operatorname{Re} k_z = 0$ or $\operatorname{Im} k_z = 0$, respectively. Thus for example, if branch cuts are chosen along the contours $\operatorname{Im} k_z = 0$, the sign of $\operatorname{Im} k_z$ is constant on either the top sheet or the bottom sheet, since the crossing of the $\operatorname{Im} k_z=0$ curve leads from one sheet to the other. It then suffices to specify $\operatorname{Re} k_z$ and $\operatorname{Im} k_z$ simultaneously at a single point on the top sheet, for example at $k_\rho=0$, since one may then deduce the sign alternations in $\operatorname{Re} k_z$ from the crossing of the $\operatorname{Re} k_z=0$ contours. In this manner, one may arrive at the designation of Figure 4.12(a). If the branch cut is chosen along the contours $\operatorname{Re} k_z = 0$, similarly we may have Figure 4.12(b).

For multilayered structure, the spectral domain Green's functions, the integrands of the Sommerfeld integrals may be expressed in terms of V and I that are given in Section 4.7. For a microstrip substrate, V_i^h and V_i^e are given in (4.7.23) and (4.7.24). Let $\gamma = jk_z$, it is seen from (4.7.23) and (4.7.24) that the functional dependence of V_i^h and V_i^e is odd for k_{z2} in region 2 and is even for k_{z1} in region 1. Consequently, in region 2 there is a double-valued problem for k_{z2}. Region 2 is a semi-infinite region, the choice of the sign for k_{z2} should ensure right phase dependence and convergence when $z \to \infty$. Therefore $\pm k_2$ are the branch points which are related to the contribution of radiation. In region 1, there is no double-valued problem for k_{z1}. This is in accordance with the fact that in this region waves bounce forth and back resulting in the arbitrary choice of the sign for k_{z1} and the appearance of the poles as seen from (4.7.23) and (4.7.24). The poles are related to the possible residue contribution and will be discussed in Section 4.8.2. The number of branch point pairs is equal to that of the semiinfinite regions. For the structure of microstrip substrate shown in Figure 4.7, this number is one. This structure may be alternatively analyzed by dividing the original problem into a series of half space problems based upon the multi-bouncing. In the equivalent problem, the branch points exist in each half space problem.

In the Sommerfeld integrals, there will be a Hankel function in the integrand with the branch point $k_\rho = 0$ associated with $\sqrt{k_\rho}$. It must be noted that an additional branch cut does not actually exist in the k_ρ plane located on the negative real axis. As we will see later it is always possible not to cross this cut and the unique value of $\sqrt{k_\rho}$ is guaranteed.

4.8.2   Poles

We still use the example in Figure 4.7 to analyze the poles. In (4.7.23), there will be poles for TE modes when

$$D_e = \gamma_1 \coth \gamma_1 d + \gamma_2 = 0 \qquad (4.8.12)$$

If $\gamma = jk_z$, we have

$$jk_{z1}(-j \cot k_{z1}d) + jk_{z2} = 0 \qquad (4.8.13)$$

Letting $k_{z2} = -jp$, p is real and larger than zero for the requirement of top Riemann sheet and

$$k_{z1} \cot k_{z1}d = -p \qquad (4.8.14)$$

Together with the relation obtained by equating the two expressions for k_ρ^2

$$k_{z1}^2 + p^2 = k_0^2(\epsilon_r - 1) \qquad (4.8.15)$$

a solution for d and p can be found. A solution may readily be obtained by graphical means. For this purpose it is convenient to rewrite (4.8.14), (4.8.15) as follows

$$k_{z1}d \cot k_{z1}d = -pd \qquad (4.8.16)$$
$$(k_{z1}d)^2 + (pd)^2 = (k_0d)^2(\epsilon_r - 1) \qquad (4.8.17)$$

In this later form the equations are independent of frequency, provided the ratio of thickness d to free space wavelength λ_0 is kept constant.

Eqn. (4.8.16) determines one relation between $k_{z1}d$ and pd, which may be plotted on a k_zd and pd plane. The other equation, (4.8.17), is a circle of radius $(\epsilon_r - 1)^{1/2}k_0d$ in the same plane. A typical plot is given in Figure 4.13. The points of intersection between the two curves such as point A, determine the eigenvalues k_{z1} and p. From the typical plot in Figure 4.13, the number N^h of the poles for $p > 0$ may be determined by

$$N^h = \begin{cases} 0, & a < \dfrac{\pi}{2} \\[2mm] n, & \left(n - \dfrac{1}{2}\right)\pi < a < \left(n + \dfrac{1}{2}\right)\pi, \quad n = 1, 2, \cdots \end{cases} \qquad (4.8.18)$$

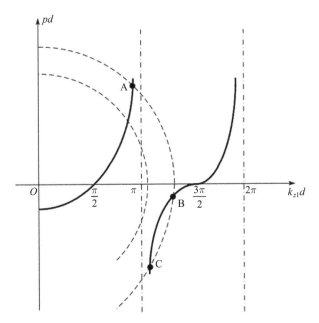

Figure 4.13   Graphical solution for eigenvalues for TE modes.

where $a = k_0d\sqrt{\epsilon_r - 1}$. The pole exists only if a is greater than $\pi/2$, that is $d > \pi/2k_0\sqrt{\epsilon_r - 1}$. As k_0d increases, the number of the poles continually increases. It can be proved that the real part of p is always negative for complex roots. The poles B, C, in Figure 4.13 are also with negative p. All these poles with negative real part of p are located on the bottom Reimann sheet, it is also called the improper sheet, and their contributions will be discussed later.

The solution for the TM modes is in the same form as that for the TE modes. In the TM case, (4.8.16) should be replaced by

$$\epsilon_r pd = k_{z1}d \tan k_{z1}d \qquad (4.8.19)$$

Combined with (4.8.17), a similar graphical solution may be used. A typical plot is given in Figure 4.14. The points of intersection between the curves, such as point A, determine the eigenvalues k_{z1} and p. Other discussions are similar to those for TE case.

An interesting property of the modes may be deduced from the graphical solution given in Figure 4.14. The two sets of curves will always have at least one point of intersection, even for d/λ_0 approaching zero. Hence the mode has no low-frequency cutoff.

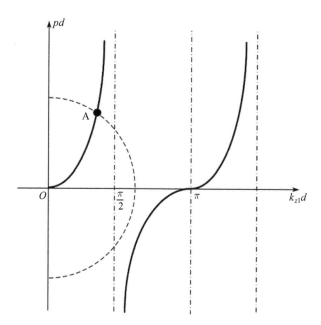

Figure 4.14   Graphical solution for eigenvalues for TM modes.

4.8.3   Integration Paths

Consider the following complex integral

$$M = \int_{-\infty}^{\infty} \widetilde{M}\, H_0^{(2)}(k_\rho \rho) e^{-jk_{\rho r}\rho} e^{k_{\rho i}\rho} e^{-jk_z z} dk_\rho \tag{4.8.20}$$

where $k_z = \sqrt{k^2 - k_\rho^2}$ and the integrand possesses both the pole and the branch-point singularities in the complex k_ρ plane. A typical integration path is shown in Figure 4.15, wherein the branch-point singularities at 0, and $\pm k$ and the pole singularities at $\pm p$ have been slightly displaced from the real axis to signify the presence of small loss. In the lossless limit, the required indentations of the integration path around the singularities are thereby evident and are shown on the left top of Figure 4.15. The branch point at the origin arises from the Hankel function in (4.8.20). The corresponding branch cut may be chosen as the negative $k_{\rho r}$ and is denoted by B_H. Branch cuts for $\pm k$ have been drawn so that $\text{Im}\, k_z$ is negative on the entire top sheet of the multisheeted Riemann surface (see Figure 4.12) and are denoted by F. The original path from $-\infty$ to ∞ parallel to $k_{\rho r}$ axis is C_0 which is slightly below the negative $k_{\rho r}$ axis on left side and slightly above the positive $k_{\rho r}$ axis on right side. Therefore C_0 never touches B_H and the branch point $k_\rho = 0$ is actually unnecessary to consider. To ensure the convergence of $\exp(k_{\rho i}\rho)$ in (4.8.20), the contour of integration should be enclosed on the lower half k_ρ plane by C_R at the infinity. In order not to intercept the branch cut F, it should be surrounded by appropriate contour C_{bf} which is in close conformity with F as is shown in Figure 4.15. The integration contour C_{bf} is along the opposite direction on two sides of F. However, due to the crossing of $\text{Re}\, k_z = 0$, the sign of $\text{Re}\, k_z$ is different, resulting in the contribution of the branch cut F, which is related to the radiation part of the solution.

In order to guarantee the convergence, that is, $\displaystyle\int_{C_R} = 0$ and from the residue theorem, we have

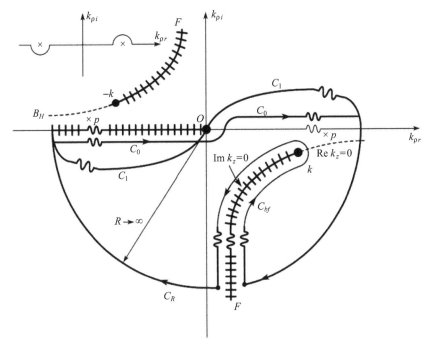

Figure 4.15   Branch cuts and integration paths.

$$\int_{C_0} + \int_{C_R} + \int_{C_{bf}} = \int_{C_0} + \int_{C_{bf}} = -2\pi j \cdot Res$$

or

$$\int_{C_0} = \left(-\int_{C_{bf}} \right) - 2\pi j \cdot Res$$

where Res denotes the residue due to the pole p. If there are multiple poles, Res should be replaced by $\sum Res$.

C_1 in Figure 4.15 is a possible deformed path. In this case, there is no pole between C_1 and C_0, thus $\int_{C_1} = \int_{C_0}$. We may write

$$\int_{C_0} = \int_{C_1} = \left(-\int_{C_{bf}} \right) - 2\pi j \sum Res \tag{4.8.21}$$

(4.8.21) indicates that $\int_{C_0}$ or $\int_{C_1}$ incorporates all the contributions from branch cuts and poles.

The hyperbolic branch cuts F may be used to identify the sign of both $\operatorname{Im} k_z$ and $\operatorname{Re} k_z$; these cuts are called the fundamental cuts or Sommerfeld cuts. The alternative cuts are straight line cuts S_1 which are in parallel with $k_{\rho i}$ axis, as is shown in Figure 4.16. It is seen that contour C_{bs} intercepts the branch cut F after turning around the branch point k. Therefore, the left side of C_{bs} is located on the bottom Reimann sheet. However, for this specific case, where point A moves to point B along C_{bs} and $R \to \infty$, $\alpha + \beta = -2\pi$, it means that $\operatorname{Im} k_z = 0$ when $k_\rho \to \infty$ and the convergence at the infinity is ensured as well. For an arbitrary straight line from k such as S_2, when point C approaches to infinity, $\alpha + \beta \neq 0, \pm 2\pi, \cdots$ and $\operatorname{Im} k_z > 0$. Obviously this straight line may not serve as a branch cut.

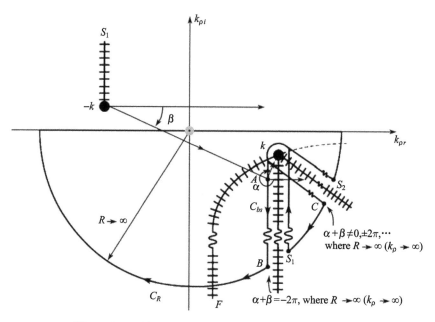

Figure 4.16　Alternative branch cuts and integration path.

Now we are going to describe the situation in which the fundamental branch cut is crossed twice, that is, the integration path is temporarily located on the bottom Reimann sheet. The convergence is ensured because the path finally returns back to the top Reimann sheet before approaching infinity. We have to examine whether the integration in this case is equivalent to that without crossing. Consider the two segments A and B of the total path as is shown in Figure 4.17. It is to prove that the integration results from segment A and segment B are equal. Both segment B and segment B' are located on the bottom Reimann sheet. In the case where there are no poles between B and B', so B and B' are equivalent in integration, it is assumed that A and B' are close to branch cut F, therefore the imaginary parts of k_z on A and B' approach zero and the real parts of k_z are not only equal in magnitude, but also have the same sign due to no crossing of $\operatorname{Re} k_z = 0$. It is evident that A

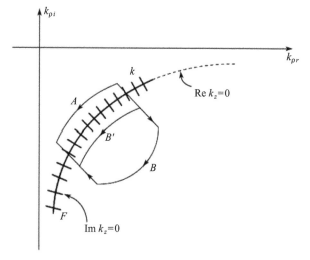

Figure 4.17　Two paths with and without crossing the branch cut.

and B are equivalent in integration.

It is clear from the above discussion that the choice of branch cut and integration path are wholly arbitrary as long as the unique value is determined on the physical base, that is, the right phase dependence required and the convergence. If there is no convergence problem, for example in the half space problem, in which case the observation point is located in the upper region, we may choose $\operatorname{Re} k_{z1} = 0$ as the branch cut, where $k_{z1} = \sqrt{k_1^2 - k_\rho^2}$, k_1 is the wavenumber of the lower region[11].

In addition to the k_ρ integration plane, there are other integration planes in the literature, for example, λ plane ($\lambda = k_\rho/k_0$), k_{z0} plane $\left(k_{z0} = \sqrt{k_0^2 - k_\rho^2}\right)$, k_{z1} plane $\left(k_{z1} = \sqrt{k_1^2 - k_\rho^2}\right)$, u_0 plane ($u_0 = jk_{z0}/k_0$), u_1 plane ($u_1 = jk_{z1}/k_0$) and θ plane ($k_\rho = k_0 \sin(\theta_r + j\theta_i)$).

The last θ plane needs more explanation. The transform $k_\rho = k_0 \sin\theta$ makes $k_z = \sqrt{k_0^2 - k_\rho^2} = 0$ a regular point in θ plane. The transcendental function $\sin\theta$ is single-valued. From its periodicity property $\sin(\theta + 2n\pi) = \sin\theta$, $n = \pm 1, \pm 2, \cdots$, it is evident that a multiplicity of θ values correspond to the same value of k_ρ. Thus, the entire k_ρ plane can be mapped into various adjacent sections of "width" 2π in the θ plane. The inverse function $\sin^{-1}(k_\rho/k_0)$ in the k_ρ plane is multi-valued, implying the existence of branch points in that plane. The correspondence between the k_ρ plane and the θ plane is defined by

$$\begin{aligned} k_\rho &= k_{\rho r} + jk_{\rho i} \\ &= k_0 \sin(\theta_r + j\theta_i) \\ &= k_0 \sin\theta_r \cosh\theta_i + jk_0 \cos\theta_r \sinh\theta_i \end{aligned} \quad (4.8.22)$$

For example $k_\rho = -\infty$ corresponds to $\theta_r = -\pi/2$, $\theta_i = \pm\infty$; $k_\rho = +\infty$ corresponds to $\theta_r = \pi/2$, $\theta_i = \pm\infty$ and so on. The mapping from the complex k_ρ plane to the complex θ plane is shown in Figure 4.18 assuming that there are two pairs of branch points $\pm k_0$ and $\pm k_1$, where T and B denote the top and bottom sheet respectively. The identification of T or B may be simply carried out according to

$$\begin{aligned} k_{0z} &= k_0 \cos(\theta_r + j\theta_i) \\ &= k_0 \cos\theta_r \cosh\theta_i - jk_0 \sin\theta_r \sinh\theta_i \end{aligned}$$

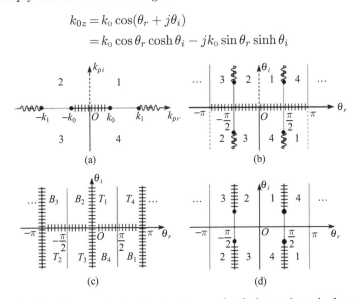

(a)

(b)

(c)

(d)

Figure 4.18 Mapping from the complex k_ρ plane to the complex θ plane, where the four quadrants 1,2,3,4 and the Reimann sheets are unwrapped. (a) k_ρ plane, (b) the corresponding segments on θ plane, (c) the branch cuts due to the branch points $\pm k_0$ which refer to the solid line, dashed line and slashed line in (a), (d) the branch cuts due to the branch points $\pm k_1$ which refers to the rippled line in (a).

and

$$\mathrm{Im}\, k_{z0} = -k_0 \sin\theta_r \sinh\theta_i \tag{4.8.23}$$

The top sheet T corresponds to $\sin\theta_r \sinh\theta_i > 0$ and the bottom sheet B corresponds to $\sin\theta_r \sinh\theta_i < 0$.

Finally, it should be pointed out that when there are poles enclosed by the integration path it is necessary to identify which Riemann sheet the poles are located on. Since the phase of k_{zi} is related to the choice of branch cut, the poles could be located on the top Reimann sheet for one choice and located on the bottom Reimann sheet for the other choice. There is no contradiction with the existence of a surface wave contribution. In fact, when the poles are located on the bottom Riemann sheet, the contribution of the surface wave is not caused by the residues but is implicit in the path integration[11].

4.9　Full Wave Discrete Image and Full Wave Analysis of Microstrip Antennas

Considering again the example in 4.7.3, according to (4.2.10), we have

$$G_{xx}^A = \frac{1}{4\pi} \int_{-\infty}^{\infty} \widetilde{G}_{xx}^A \, \mathrm{H}_0^{(2)}(k_\rho\rho) k_\rho dk_\rho$$

$$= \frac{1}{4\pi} \int_{-\infty}^{\infty} \frac{V_i^h}{j\omega\mu_0} \, \mathrm{H}_0^{(2)}(k_\rho\rho) k_\rho dk_\rho \tag{4.9.1}$$

$$G^\phi = \frac{1}{4\pi} \int_{-\infty}^{\infty} \widetilde{G}^\phi \, \mathrm{H}_0^{(2)}(k_\rho\rho) k_\rho dk_\rho$$

$$= \frac{1}{4\pi} \int_{-\infty}^{\infty} \frac{j\omega\epsilon_0}{k_\rho^2}(V_i^e - V_i^h) \, \mathrm{H}_0^{(2)}(k_\rho\rho) k_\rho dk_\rho \tag{4.9.2}$$

where

$$V_i^h = j\omega\mu_0/D_{TE}, \qquad\qquad V_i^e = -jk_{z1}k_{z0}\tan k_{z1}d/(j\omega\epsilon_0 D_{TM})$$

$$D_{TE} = k_{z1}\cot k_{z1}d + jk_{z0}, \quad D_{TM} = -k_{z1}\tan k_{z1}d + j\epsilon_r k_{z0}$$

4.9.1　Extraction of Quasi-Static Images

Assume $\omega \to 0$ but not the case of direct current, that is the charges still oscillate and the alternative currents exist. This case is referred to as the quasi-static. Under this approximation, $k_{z1} = k_{z0}$, $k_0 \approx 0$ and we have

$$V_i^h = \frac{j\omega\mu_0}{k_{z1}\cot k_{z1}d + jk_{z0}}$$

$$\approx \frac{\omega\mu_0\left(1 - e^{-j2k_{z0}d}\right)}{2k_{z0}} = V_{i,qs}^h \tag{4.9.3}$$

$$V_i^e = \frac{-jk_{z1}k_{z0}\tan k_{z1}d}{j\omega\epsilon_0(-k_{z1}\tan k_{z1}d + j\epsilon_r k_{z0})}$$

$$\approx -\frac{k_{z0}\tan k_{z0}d}{\omega\epsilon_0(-\tan k_{z0}d + j\epsilon_r)}$$

$$= \frac{k_{z0}\left(1 - e^{-j4k_{z0}d}\right)}{(1 + e^{-j2k_{z0}d})\,\omega\epsilon_0\,(1 + \epsilon_r)\left(1 - Ke^{-j2k_{z0}d}\right)} = V_{i,qs}^e \tag{4.9.4}$$

where $K = (1 - \epsilon_r)/(1 + \epsilon_r)$, the subscript qs is used to denote quasi-static. Now we have

$$\widetilde{G}^A_{xx,qs} = \frac{V^h_{i,qs}}{j\omega\mu_0} = \frac{1}{2jk_{z0}}\left(1 - e^{-j2k_{z0}d}\right) \qquad (4.9.5)$$

$$\widetilde{G}^\phi_{qs} = \frac{j\omega\epsilon_0}{k_\rho^2}\left(V^e_{i,qs} - V^h_{i,qs}\right)$$

$$= j\omega\epsilon_0\left[\frac{V^e_{i,qs}}{k_\rho^2} - \frac{k_{z0}^2 + k_\rho^2}{k_0^2 k_\rho^2}V^h_{i,qs}\right]$$

$$= -j\omega\epsilon_0\left[\frac{V^h_{i,qs}}{k_0^2} + \left(\frac{k_{z0}^2}{k_\rho^2}\right)\left(\frac{V^h_{i,qs}}{k_0^2} - \frac{V^e_{i,qs}}{k_{z0}^2}\right)\right]$$

$$= -j\omega\epsilon_0\left[\frac{V^h_{i,qs}}{k_0^2} - \left(\frac{V^h_{i,qs}}{k_0^2} - \frac{V^e_{i,qs}}{k_{z0}^2}\right)\right] \qquad (4.9.6)$$

where the approximation of $k_{z0}^2/k_\rho^2 = -1$ is used. The application of (4.9.3) and (4.9.4) yields

$$\frac{V^h_{i,qs}}{k_0^2} - \frac{V^e_{i,qs}}{k_{z0}^2}$$

$$= \frac{\omega\mu_0\left(1 - e^{-j2k_{z0}d}\right)}{2k_0^2 k_{z0}} - \frac{1 - e^{-j4k_{z0}d}}{\left(1 + e^{-j2k_{z0}d}\right)\omega\epsilon_0\left(1 + \epsilon_r\right)\left(1 - Ke^{-j2k_{z0}d}\right)k_{z0}}$$

$$= \frac{1 - e^{-j4k_{z0}d}}{2\omega\epsilon_0 k_{z0}}\left[\frac{1}{1 + e^{-j2k_{z0}d}} - \frac{2}{(1 + \epsilon_r)(1 + e^{-j2k_{z0}d})(1 - Ke^{-j2k_{z0}d})}\right]$$

$$= \frac{1 - e^{-j4k_{z0}d}}{2\omega\epsilon_0 k_{z0}}\cdot\left(\frac{-K}{1 - Ke^{-j2k_{z0}d}}\right)$$

$$= \frac{-1}{2\omega\epsilon_0 k_{z0}}R_{q,qs} \qquad (4.9.7)$$

where

$$R_{q,qs} = \frac{K\left(1 - e^{-j4k_{z0}d}\right)}{1 - Ke^{-j2k_{z0}d}}$$

$$= K + K^2 e^{-j2k_{z0}d} + K(K^2 - 1)e^{-j4k_{z0}d} + \cdots \qquad (4.9.8)$$

Then (4.9.6) may be written as

$$\widetilde{G}^\phi_{qs} = \frac{1}{2jk_{z0}}\left[1 - e^{-j2k_{z0}d} + R_{q,qs}\right]$$

$$= \frac{1}{2jk_{z0}}\left[(1 + K) + (K^2 - 1)e^{-j2k_{z0}d} + K(K^2 - 1)e^{-j4k_{z0}d} + \cdots\right] \qquad (4.9.9)$$

Substituting (4.9.5) and (4.9.9) into (4.9.1) and (4.9.2) and making use of the Sommerfeld identity in (4.2.12) result in

$$G^A_{xx,qs} = \frac{1}{4\pi}\left(\frac{e^{-jk_0 r_0}}{r_0} - \frac{e^{-jk_0 r_1}}{r_1}\right) \qquad (4.9.10)$$

$$G^\phi_{qs} = \frac{1 + K}{4\pi r_0}e^{-jk_0 r_0} + \frac{K^2 - 1}{4\pi r_1}e^{-jk_0 r_1} + \frac{K(K^2 - 1)}{4\pi r_2}e^{-jk_0 r_2} + \cdots \qquad (4.9.11)$$

where $r_i = \sqrt{\rho^2 + (z + 2id)^2}$, $i = 0,1,2,\cdots$, d is the thickness of the microstrip substrate as is shown in Figure 4.7.

From (4.6.24), it is seen that G_{qs}^ϕ is the potential produced by unit $\nabla' \cdot \mathbf{J}$. Notice that $\nabla' \cdot \mathbf{J} = -\nabla \cdot \mathbf{J} = j\omega\rho$, so the potential produced by unit q is then given by

$$
\begin{aligned}
\phi_q &= \frac{1+K}{4\pi\omega\epsilon_0 r_0} e^{-jk_0 r_0} + \frac{K^2 - 1}{4\pi\omega\epsilon_0 r_1} e^{-jk_0 r_1} + \frac{K(K^2 - 1)}{4\pi\omega\epsilon_0 r_2} e^{-jk_0 r_2} + \cdots \\
&= \frac{1+K}{4\pi\omega\epsilon_0} \left[\frac{e^{-jk_0 r_0}}{r_0} - (1 - K) \sum_{i=1}^\infty K^{i-1} \frac{e^{-jk_0 r_i}}{r_i} \right]
\end{aligned}
\tag{4.9.12}
$$

If the unit source is assumed to be current I, according to the relationship between q and I: $q = I/j\omega$, the potential ϕ_I produced by unit I is given as

$$
\phi_I = \frac{1}{j\omega} \phi_q
\tag{4.9.13}
$$

The same result was derived on purely physical bases in [25] and used for the dynamic study of the microstrip line[26]. It is pointed out in [25] that the Faraday field related to **A** is not affected by the dielectric interfaces, whereas the Coloumb field is affected by the dielectric interfaces through an infinite number of images. From the viewpoint of spectral domain analysis, this quasi-static model is a low frequency approximation of the full-wave solution.

When ϵ_r is large, G_{qs}^ϕ in (4.9.11) is a slowly convergent series. For instance, if $\epsilon_r = 12.9$, $h/\lambda_0 = 0.05$, as many as 80 terms should be taken to ensure the convergence. In [27], a reduced image model was presented, in which only four images are necessary. Such reduction is not necessarily unique, a simpler alternative reduced image scheme is given below.

Expanding the $R_{q,qs}$ of (4.9.8) in a Taylor series and taking only the two leading terms, we have

$$
R_{q,qs} \approx K \left(1 - e^{-j4k_z 0 d} \right) \left(1 + Ke^{-j2k_z 0 d} \right)
\tag{4.9.14}
$$

Corresponding to (4.9.14), the quasi-static Green's function for the scalar potential should be as follows

$$
G_{qs}^\phi = \frac{1}{4\pi} \left[\frac{(1+K)e^{-jk_0 r_0}}{r_0} + (K^2 - 1)\frac{e^{-jk_0 r_1}}{r_1} - \frac{Ke^{-jk_0 r_2}}{r_2} - K^2 \frac{e^{-jk_0 r_3}}{r_3} \right]
\tag{4.9.15}
$$

After extracting the quasi-static images, (4.9.1) and (4.9.2) may be rewritten as

$$
G_{xx}^A = G_{xx,qs}^A + \frac{1}{4\pi} \int_{-\infty}^\infty \left(\widetilde{G}_{xx}^A - \widetilde{G}_{xx,qs}^A \right) \mathrm{H}_0^{(2)}(k_\rho\rho) k_\rho dk_\rho
\tag{4.9.16}
$$

$$
G^\phi = G_{qs}^\phi + \frac{1}{4\pi} \int_{-\infty}^\infty \left(\widetilde{G}^\phi - \widetilde{G}_{qs}^\phi \right) \mathrm{H}_0^{(2)}(k_\rho\rho) k_\rho dk_\rho
\tag{4.9.17}
$$

Since the quasi-static fields are defined in the range in which the observation distance is much smaller than the free-space wavelength ($\rho \ll \lambda_0$), they correspond to $k_\rho \to \infty$. Therefore, the subtraction of these quasi-static terms $\widetilde{G}_{xx,qs}^A$ and $\widetilde{G}_{qs}^\phi$ in spectral domain makes the remaining integrands of (4.9.16) and (4.9.17) decay faster for large k_ρ.

4.9.2   Extraction of Surface Waves

The surface waves play a rather significant role as they are guided along the interface without leaking energy. The corresponding pole singularities are located on the real axis of the k_ρ plane. Even if the integration path is deformed such that it is not too close to these poles' singularities, their presence still affects the value of the integral for small values of k_ρ.

Since the surface-wave poles always occur in complex conjugate pairs, a typical pair can be represented mathematically as

$$\frac{2k_{\rho p}(Residue \; at \; k_{\rho p})}{k_\rho^2 - k_{\rho p}^2}$$

where $k_{\rho p}$ is the surface-wave pole. By subtracting these poles from the integrands of (4.9.16) and (4.9.17), and analytically evaluating their contributions via the residue calculus technique, we can derive the following representation for the spatial domain Green's functions G_{xx}^A and G^ϕ:

$$G_{xx}^A = G_{xx,qs}^A + G_{xx,sw}^A + \frac{1}{4\pi} \int_{-\infty}^{\infty} \frac{1}{2jk_{z0}} F_1(k_\rho) \, \mathrm{H}_0^{(2)}(k_\rho \rho) k_\rho dk_\rho \qquad (4.9.18)$$

$$G^\phi = G_{qs}^\phi + G_{sw}^\phi + \frac{1}{4\pi} \int_{-\infty}^{\infty} \frac{1}{2jk_{z0}} F_2(k_\rho) \, \mathrm{H}_0^{(2)}(k_\rho \rho) k_\rho dk_\rho \qquad (4.9.19)$$

where

$$\frac{1}{2jk_{z0}} F_1(k_\rho) = \widetilde{G}_{xx}^A - \widetilde{G}_{xx,qs}^A - \widetilde{G}_{xx,sw}^A \qquad (4.9.20)$$

$$\frac{1}{2jk_{z0}} F_2(k_\rho) = \widetilde{G}^\phi - \widetilde{G}_{qs}^\phi - \widetilde{G}_{sw}^\phi \qquad (4.9.21)$$

$$G_{xx,sw}^A = \frac{1}{4\pi} \sum_{i=1}^{N^h} \int_{-\infty}^{\infty} \frac{2k_{\rho p(i)} Res1^{(i)}}{k_\rho^2 - k_{\rho p}^2} \, \mathrm{H}_0^{(2)}(k_\rho \rho) k_\rho dk_\rho$$

$$= \frac{1}{4\pi}(-j2\pi) \sum_{i=1}^{N^h} k_{\rho p(i)} \, \mathrm{H}_0^{(2)}\left(k_{\rho p(i)}\rho\right) Res1^{(i)} \qquad (4.9.22)$$

$$\widetilde{G}_{xx,sw}^A = \sum_{i=1}^{N^h} \frac{2k_{\rho p(i)} Res1^{(i)}}{k_\rho^2 - k_{\rho p(i)}^2} \qquad (4.9.23)$$

$$Res1^{(i)} = \lim_{k_\rho \to k_{\rho p(i)}} \left(k_\rho - k_{\rho p(i)}\right) \cdot \widetilde{G}_{xx}^A \qquad (4.9.24)$$

$$G_{sw}^\phi = \frac{1}{4\pi} \sum_{i=1}^{N^e} \int_{-\infty}^{\infty} \frac{2k_{\rho p(i)} Res2^{(i)}}{k_\rho^2 - k_{\rho p}^2} \, \mathrm{H}_0^{(2)}(k_\rho \rho) k_\rho dk_\rho$$

$$= \frac{1}{4\pi}(-j2\pi) \sum_{i=1}^{N^e} k_{\rho p(i)} \, \mathrm{H}_0^{(2)}\left(k_{\rho p(i)}\rho\right) Res2^{(i)} \qquad (4.9.25)$$

$$\widetilde{G}_{sw}^\phi = \sum_{i=1}^{N^e} \frac{2k_{\rho p(i)} \cdot Res2^{(i)}}{k_\rho^2 - k_{\rho p(i)}^2} \qquad (4.9.26)$$

$$Res2^{(i)} = \lim_{k_\rho \to k_{\rho p(i)}} \left(k_\rho - k_{\rho p(i)}\right) \widetilde{G}^\phi \qquad (4.9.27)$$

where N^h and N^e denote the number of poles for TE modes and TM modes. In calculating residues of (4.9.24) and (4.9.27), we may use the following theorem:

$$\lim_{z \to z_0} (z - z_0) \frac{g(z)}{h(z)} = Res \left[\frac{g(z)}{h(z)} \bigg|_{z=z_0} \right] = \lim_{z \to z_0} \frac{g(z)}{[h(z) - h(z_0)]/(z - z_0)}$$

$$= \frac{g(z_0)}{h'(z_0)} \tag{4.9.28}$$

where $f(z) = g(z)/h(z)$ is a holomorphic function over a domain of definition, $\widetilde{G}$ and h are holomorphic functions in the same domain. $h(z_0) = 0$ but $h'(z_0) \neq 0$.

Since the surface waves dominate in the range in which the observation distance is much larger than the free-space wavelength ($\rho \gg \lambda_0$), they correspond to the values of small k_ρ. The subtraction of these surface wave terms $\widetilde{G}^A_{xx,sw}$ and $\widetilde{G}^\phi_{sw}$ in spectral domain makes the remaining integrands of (4.9.18) and (4.9.19) smooth to the values of small k_ρ.

4.9.3   Approximation for the Remaining Integrands

In this step $F_1(k_\rho)$ and $F_2(k_\rho)$ given in (4.9.20) and (4.9.21), respectively, are approximated in terms of complex exponentials by Prony's method[28, 29]. In order to be able to use the Sommerfeld identity for the exponentials obtained from the approximations of $F_1(k_\rho)$ and $F_2(k_\rho)$, these exponentials should be functions of k_{z0} which is generally a complex number. In Figure 4.19(a) and Figure 4.19(b), we choose a finite path C_1. Since Prony's method

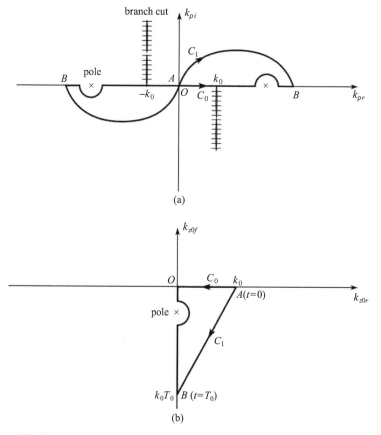

Figure 4.19   (a) The integration paths C_0 and C_1 on the k_ρ plane which is a mapping of k_z plane according to $k_\rho^2 = k_0^2 - k_z^2$, (b) The integration paths C_0 and C_1 on the k_{z0} plane.

is applicable to complex functions with real variables, we need to transform the complex variable k_{z0} into a real variable t by a parametric function defined as

$$C_1 : k_{z0} = k_0 \left[-jt + \left(1 - \frac{t}{T_0} \right) \right], \quad 0 \leqslant t \leqslant T_0 \tag{4.9.29}$$

which maps $t \in [0, T_0]$ into $k_{z0} \in [k_0, -jk_0T_0]$. When $t = T_0$, then $k_{z0} = -jk_0T_0$ and $k_\rho = k_0\sqrt{1 + T_0^2}$. The choice of path is dependent upon the behavior of the integrands to the approximated for large k_ρ. Both $F_1(k_\rho)$ and $F_2(k_\rho)$ are uniformly sampled along the integration path C_1, which corresponds to the real variable t, and approximated in terms of exponentials of variable t, or k_{z0} as follows:

$$F_1(k_\rho) = \sum_{i=1}^{N} a_i e^{-b_i k_{z0}} = \sum_{i=1}^{N} A_i e^{B_i t} \tag{4.9.30}$$

$$F_2(k_\rho) = \sum_{i=1}^{N} a_i' e^{-b_i' k_{z0}} = \sum_{i=1}^{N} A_i' e^{B_i' t} \tag{4.9.31}$$

where

$$a_i = A_i e^{B_i T_0/(1+jT_0)}, \quad b_i = B_i \frac{T_0}{k_0(1+jT_0)} \tag{4.9.32}$$

and similarly for a_i' and b_i'.

Substituting (4.9.30) and (4.9.31) into (4.9.18) and (4.9.19) and using the Sommerfeld identity, we have

$$G_{xx}^A = G_{xx,qs}^A + G_{xx,sw}^A + G_{xx,ci}^A \tag{4.9.33}$$

$$G^\phi = G_{qs}^\phi + G_{sw}^\phi + G_{ci}^\phi \tag{4.9.34}$$

where

$$G_{xx,ci}^A = \frac{1}{4\pi} \sum_{i=1}^{N} a_i \frac{e^{-jk_0 r_i}}{r_i} \tag{4.9.35}$$

$$G_{ci}^\phi = \frac{1}{4\pi} \sum_{i=1}^{N} a_i' \frac{e^{-jk_0 r_i'}}{r_i'} \tag{4.9.36}$$

$$r_i = \sqrt{\rho^2 + (z - jb_i)^2}, \quad r_i' = \sqrt{\rho^2 + (z - jb_i')^2} \tag{4.9.37}$$

The subscript ci means complex image; r_i and r_i' are complex distances; and a_i and a_i' are complex amplitudes. Each term in (4.9.35) and (4.9.36) represents a complex image.

In Prony's method, the required number of sampling points is at least twice the number of exponentials to be used in the approximation. If the number of sample points is chosen to be exactly twice the number of exponentials, the approximation will be exact only at the sampling points, and there will be no guarantee that it would be accurate elsewhere. It is therefore essential to take as many samples as necessary to ensure the capture of any rapid variation of the function being sampled. Consequently, the number of sampling points is usually taken to be much higher than twice the number of exponentials and prompts the using of the least square Prony's method proposed in [30].

To distinguish the full-wave continuous image method[31] from approximate discrete image method[32] of the multilayer planar structures, this method is called the full-wave discrete image method or the complex image method (CIM), and the name discrete complex image method (DCIM) is commonly used by most of the authors.

Like most inverse problems, the set of the image is nonunique. The nonuniqueness shows up in the complex image method as unsteadiness of the image locations and amplitudes over frequency. The nonuniqueness also points to the possibility of perfect steadiness by fixing the images at real, instead of complex locations. These images are called the simulated images. To satisfy the low-frequency asymptotes, the real locations of the images can be the first few of the classical quasi-static images. With the location fixed, in spectral domain instead of Prony's method, the moment method matching is used to find the images' amplitudes. The amplitudes found are indeed smoothly varying functions of frequency[33]. The further simplification of DCIM to arrive at formulas with insight from the Green's functions is done in [34]. The extension of DCIM to time domain has been developed in [35].

The DCIM introduced above can not be made fully robust and suitable for the development of CAD software. This is because in general cases, it requires users first to investigate the spectral domain behavior of the Green's function and then to perform a few iterations to find the best possible combination of the approximation. To circumvent these difficulties, a two-level approximation scheme has been developed in conjunction with the use of the generalized pencil of function (GPOF) method, which is superior to either Prony's method or least-square Prony's method. In addition, in this new scheme, neither surface wave poles nor the real images need to be extracted[36]. Now consider the general form of Sommerfeld integral

$$G = \int_{-\infty}^{\infty} \frac{1}{2jk_z} F(k_\rho) \, \mathrm{H}_0^{(2)}(k_\rho\rho) k_\rho dk_\rho \qquad (4.9.38)$$

where $F(k_\rho)$ could be either $F_1(k_\rho)$ in (4.9.20) or $F_2(k_\rho)$ in (4.9.21). The parametric equations are defined as

$$C_1: \ k_z = -jkt, \qquad\qquad T_{02} \leqslant t \leqslant T_{01} + T_{02} \qquad (4.9.39)$$

$$C_2: \ k_z = k\left[-jt + \left(1 - \frac{t}{T_{02}}\right)\right], \quad 0 \leqslant t \leqslant T_{02} \qquad (4.9.40)$$

According to the relationship $k_\rho^2 = k^2 - k_z^2$, C_1, C_2 in k_z the plane may be mapped to the k_ρ plane as shown in Figure 4.20. From (4.9.39), when $t = T_{02}$, $k_z = -jkT_{02}$,

$$k_{\rho\max 2} = \sqrt{k^2 - k_z^2} = k\sqrt{1 + T_{02}^2} \qquad (4.9.41)$$

or

$$T_{02} = \sqrt{(k_{\rho\max 2}/k)^2 - 1} \qquad (4.9.42)$$

To choose T_{02}, it is necessary to satisfy $k_{\rho\max} \geqslant k_m$, where k_m is the maximum value of the wavenumber involved in the problem to be analyzed. For example if the highest relative dielectric constant $\epsilon_r = 12.5$, then $k_m = \sqrt{12.5}\,k \approx 3.5k$, and T_{02} can be safely chosen to be five. From (4.9.39), when $t = T_{01} + T_{02}$, $k_z = -jk(T_{01} + T_{02})$,

$$k_{\rho\max 1} = k\sqrt{1 + (T_{01} + T_{02})^2} \qquad (4.9.43)$$

or

$$T_{01} = \sqrt{(k_{\rho\max 1}/k)^2 - 1} - T_{02} \qquad (4.9.44)$$

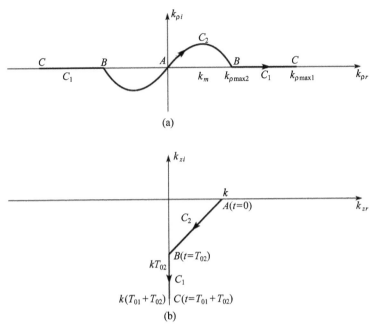

Figure 4.20   (a) The integration paths C_1 and C_2 on the k_ρ plane which is a mapping of the k_z plane according to $k_\rho^2 = k_0^2 - k_z^2$, (b) The integration paths C_1 and C_2 on the k_z plane.

The choice of T_{01} is not very critical as long as one chooses $k_{\rho max\,1}$ large enough to pick up the behavior of the spectral domain Green's function for large k_ρ. Since the spectral domain behaviors of the Green's functions are always smooth beyond $k_{\rho max\,2}$, it is not necessary to have a large number of samples on $[k_{\rho max\,2}, k_{\rho max\,1}]$. Typical value for T_{01} is $300 \sim 500$. The sampling interval Δt for both C_1 and C_2 is typically 0.025.

On C_1 shown in Figure 4.20(b), approximating $F(k_\rho)$ by using the matrix pencil method subroutine in [37], we have

$$F_1(k_\rho) = \sum_{k=1}^{M_1} b_{1k} e^{B_{1k}t} = \sum_{k=1}^{M_1} a_{1k} e^{-A_{1k}k_z}$$

where

$$A_{1k} = B_{1k}/jk, \quad a_{1k} = b_{1k} e^{-jkT_0 A_{1k}}$$

then, G in (4.9.38) may be rewritten as

$$G = \int_{-\infty}^{\infty} \frac{1}{2jk_z} F_2(k_\rho)\, \mathrm{H}_0^{(2)}(k_\rho\rho)k_\rho dk_\rho$$

$$+ \int_{-\infty}^{\infty} \frac{1}{2jk_z} \left[\sum_{k=1}^{M_1} a_{1k} \exp(-A_{1k}k_z) \right] \mathrm{H}_0^{(2)}(k_\rho\rho)k_\rho dk_\rho \qquad (4.9.45)$$

where

$$F_2(k_\rho) = F(k_\rho) - \sum_{k=1}^{M_1} a_{1k} e^{-A_{1k}k_z} \qquad (4.9.46)$$

The above is the first level approximation. The second level one is to approximate $F_2(k_\rho)$ on C_2 shown in Figure 4.20(b) by using the same subroutine; then we have

$$F_2(k_\rho) = \sum_{k=1}^{M_2} b_{2k} e^{B_{2k}t} = \sum_{k=1}^{M_2} a_{2k} e^{-A_{2k}k_z} \qquad (4.9.47)$$

The typical number of M_1 or M_2 is around 10. Finally we have the closed form for G

$$
\begin{aligned}
G &= \int_{-\infty}^{\infty} \frac{1}{2jk_z} \left[\sum_{k=1}^{M_1} a_{1k} e^{-A_{1k}k_z} + \sum_{k=1}^{M_2} a_{2k} e^{-A_{2k}k_z} \right] \mathrm{H}_0^{(2)}(k_\rho \rho) k_\rho dk_\rho \\
&= \sum_{k=1}^{M_1} \int_{-\infty}^{\infty} \frac{1}{2jk_z} a_{1k} e^{-A_{1k}k_z} \, \mathrm{H}_0^{(2)}(k_\rho \rho) k_\rho dk_\rho \\
&\quad + \sum_{k=1}^{M_2} \int_{-\infty}^{\infty} \frac{1}{2jk_z} a_{2k} e^{-A_{2k}k_z} \, \mathrm{H}_0^{(2)}(k_\rho \rho) k_\rho dk_\rho \\
&= \sum_{k=1}^{M_1} a_{1k} \frac{e^{-jkr_{1k}}}{r_{1k}} + \sum_{k=1}^{M_2} a_{2k} \frac{e^{-jkr_{2k}}}{r_{2k}}
\end{aligned}
\tag{4.9.48}
$$

where

$$
r_{nk} = \sqrt{\rho^2 + (A_{nk}/j)^2} = \sqrt{\rho^2 - A_{nk}^2} \quad n = 1 \quad \text{or} \quad 2
$$

When sources are in the bounded regions of multilayer structures shown in Figure 4.21, after extracting the quasi-static images, in addition the branch cut for k_{z0}, we introduce a branch cut for k_{zs}. Therefore function $\widetilde{G}$ should be rewritten as

$$
\widetilde{G} = \frac{F(k_\rho)}{j2k_{zs}} = \frac{F'(k_\rho)}{j2k_{z0}}
$$

where $F'(k_\rho) = F(k_\rho)k_{z0}/k_{zs}$. Note that F' only has a branch cut for k_{z0} and not for k_{zs}. This important correction in [38] makes original CIM available to the general multilayer structures.

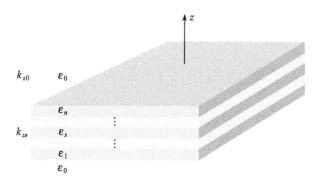

Figure 4.21   Multilayer planar structure with embedded source.

In (4.9.24) and (4.9.27), the surface-wave contributions are analytically extracted by using residue calculations. For multilayer cases these calculations are always very difficult to carry out. To avoid this difficulty, $Res(i)$ are obtained by recursively evaluating the contour integral numerically in the complex k_ρ plane. The integral begins with the contour enclosing the region we are interested in. If it is nonzero, then subdivide it into four pieces and evaluate the contour integral of each piece. Repeat this process until the location k_{ρ_i} and residue $Res(i)$ for all the poles are found[39].

As an alternative to [38] and [39], in [36], there is no need to extract the quasi-static images and to calculate the residues.

The DCIM has been considered as a prominent approach[9] and has received significant attention in recent years[40]. However, there are still some problems in using it. The approach proposed in [36] was proved to be very efficient and robust, but only for the case of thin layers where the propagating wave contribution is insignificant. Further investigations[23] showed that for thick layers, a large error occurs in the far field if the propagating modes are not extracted. The physical explanation is that use of complex images is in fact an attempt to approximate the cylindrical nature of propagating waves in terms of spherical ones. In [29], this difficulty was overcome by implementing Hankel functions to describe propagating modes and complex images for the decaying part of the field. In [40], an alternative to DCIM is presented and is combined with the DCIM process. In conclusion, like other methods, DCIM is still under development and competes with other methods.

4.9.4 Application of Full Wave Discrete Image Method in Microstrip Structures

The full-wave discrete images are actually the spatial Green's functions for microstrip structures. Combined with the method of moments (MoM)[41], the full-wave analysis of these structures, such as microstrip antennas and their feed networks, can be carried out. The integral equation is formed according to the boundary condition matching, that is, the tangential electric field on the microstrips should be zero and is given by

$$\hat{\mathbf{n}} \times \mathbf{E}^s + \hat{\mathbf{n}} \times \mathbf{E}^i = 0 \qquad (4.9.49)$$

where $\mathbf{E}^i$ and $\mathbf{E}^s$ denote the impressed and scattered fields respectively. Using $\mathbf{E}^s$ given in (4.6.24), (4.9.49)may be written as

$$\hat{\mathbf{n}} \times \left[-j\omega\mu_0 \langle \underline{\underline{\mathbf{K}}}^A; \mathbf{J} \rangle + \frac{1}{j\omega\epsilon_0} \nabla(\langle G^\phi, \nabla' \cdot \mathbf{J} \rangle) \right] = -\hat{\mathbf{n}} \times \mathbf{E}^i \qquad (4.9.50)$$

The induced current on the microstrip can be found by solving the above integral equation (4.9.50).

A critical factor for an efficient and accurate MoM analysis is the choice of basis functions. Traditional numerical modelling employs root-top functions for rectangular discretization or Rao-Wilton-Glisson(RWG) functions[42] for triangular discretization. For these functions, a very fine discretization is often required to yield an accurate solution. This leads to a large matrix equation, which is computationally expensive to solve. In addition, the numerical solution converges slowly to the exact one when the discretization is made finer. A solution to this problem is to employ higher-order basis functions, which have a better convergence rate and can yield an accurate solution with a rather coarse discretization. The higher-order interpolatory basis functions developed by Graglia et al[43] are employed in [39]. Also, in [39] the curvilinear discretization is used, which provides more flexibility to model arbitrary shapes.

4.10 Asymptotic Integration Techniques and Their Applications

Once the current distribution on the microstrips are obtained, all the information may be extracted, such as the input impedance of microstrip antenna, the parameters of the network and the radiation pattern. The radiation pattern is related to the far fields, which in particular can be calculated by the asymptotic integration techniques which will be introduced and then will be used for the calculation of the radiation pattern of microstrip antennas. In addition, these techniques are also useful in the analysis of multilayer problems by using the geometrical optics method (saddle point method)[11].

4.10.1   The Saddle Point Method

Consider a complex integral

$$A(\alpha) = \int_c e^{\alpha g(\theta)} d\theta \qquad (4.10.1)$$

where α is a large real number, $g(\theta)$ is an analytic function and $\theta = \theta_r + j\theta_i$, with the subscripts r and i denoting the real and imaginary parts respectively. The point at which the first derivative of $g(\theta)$ vanishes with respect to θ is the point θ_s such that $g'(\theta) = 0$ and is called the saddle point. In the neighborhood of θ_s, $g(\theta)$ can be approximated by using Taylor's series up to second order

$$g(\theta) \approx g(\theta_s) + \frac{1}{2}g''(\theta_s)(\theta - \theta_s)^2 \qquad (4.10.2)$$

If higher accuracy is required, still higher order terms may be invoked in Taylor's series (4.6.24).

Introduce

$$g(\theta_s) = u(\theta_s) + jv(\theta_s), \quad g''(\theta_s) = \eta e^{j\beta}, \quad (\theta - \theta_s) = se^{j\phi} \qquad (4.10.3)$$

then

$$g(\theta) = u(\theta_s) + jv(\theta_s) + \frac{1}{2}\eta s^2 e^{j(\beta+2\phi)} = u + jv \qquad (4.10.4)$$

where

$$u = u(\theta_s) + \frac{1}{2}\eta s^2 \cos(\beta + 2\phi) \qquad (4.10.5)$$

$$v = v(\theta_s) + \frac{1}{2}\eta s^2 \sin(\beta + 2\phi) \qquad (4.10.6)$$

If $\beta + 2\phi = \pm\pi$, then along this direction of ϕ, $e^{\alpha g(\theta)}$ decays rapidly because of the large α, but the phase remains constant. This path defined by $v = v(\theta_s)$ is called the steepest descent path and is denoted by C' in Figure 4.22. It is beneficial to carry out the integral $A(\alpha)$ in (4.10.1) along this deformed path C'. Then we have

$$A(\alpha) \approx \int_{c'} e^{\alpha g(\theta_s)} e^{-\alpha \eta s^2/2} d\theta = e^{\alpha g(\theta_s)} e^{j\phi} \int_{c'} e^{-\alpha \eta s^2/2} ds \qquad (4.10.7)$$

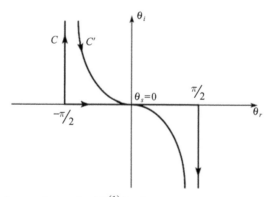

Figure 4.22   Integration path of $H_0^{(1)}(k_\rho\rho)$ showing the steepest descent path C'.

Because of the rapid decay of the integrand, the limits of the integral may be approximately taken as $-\infty$ to $+\infty$, enabling us to make use of the following formula:

$$\int_{-\infty}^{+\infty} e^{-\xi s^2} ds = \sqrt{\frac{\pi}{\xi}} \tag{4.10.8}$$

Using (4.10.8), (4.10.7) may be written as

$$A(\alpha) = e^{\alpha g(\theta_s)} e^{j\phi} \sqrt{\frac{2\pi}{\alpha \eta}} \tag{4.10.9}$$

(4.10.9) is the asymptotic expression of (4.10.1).

Example: To find the asymptotic expression of $H_0^{(1)}(k_\rho \rho)$ with $k_\rho \rho \gg 1$, where

$$H_0^{(1)}(k_\rho \rho) = \frac{1}{\pi} \int_{-\frac{\pi}{2}+j\infty}^{\frac{\pi}{2}-j\infty} e^{jk_\rho \rho \cos\theta} d\theta \tag{4.10.10}$$

According to (4.10.1) and keeping α real, we have $\alpha = k_\rho \rho$, $g(\theta) = j\cos(\theta)$. From $g'(\theta_s) = 0$, we find $\theta_s = 0°$. From

$$g(\theta) = j\cos\theta = \sin\theta_r \sinh\theta_i + j\cos\theta_r \cosh\theta_i = u + jv \tag{4.10.11}$$

and $v(\theta_s) = 1$, the equation of steepest descent path step is given by $\cos\theta_r \cosh\theta_i = 1$. The final step is to find η and φ. From $g''(\theta_s) = -j = \eta e^{j\beta}$, we have $\eta = 1$, $\beta = -\pi/2$. The reasonable solution of φ may be determined by $\varphi = (-\pi + \pi/2)/2 = -\pi/4$.

From (4.10.9), the asymptotic solution of $H_0^{(1)}(k_\rho \rho)$ will be

$$H_0^{(1)}(k_\rho \rho) \approx \frac{1}{\pi} e^{jk_\rho \rho} e^{-j\pi/4} \sqrt{\frac{2\pi}{k_\rho \rho}} = \sqrt{\frac{2}{\pi k_\rho \rho}} e^{jk_\rho \rho} e^{-j\pi/4} \tag{4.10.12}$$

4.10.2   The Steepest Descent Method

The essence of this method is the same as that of the saddle point method. The only difference is that in the steepest descent method, (4.10.2) is alternatively expressed as

$$g(\theta) = g(\theta_s) + \frac{1}{2} g''(\theta_s)(\theta - \theta_s)^2 = g(\theta_s) - s^2 \tag{4.10.13}$$

Then

$$\theta - \theta_s = \pm s \sqrt{\frac{-2}{g''(\theta_s)}} = \pm s \sqrt{\frac{-2}{\eta \exp(j\beta)}} \tag{4.10.14}$$

Similar to (4.10.7), we have

$$A(\alpha) \approx \pm e^{\alpha g(\theta_s)} \sqrt{\frac{-2}{\eta \exp(j\beta)}} \int_{-\infty}^{\infty} e^{-\alpha s^2} ds \tag{4.10.15}$$

Substituting (4.10.8) into (4.10.15) yields

$$A(\alpha) \approx e^{\alpha g(\theta_s)} e^{\pm j\frac{\pi}{2}} e^{-j\frac{\beta}{2}} \sqrt{\frac{2\pi}{\alpha \eta}} \tag{4.10.16}$$

The integral path C' is the same as that in Figure 4.22. And we also have $\beta + 2\phi = \pm \pi$, that is, $\phi = (\pm \pi - \beta)/2$. Evidently (4.10.16) is exactly the same as (4.10.9).

If higher accuracy is required, higher order terms may be invoked in Taylor's series of (4.10.13)[11].

4.10.3   The Stationary Phase Method

Consider a complex integral

$$A(\alpha) = \int_c e^{j\alpha g(\theta)} d\theta \qquad (4.10.17)$$

where α is the large real number. In this case, we simply modify (4.10.7) and (4.10.8) into the following expressions

$$A(\alpha) = \int_c e^{j\alpha g(\theta)} e^{j\alpha \eta s^2/2} ds \qquad (4.10.18)$$

$$\int_{-\infty}^{\infty} e^{j\xi s^2} ds = \sqrt{\frac{\pi}{j\xi}} = \frac{1-j}{\sqrt{2}}\sqrt{\frac{\pi}{\xi}} \qquad (4.10.19)$$

Substituting (4.10.19) into (4.10.18) results in

$$A(\alpha) = e^{j\alpha g(\theta_s)} e^{j\phi}\frac{1-j}{\sqrt{2}}\sqrt{\frac{2\pi}{\alpha\xi}} = e^{j\alpha g(\theta_s)} e^{j\phi} e^{-j\pi/4}\sqrt{\frac{2\pi}{\alpha\xi}} \qquad (4.10.20)$$

(4.10.20) is another asymptotic expression obtained from the stationary phase method for (4.10.1).

Now we recalculate $H_0^{(1)}(k_\rho\rho)$ with (4.10.20). In this case, $\alpha = k_\rho\rho$, $g(\theta) = \cos\theta$, $\theta_s = 0°$, $g(\theta_s) = 1$, $g''(\theta_s) = -\cos\theta_s = -1$. From $g''(\theta_s) = \eta\exp(j\beta)$, we have $\eta = 1$, $\beta = \pm\pi$. Because $\beta + 2\phi = \pm\pi$, we have $\phi = 0°$. From (4.10.20) we obtain the same result as that in (4.10.12).

Compared with the saddle point method and steepest descent method, we see that, in the stationary phase method, the integration path near the saddle point is no longer along the direction of $\phi = 45°$ but $\phi = 0°$ at which the phase changes rapidly and the amplitude remains constant. This path is defined by $\sin\theta_r\sinh\theta_i = 1$. The contribution of the integral is still confined in a very small region. However, the reason is not because of the rapid decay of the integrand but is due to the rapid change of the phase, which results in the destructive interference away from θ_s. θ_s is recognized as the stationary phase point.

4.10.4   Extensions of the Above Asymptotic Formulas

All the above asymptotic formulas may contain a slowly varying function $f(\theta)$ in the integrand. For example

$$A(\alpha) = \int_c f(\theta) e^{\alpha g(\theta)} d\theta \qquad (4.10.21)$$

In this case (4.10.9) becomes

$$A(\alpha) = f(\theta_s) e^{\alpha g(\theta_s)} e^{j\phi}\sqrt{\frac{2\pi}{\alpha\eta}} \qquad (4.10.22)$$

The reason is that since the contribution of the integral is confined nearby θ_s, there will be no significant error to replace $f(\theta)$ by $f(\theta_s)$.

The asymptotic technique may be extended to the double variable problems, such as in [44]

$$A(\alpha, \beta) = \iint f(\alpha, \beta) e^{-jkr g(\alpha, \beta)} d\alpha d\beta, \qquad kr \gg 1 \qquad (4.10.23)$$

For the stationary phase method, the stationary phase point is located at $\alpha = \alpha_0$, $\beta = \beta_0$ from the solution of

$$\frac{\partial g}{\partial \alpha} = \frac{\partial g}{\partial \beta} = 0 \qquad (4.10.24)$$

Expanding $g(\alpha, \beta)$ into Taylor's series around α_0, β_0 up to second order

$$g(\alpha, \beta) \approx g(\alpha_0, \beta_0) + \frac{1}{2}a(\alpha - \alpha_0)^2 + \frac{1}{2}b(\beta - \beta_0)^2 + c(\alpha - \alpha_0)(\beta - \beta_0) \qquad (4.10.25)$$

and introducing $\xi = \alpha - \alpha_0, \eta = \beta - \beta_0$, we have

$$g(\alpha, \beta) \approx g(\alpha_0, \beta_0) + \frac{1}{2}a\xi^2 + \frac{1}{2}b\eta^2 + c\eta\xi \qquad (4.10.26)$$

Finally, we have

$$A(\alpha, \beta) \approx f(\alpha_0, \beta_0)e^{-jkrg(\alpha_0,\beta_0)} \iint e^{-jkr(a\xi^2+b\eta^2+2c\xi\eta)/2}d\xi d\eta$$

$$= \frac{-2\pi j\sigma}{kr\sqrt{|ab - c^2|}}f(\alpha_0, \beta_0)e^{-jkrg(\alpha_0,\beta_0)} \qquad (4.10.27)$$

where

$$\sigma = \begin{cases} 1, & ab > c^2, \quad a > 0 \\ -1, & ab > c^2, \quad a < 0 \\ j, & ab < c^2 \end{cases} \qquad (4.10.28)$$

4.10.5   Radiation Patterns of Microstrip Antennas

To derive the formulas of radiation patterns of microstrip antennas, we need to use the following Fourier Transform pair for E_x and $\widetilde{E}_x$[44] which is similar to (4.2.1), (4.2.2)

$$E_x(x, y, z) = \frac{1}{4\pi^2} \int_{-\infty}^{\infty} \int_{-\infty}^{\infty} \widetilde{E}_x e^{-j(k_x x + k_y y + k_z z)} dk_x dk_y \qquad (4.10.29)$$

where $k_x = k\sin\alpha\cos\beta$, $k_y = k\sin\alpha\sin\beta$ and $k_z = k\cos\alpha$ (see Figure 4.23). k_z may also be expressed in the same form as that in (4.2.4).

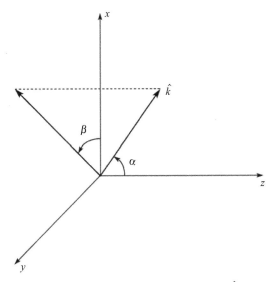

Figure 4.23   3D coordinates related to $\hat{\mathbf{k}}$.

The aperture field E_x is given by

$$E_x(x, y, 0) = \frac{1}{4\pi^2} \int_{-\infty}^{\infty} \int_{-\infty}^{\infty} \tilde{E}_x(k_x, k_y) e^{-j(k_x x + k_y y)} dk_x dk_y \tag{4.10.30}$$

where $\tilde{E}_x(k_x, k_y)$ is the spectral aperture field and is given by

$$\tilde{E}_x(k_x, k_y) = \int_{-\infty}^{\infty} \int_{-\infty}^{\infty} E_x(x, y, 0) e^{j(k_x x + k_y y)} dx dy \tag{4.10.31}$$

The z-component of E in each plane wave follows from

$$\mathbf{k} \cdot \tilde{\mathbf{E}} = (\hat{\mathbf{x}} k_x + \hat{\mathbf{y}} k_y + \hat{\mathbf{z}} k_z) \cdot \left(\hat{\mathbf{x}} \tilde{E}_x + \hat{\mathbf{z}} \tilde{E}_z \right) = 0$$

or

$$\tilde{E}_z(x, y, 0) = -\frac{k_x}{k_z} \tilde{E}_x(x, y, 0)$$

and the complete $\mathbf{E}$ field is

$$\mathbf{E}(x, y, z) = \frac{1}{4\pi^2} \int_{-\infty}^{\infty} \int_{-\infty}^{\infty} (\hat{\mathbf{x}} k_z - \hat{\mathbf{z}} k_x) \tilde{E}_x(k_x, k_y) e^{-j(k_x x + k_y y + k_z z)} \frac{dk_x dk_y}{k_z} \tag{4.10.32}$$

To evaluate (4.10.32) for large kr, the asymptotic formula in (4.10.27) can be used. To use this formula, we first need convert the integral into spherical coordinates by the following identities

$$\hat{\mathbf{x}} k_z - \hat{\mathbf{z}} k_x$$
$$= k \cos \alpha [\hat{\mathbf{r}} \sin \theta \cos \varphi + \hat{\boldsymbol{\theta}} \cos \theta \cos \varphi - \hat{\boldsymbol{\varphi}} \sin \varphi] - k \sin \alpha \cos \beta [\hat{\mathbf{r}} \cos \theta - \hat{\boldsymbol{\theta}} \sin \theta]$$
$$= \hat{\boldsymbol{\theta}}(k \cos \alpha \cos \theta \cos \varphi + k \sin \alpha \cos \beta \sin \theta) - \hat{\boldsymbol{\varphi}} k \cos \alpha \sin \varphi$$
$$+ \hat{\mathbf{r}}(k \cos \alpha \sin \theta \cos \varphi - k \cos \theta \sin \alpha \cos \beta)$$
$$\overset{\substack{\alpha = \alpha_0 = \theta \\ \beta = \beta_0 = \varphi}}{=} k(\hat{\boldsymbol{\theta}} \cos \varphi - \hat{\boldsymbol{\varphi}} \sin \varphi \cos \theta) \tag{4.10.33}$$

and the following transform from Jacobian

$$\frac{dk_x dk_y}{k_z} = k \sin \alpha \, d\alpha \, d\beta \tag{4.10.34}$$

The function g, parameters a, b, c in (4.10.27) then may be obtained as

$$g = \sin \alpha \cos \beta \sin \theta \cos \varphi + \sin \alpha \sin \beta \sin \theta \sin \varphi + \cos \alpha \cos \theta$$

$$a = \left. \frac{\partial^2 g}{\partial \alpha^2} \right|_{\substack{\alpha = \alpha_0 = \theta \\ \beta = \beta_0 = \varphi}} = -1$$

$$b = \left. \frac{\partial^2 g}{\partial \beta^2} \right|_{\substack{\alpha = \alpha_0 = \theta \\ \beta = \beta_0 = \varphi}} = -\sin^2 \theta$$

$$c = \left. \frac{\partial^2 g}{\partial \alpha \partial \beta} \right|_{\substack{\alpha = \alpha_0 = \theta \\ \beta = \beta_0 = \varphi}} = 0$$

σ is determined by (4.10.28) to be -1. The final result from (4.10.27), (4.10.33) and (4.10.34) is

$$\mathbf{E}\left(r,\theta,\varphi\right)\approx\frac{je^{-jkr}}{\lambda r}\left(\hat{\boldsymbol{\theta}}\cos\varphi-\hat{\boldsymbol{\phi}}\sin\varphi\cos\theta\right)\widetilde{E}_x\left(k\sin\theta\cos\varphi,k\sin\theta\sin\varphi\right)$$

$$=\frac{jk_z e^{-jkr}}{k\lambda r}\left[\left(\hat{\mathbf{x}}-\hat{\mathbf{z}}\frac{k_x}{k_z}\right)\widetilde{E}_x\left(k\sin\theta\cos\varphi,k\sin\theta\sin\varphi\right)\right] \tag{4.10.35}$$

Similarly we have the following formula for $\widetilde{E}_y$

$$\mathbf{E}\left(r,\theta,\varphi\right)\approx\frac{je^{-jkr}}{\lambda r}\left(\hat{\boldsymbol{\theta}}\sin\varphi+\hat{\boldsymbol{\phi}}\cos\varphi\cos\theta\right)\widetilde{E}_y\left(k\sin\theta\cos\varphi,k\sin\theta\sin\varphi\right)$$

$$=\frac{jk_z e^{-jkr}}{k\lambda r}\left[\left(\hat{\mathbf{y}}-\hat{\mathbf{z}}\frac{k_y}{k_z}\right)\widetilde{E}_y\left(k\sin\theta\cos\varphi,k\sin\theta\sin\varphi\right)\right] \tag{4.10.36}$$

where $\widetilde{E}_y(k_x,k_y)$ is the spectral aperture field and is given by

$$\widetilde{E}_y\left(k_x,k_y\right)=\int_{-\infty}^{\infty}\int_{-\infty}^{\infty}E_y\left(x,y,0\right)e^{j(k_x x+k_y y)}dk_x dk_y \tag{4.10.37}$$

$$\mathbf{E}(r,\theta,\varphi)\approx\frac{je^{-jkr}}{\lambda r}\left[\left(\hat{\boldsymbol{\theta}}\cos\varphi-\hat{\boldsymbol{\varphi}}\sin\varphi\cos\theta\right)\widetilde{E}_x\left(k\sin\theta\cos\varphi,k\sin\theta\sin\varphi\right)\right.$$

$$\left.+\left(\hat{\boldsymbol{\theta}}\sin\varphi+\hat{\boldsymbol{\varphi}}\cos\varphi\cos\theta\right)\widetilde{E}_y\left(k\sin\theta\cos\varphi,k\sin\theta\sin\varphi\right)\right]$$

$$=\frac{jk_z e^{-jkr}}{k\lambda r}\left[\left(\hat{\mathbf{x}}-\hat{\mathbf{z}}\frac{k_x}{k_z}\right)\widetilde{E}_x\left(k\sin\theta\cos\varphi,k\sin\theta\sin\varphi\right)\right.$$

$$\left.+\left(\hat{\mathbf{y}}-\hat{\mathbf{z}}\frac{k_y}{k_z}\right)\widetilde{E}_y\left(k\sin\theta\cos\varphi,k\sin\theta\sin\varphi\right)\right]$$

$$=\frac{jk_z e^{-jkr}}{2\pi r}\widetilde{\mathbf{E}}\left(k\sin\theta\cos\varphi,k\sin\theta\sin\varphi\right) \tag{4.10.38}$$

where $\widetilde{\mathbf{E}}=\hat{\mathbf{x}}\widetilde{E}_x+\hat{\mathbf{y}}\widetilde{E}_y+\hat{\mathbf{z}}\widetilde{E}_z$. In deriving (4.10.38), the relationship $\mathbf{k}\cdot\widetilde{\mathbf{E}}=k_x\widetilde{E}_x+k_y\widetilde{E}_y+k_z\widetilde{E}_z=0$ is used. From this relationship, the Fourier transform for an arbitrarily polarized aperture field may also be easily derived as

$$\mathbf{E}(x,y,z)=\frac{1}{4\pi^2}\int_{-\infty}^{\infty}\int_{-\infty}^{\infty}\widetilde{\mathbf{E}}e^{-j(k_x x+k_y y+k_z z)}dk_z dk_y \tag{4.10.39}$$

For an arbitrarily polarized aperture field, the radiation field is the sum of (4.10.35) and (4.10.36).

The radiation pattern for a rectangular patch microstrip antenna may be obtained from (4.10.35) and (4.10.36)

$$E_\theta=K\left[\cos\varphi\widetilde{E}_x+\sin\varphi\widetilde{E}_y\right] \tag{4.10.40}$$

$$E_\varphi=K\left[-\sin\varphi\cos\theta\widetilde{E}_x+\cos\varphi\cos\theta\widetilde{E}_y\right] \tag{4.10.41}$$

where $K=j\exp(-jkr)/\lambda r$. In the simple case when $\widetilde{J}_y=0,\widetilde{J}_x\neq0$

$$\widetilde{\mathbf{E}}=\widetilde{\underline{\mathbf{G}}}^{EJ}\cdot\widetilde{J}_x\hat{\mathbf{x}} \tag{4.10.42}$$

From (4.5.15), (4.4.19) and (4.4.20) we have

$$\tilde{E}_x = \frac{-1}{k_x^2 + k_y^2} \left(k_x^2 V_i^e + k_y^2 V_i^h \right) \tilde{J}_x \tag{4.10.43}$$

$$\tilde{E}_y = \frac{k_x k_y}{k_x^2 + k_y^2} \left(-V_i^e + V_i^h \right) \tilde{J}_x \tag{4.10.44}$$

Consider the covered microstrip antenna with the feed location at $y_s = W_y/2$ as shown in Figure 4.24. In this case, only the currents in the x direction are excited. The transversal dependence is taken to be constant. The longitudinal distribution is assumed equal to that of a resonant end fed half-wavelength of a microstrip transmission line and is given by [45]

$$\mathbf{J}(x,y) = \begin{cases} \hat{\mathbf{x}} I_1 \sin \dfrac{\pi x}{W_x}, & 0 < x < W_x, 0 < y < W_y \\ 0, & \text{otherwise} \end{cases} \tag{4.10.45}$$

where I_1 is a constant. Consequently

$$\tilde{J}_x(k_x, k_y) = \iint_{patch} J(x,y) e^{-j(k_x x + k_y y)} dx dy$$

$$= I_1 \left[e^{jk_y W_y/2} \frac{W_y \sin k_y W_y/2}{W_y k_y/2} \right] \left[e^{jk_x W_x/2} \frac{2\pi W_x \cos(k_x W_x/2)}{\pi^2 - (k_x W_x)^2} \right] \tag{4.10.46}$$

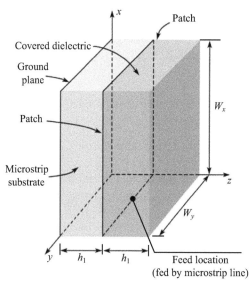

Figure 4.24   Covered rectangular microstrip antenna
(the ground plane and the dielectric are partly shown).

When the dielectric constants and the thicknesses in Figure 4.24 are given, V_i^e and V_i^h may be easily found from the formulas in Section 4.7. It is seen that the solution of the radiation pattern for covered microstrip patch antenna may be simply obtained by solving the transmission line problems.

The analytical formula of the radiation pattern for the covered microstrip patch antenna is very useful. One example of using it is to make the optimization of enhancing the directivity. In the optimization procedure, we may choose a certain applicable dielectric as the cover, and change the thickness to ensure the radiation pattern is as narrow as possible in order to enhance the directivity. In the case of a single layer, the enhancing of the directivity is more than 1dB[46]. Another example is to make the optimization to equalize the beam-widths of radiation pattern of E and H planes. The parameters of the covered microstrip patch are as follows: the width along E plane is 2.6mm, along H plane is 4mm, the relative dielectric constant and thickness of the microstrip substrate are 2.2 and 0.254mms, and for the cover are 2.2 and 1.018 respectively. The working frequency is 34.5GHz. The calculated patterns are shown in Figure 4.25 which are in good agreement with those from IE3D^{TM}. In Figure 4.24(a), it is seen that the beam-widths of two patterns are not exactly equal. This is because the dielectric constant of the applicable material is not exactly the same as it is for the optimized one. When the microstrip patch is covered by dielectric, its properties like resonant frequency, Q factor, may change and modification of the design is needed.

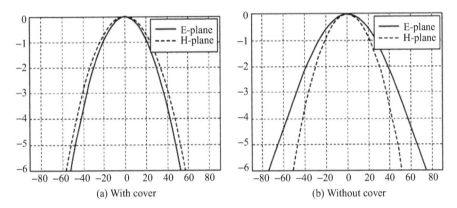

(a) With cover   (b) Without cover

Figure 4.25   Radiation patterns of E and H-planes calculated by (4.10.40)–(4.10.44).

Bibliography

[1] Y. R. Samii and E. Michielssen, *Electromagnetic Optimization by Genetic Algorithm*, John Wiley & Sons, Inc., 1999.

[2] J. W. Bandler et al, "Neural space mapping EM optimization of microwave structures," *IEEE MTT-S Int. Microwave Symp. Dig.*, vol. 2, pp. 879–882, 2000.

[3] D. G. Fang, *Spectral Domain Method in Elecromagnetics* (in Chinese), Anhui Education Press, 1995.

[4] D. G. Fang, "Spectral domain method," "Image theory" in *Dictionary of Review on Modern Science and Technology* (in Chinese), pp. 229–231, Beijing Press, 1998.

[5] L. M. Brekhovskikh, *Waves in Layered Media* (second edition), Academic Press, 1980.

[6] J. W. Harris and H. Stocker, *Handbook of Mathematics and Computational Science*, Springer, 1998.

[7] K. A. Michalski and D. Zheng, "Electromagnetic scattering and radiation by surfaces of arbitrary shape in layered media, Part I: theory," *IEEE Trans. Antennas Propagat.*, vol. 38, no. 3, pp. 335–344, Mar., 1990.

[8] C. A. Balanis, *Antenna Theory–Analysis and Design* (second edition), John Wiley & Sons, Inc., 1997.

[9] K. A. Michalski and J. R.Mosig, "Mutilayered media Green's function in integral equation formulations," *IEEE Trans. Antennas Propagat.*, vol. 45, no. 3, pp. 508–519, Mar., 1997.

[10] R. A. Kipp and C. H. Chan, "A complex image method for the vertical component of the magnetic potential of a horizontal dipole in layered media," *IEEE AP-S Int. Antennas and Propagation Symp. Dig.*, vol. 2, pp. 1366–1369, 1994.

[11] J. A. Kong, *Theory of Electromagnetic Waves*, John Wiley & Sons, 1975.

[12] W. C. Chew and T. M. Halashy, "The use of vector transforms in solving some electromagnetic scattering problems," *IEEE Trans. Antennas Propagat.*, vol. 34, no. 7, pp. 871–879, Jul., 1986.

[13] G. Z. Jiang and W. X. Zhang, "The focused fields of Fresnel zone plate lens," *Proceeding of Inter. Symp. on Antennas and Propagation*, pp. 57–60, 1996.

[14] J. C. Chao, Y. J. Liu, F. J. Rizzo, P. A. Martin, and L. Udpa, "Regularized integral equations for curvilinear boundary elements for electromagnetic wave scattering in three dimensions," *IEEE Trans. Antennas Propagat.*, vol. 45, no. 12, pp. 1416–1422, Dec., 1995.

[15] S. M. Rao, D. R. Wilton, and A. W. Glisson, "Electromagnetic scattering by surfaces of arbitrary shape," *IEEE Trans. Antennas Propagat.*, vol. 30, no. 5, pp. 409–418, May, 1982.

[16] D. C. Stinson, *Intermediate Mathematics of Electromagnetics*, Prentice-Hall, Inc., 1976.

[17] R. E. Collin, *Field Theory of Guided Waves*(second edition), IEEE Press, 1991.

[18] R. F. Harrington, *Time-Harmonic Electromagnetic Field*, McGraw-Hill Book Co., 1961.

[19] D. M. Pozar, *Microwave Engineering*(second edition), John Wiley & Sons, Inc., 1998.

[20] D. G. Fang and C. Y. Liu, *Microwave Theory and Technique* (in Chinese), Ordnance Industry Press, 1987.

[21] A. Ishimaru, *Electromagnetic Wave Propagation, Radiation, and Scattering*, Prentice-Hall, Inc., 1991.

[22] C. E. Pearson, *Handbook of Applied Mathematics*, Van Nostrand Reinhold Company, 1974.

[23] F. Ling, *Fast Electromagnetic Modeling of Multilayer Microstrip Antennas and Circuits*, PhD. Dissertation University of Illinois at Urbana-Champaign, 2000.

[24] L. B. Felsen and N. Marcuvitz, *Radiation and Scattering of Waves*, Prentice-Hall Inc., 1973.

[25] Y. L. Chow and I. N. El-behery, "An approximate dynamic spatial Green's function for microstriplines," *IEEE Trans. Microwave Theory Tech.*, vol. 26, no. 12, pp. 978–983, Dec., 1978.

[26] Y. L. Chow, "An approximate dynamic Green's function in three dimensions for finite length microstrip line," *IEEE Trans. Microwave Theory Tech.*, vol. 28, no. 4, pp. 393–397, Apr., 1980.

[27] J. Dai and Y. L. Chow, "A reduced model of a series of image charges for study MMIC's," *Proc. Second Asia-Pacific Microwave Conf.* (Beijing China), pp. 26–28, 1988.

[28] D. G. Fang, J. J. Yang and G. Y. Delisle, "Discrete image theory for horizontal electric dipoles in a multilayered medium," *Proc Inst. Elec. Eng.*, vol. 135, Pt. H., pp. 297–303, 1988.

[29] Y. L. Chow, J. J. Yang, D. G. Fang and G. E. Howard,"Closed-form spatial Green's function for the thick substrate," *IEEE Trans. Microwave Theory Tech.*, vol. 39, no. 3, pp. 588–592, Mar., 1991.

[30] M. I. Aksun and R. Mittra, "Derivation of closed-form Green's functions for a general microstrip geometry," *IEEE Trans. Microwave Theory Tech.*, vol. 40, no. 11, pp. 2055–2062, Nov., 1992.

[31] I. V. Lindell, E. Alanen and H. V. Bagh,"Exact image theory for the calculation of fields transmitted through a planar interface of two media," *IEEE Trans. Antennas Propagat.*, vol. 34, no. 2, pp. 129–137, Feb., 1986.

[32] S. F. Mahmoud, "Image theory for electric dipoles above a conducting anisotropic earth," *IEEE Trans. Antennas Propagat.*, vol. 32, no. 7, pp. 679–683, Jul., 1984.

[33] A. Torabian and Y. L. Chow, "Simulated image method for Green's function of multilayer media," *IEEE Trans. Microwave Theory Tech.*, vol. 47, no. 9, pp. 1777–1781, Sept., 1999.

[34] Y. L. Chow and W. C. Tang, "3-D Green's functions of microstrip separated into simpler terms– behavior, mutual interaction and formulas of the terms," *IEEE Trans. Microwave Theory Tech.*, vol. 49, no. 8, pp. 1483–1491, Aug., 2001.

[35] Y. Xu, D. G. Fang, M. Y. Xia and C. H. Chan, "Speedy computation of the time-domain Green's function for microstrip structures," *Electronics Letters*, vol. 36, no. 22, 26th, pp. 1855–1857, Oct., 2000.

[36] M. I. Aksun,"A robust approach for the derivation of closed-form Green's functions," *IEEE Trans. Microwave Theory Tech.*, vol. 44, no. 5, pp. 651–658, May, 1996.

[37] T. K. Sarkar and O. Pereira, "Using the matrix pencil method to estimate the parameters of a sum of complex exponenetials," *IEEE Antennas and Propagation Magazine*, vol. 37, no. 1, pp. 48–55, 1995.

[38] R. A. Kipp and C. H. Chan, "Complex image method for sources in bounded regions of multilayer structures," *IEEE Trans. Microwave Theory Tech.*, vol. 42, no. 5, pp. 860–865, May, 1994.

[39] F. Ling and J. M. Jin,"Full-wave analysis of multilayer microstrip problems," Chapter 16 in a book :*Fast and Efficient Algorithms in Computational Electromagnetics*, edited by W. C. Chew, J. M. Jin, E. Michielssen, and J. Song, Artech House, 2001.

[40] V. I. Okhmatovslei and A. C. Cangellaris, "A new technique for the derivation of closed-form electromagnetic Green's functions for unbounded planar layered media," *IEEE Trans. Antennas Propagat.*, vol. 50, pp. 1005–1015, Jul., 2002.

[41] R. F. Harrington, *Field Computation by Moment Method*, The Macmillan Company, 1968.

[42] S. M. Rao, D. R. Wiltion and A. W. Glisson, "Electromagnetic scattering by surface of arbitrary shape," *IEEE Trans. Antennas Propagat.*, vol. 30, pp. 409–418, May, 1982.

[43] R. D. Graglia, D. R. Wilton and A. F. Peterson, "Higher order interpolatory vector bases for computational eclctromagnetics," *IEEE Trans. Antennas Propagat.*, vol. 45, pp. 329–342, Mar., 1997.

[44] E. V. Jull, *Aperture Antennas and Diffraction Theory*, Peter Peregrinus Ltd., 1981.

[45] P. Perlmutter, S. Shtrikman and D. Treves, "Electric surface current model for the analysis of microstrip antennas with application to rectangular elements," *IEEE Trans. Antennas Propagat.*, vol. 33, pp. 301–311, Mar., 1985.

[46] D. G. Fang, J. J. Yang and K. Sha, "Optimization of directivity of the covered rectangular microstrip antennas." *Proceedings of 1985 Inter. Symp. in Antenna and EM Theory*, Beijing, China, pp. 18–21, Aug., 1985.

[47] D. J. Hoppe and Y. R. Samii, "Scattering by superquadric dielectric-coated cylinders using higher order impedance boundary conditions," *IEEE Trans. Antennas Propagat.*, vol. 40, no. 12, pp. 1513–1523, Dec., 1992.

[48] D. J. Hoppe and Y. R. Samii, "Higher order impedance boundary conditions applied to scattering by coated bodies of revolution," *IEEE Trans. Antennas Propagat.*, vol. 42, no. 12, Dec., 1994.

Problems

4.1 Prove (4.6.28).

4.2 For a lossless microstrip substrate with $\epsilon_r = 2.2, \lambda = 3$cm

1. Find the branch cut contribution from the paths along the branch cuts shown in Figures (a)-(c);

2. Find the branch cut contribution from the path that cuts through the branch cut twice as shown in Figure (d), where C_{SD} is the steepest descent path.

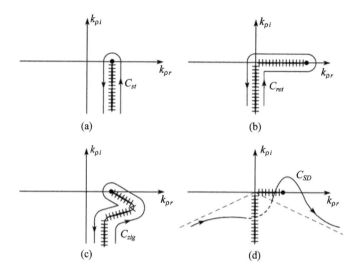

4.3 Write a program for the full-wave analysis of a rectangular patch antenna.

4.4 Write a program for calculation of the radiation pattern of dielectric loaded rectangular patch antenna. The direction of the currents on the patch is assumed to be along x direction and the distribution of currents is assumed to be constant along y direction and is of triangular shape function along x direction with zero value at the ends[46].

4.5 Check the results in Figure 4.5(a)(b) of the paper in [29] by using the two-level approximation scheme shown in Figure 4.20.

4.6 For Figure 4.7, prove that the poles of surface wave $k_{\rho p(i)}$ in (4.8.13) are located within k_0 and $\sqrt{\epsilon_r}\,k_0$.

Hints: For the surface wave which propagates along a zigzag path in dielectric region 1 along ρ direction, k_ρ and k_{z1} should be positive real, k_{z2} should be negative imaginary. In region 2, the radiation wave propagates along z direction, therefore $k_\rho < k_0$.

4.7 Use the relationship given in (4.7.43) to prove (4.3.11).

4.8 From Maxwell's equations, prove

$$\begin{bmatrix} \tilde{E}_x \\ \tilde{E}_y \end{bmatrix} = \frac{1}{\omega\varepsilon k_z}\mathbf{A}\begin{bmatrix} \tilde{H}_y \\ -\tilde{H}_x \end{bmatrix}$$

where $\mathbf{A} = \begin{bmatrix} (k^2 - k_x^2) & -k_x k_y \\ -k_x k_y & (k^2 - k_y^2) \end{bmatrix}$

4.9 To prove the eigenvalues of $\mathbf{A}$ are

$$\lambda_1 = k^2 - (k_x^2 + k_y^2), \quad \lambda_2 = k^2$$

and their related eigenvectors are

$$[k_x \quad k_y]^{\mathrm{T}} \qquad [k_y \quad -k_x]^{\mathrm{T}}$$

the normalized orthogonal transform matrix is

$$\psi = \frac{1}{\sqrt{k_x^2 + k_y^2}}\begin{bmatrix} k_x & k_y \\ k_y & -k_x \end{bmatrix} = \psi^{-1}$$

4.10 Using the coordinate transform

$$\begin{bmatrix} x \\ y \end{bmatrix} = \psi\begin{bmatrix} u \\ -v \end{bmatrix}, \quad \begin{bmatrix} y \\ -x \end{bmatrix} = \psi\begin{bmatrix} v \\ u \end{bmatrix}$$

prove

$$\begin{bmatrix} \widetilde{E}_u \\ -\widetilde{E}_v \end{bmatrix} = \frac{1}{\omega\varepsilon k_z} \begin{bmatrix} k_z^2 & 0 \\ 0 & k^2 \end{bmatrix} \begin{bmatrix} \widetilde{H}_v \\ \widetilde{H}_u \end{bmatrix}$$

$$= \begin{bmatrix} \dfrac{k_z}{\omega\varepsilon} & 0 \\ 0 & \dfrac{\omega\mu}{k_z} \end{bmatrix} \begin{bmatrix} \widetilde{H}_v \\ \widetilde{H}_u \end{bmatrix}$$

and

$$\frac{\widetilde{E}_u}{\widetilde{H}_v} = \frac{k_z}{\omega\varepsilon} = Z_{\text{TM}}, \qquad \frac{-\widetilde{E}_v}{\widetilde{H}_u} = \frac{\omega\mu}{k_z} = Z_{\text{TE}}$$

where Z_{TM} and Z_{TE} are the characteristic impedance for TM mode and for TE mode respectively.

CHAPTER 5

Effective Methods in Using Commercial
Software for Antenna Design

5.1 Introduction

Computational electromagnetics is closely related to antenna design and other research areas such as electromagnetic compatibility (EMC) analysis, microstrip circuit including the 3D multichip module (MCM), and electromagnetic scattering problems. Advances in CAD commercial software have been a major impact on it. Yet, not all effective numerical methods recently developed have been involved in those software packages. Efforts are still required to make full use of these new methods in different ways. For example, one may create formulas through synthetic asymptote or train formulas through artificial neural networks (ANN) by using commercial software. After that is done, the software may be replaced by formulas in the specified range for the consequent computation, resulting in the dramatic reduction of the computation burden. In using these new methods, it is also possible to communicate with the software if it is not completely canned, otherwise it is necessary to write the program on one's own.

One way to improve the computational efficiency is to put efforts on the algorithm or method itself, both in the frequency domain and in the time domain. Some examples are the fast multipole method (FMM) algorithm[1], the unconditionally stable scheme without time derivatives for FDTD[2], the wavelet-based algebraic multi-grid preconditioned CG method[3] and the time-domain image method[4]. However, when the number of sampling points increases looking at wide frequency band response or during optimization, the computer burden is still a problem. The other way is then to put some efforts beyond the software and related algorithms. This will be discussed in this chapter.

5.2 The Space Mapping (SM) Technique

The SM technique was proposed by J. Bandler et al. in 1994[5]. This technique establishes a mapping between two spaces, the coarse model space and the fine model space. An accurate but computationally intensive EM fine model is sparingly used to calibrate a less accurate but efficient coarse model. The SM technique takes full advantage of the high efficiency of the coarse model and the high accuracy of the fine model. Thus, the fine model design is reduced to the inverse mapping of the optimal coarse model design. Consequently, intensive computation of the fine model during the optimization process is avoided.

A crucial step for the SM is the parameter extraction (PE), in which a coarse model point corresponding to a given fine model response is obtained through an optimization process. The nonuniqueness of the PE procedure can lead to divergence or oscillation of the optimization iterations. In order to improve the convergence, the aggressive space mapping (ASM) algorithm was proposed[6]. The ASM utilizes the linear approximation to construct the mapping, which is iteratively updated in each step. It is more efficient than the original SM algorithm for it aggressively exploits every available fine model analysis, producing dramatic results right from the first step.

In this section, we start with the original SM in order to get a better understanding of this technique. Then we will focus on the ASM and its applications in antenna design. Some recent development in SM will be briefly introduced later.

5.2.1   Original Space Mapping Algorithm

The coarse model in SM could be coarse discretization or some approximation model such as the cavity model in the microstrip antenna. The fine model is the fine discretization in the electromagnetic model (EM).

We use l-dimensional vector $\mathbf{x}_c = [x_{c1}, x_{c2}, \cdots, x_{cl}]^T$ to describe the parameters of the coarse model and use k-dimensional vector $\mathbf{x}_f = [x_{f1}, x_{f2}, \cdots, x_{fk}]^T$ to describe the parameters of the fine model. For simple illustration, we assume that $l = k$. We use $\mathbf{R}_c(\mathbf{x}_c)$ and $\mathbf{R}_f(\mathbf{x}_f)$ to denote the coarse model response and the fine model response respectively. The key idea behind the SM is the generation of an appropriate transformation:

$$\mathbf{x}_c = \mathbf{P}\mathbf{x}_f \tag{5.2.1}$$

Mapping the fine model parameter space to the coarse model parameter space such that

$$\| \mathbf{R}_f(\mathbf{x}_f) - \mathbf{R}_c(\mathbf{x}_c) \| \leqslant \epsilon \tag{5.2.2}$$

within some local modeling region around optimal coarse model solution, where $\| \cdot \|$ indicates a suitable norm and ϵ is a small positive constant.

Finding $\mathbf{P}$ is an iterative process. After each iteration, the parameters obtained will be supplemented into $\mathbf{x}_c$ and $\mathbf{x}_f$ to generate the updated $\mathbf{P}$. We begin with a set of fine model base points $\mathbf{B}_f = \{\mathbf{x}_f^1, \mathbf{x}_f^2, \cdots, \mathbf{x}_f^m\}$. The initial m base points are selected in the vicinity of a reasonable candidate for the fine model solution. Because we have assumed that $k = l$, then the set $\mathbf{B}_f$ can be chosen as $\mathbf{x}_f^1 = \mathbf{x}_c^*$ and some local perturbations around $\mathbf{x}_f^1$. Once the set $\mathbf{B}_f$ is chosen, we evaluate the fine model responses $\mathbf{R}_f(\mathbf{x}_f^i), i = 1, 2, \cdots, m$. Next we find, by parameter extraction(PE), the coarse model set $\mathbf{B}_c = \{\mathbf{x}_c^1, \mathbf{x}_c^2, \cdots, \mathbf{x}_c^m\}$ so that (5.2.3) holds for each pair of corresponding base points in $\mathbf{B}_f$ and $\mathbf{B}_c$.

$$\min_{\mathbf{x}_c^i} \| \mathbf{R}_f(\mathbf{x}_f^i) - \mathbf{R}_c(\mathbf{x}_c^i) \| \tag{5.2.3}$$

Using these initial sets, we establish $\mathbf{P}_1$.

The base points will be increased as the iteration step goes on. At the jth iteration step both sets contain m_j base points which are used to update $\mathbf{P}_j$; then we have

$$\mathbf{x}_f^{m_j+1} = \mathbf{P}_j^{-1}(\mathbf{x}_c^*) \tag{5.2.4}$$

This procedure is repeated until (5.2.5) holds.

$$\|\mathbf{R}_f(\mathbf{x}_f^{m_j+1}) - \mathbf{R}_c(\mathbf{x}_c^*)\| \leqslant \epsilon \tag{5.2.5}$$

Finally, the solution in fine model will be $\bar{\mathbf{x}}_f = \mathbf{x}_f^{m_j+1}$.

In the following, we will take the optimization design for microstrip patch antenna given in Figure 5.1 as an example to demonstrate the application of SM. Commercial software is chosen as the EM simulator. The coarse discretization and the fine discretization correspond to the coarse model and fine model respectively.

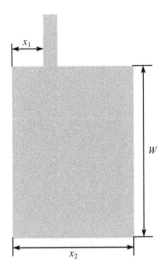

Figure 5.1   Microstrip geometry.

The parameters to be optimized are x_1, x_2. W and ϵ_r are fixed ($W = 315$mil $= 8$mm, $\epsilon_r = 2.7$). In the coarse model, the discretization is 10 grids per wavelength, and the computer time for each frequency is 1 second. Whereas in the fine model, the discretization is 30 grids per wavelength, and the computer time for each frequency is 40 seconds. The result for the initial optimization in coarse model is $x_c^* = [16.00 \quad 213.00]^T$.

The final results are given in Table 5.1, where i is the number of base points (BP).

Table 5.1   Model parameters (Unit: mils).

i	1	2	3	4	5	6	7	8	9	10	11
x_{c1}^i	16.00	15.70	15.33	13.60	17.00	15.40	16.54	16.41	16.35	16.30	
x_{c2}^i	216.75	215.84	215.74	216.99	215.80	216.99	213.37	213.28	213.24	213.19	
x_{f1}^i	16.00	16.00	15.00	15.00	17.00	17.00	16.54	16.41	16.35	16.30	16.27
x_{f2}^i	213.00	212.00	212.00	214.00	212.00	214.00	209.50	209.42	209.38	209.36	209.34

The steps are summarized as follows:

1. Initialize $\mathbf{x}_f^1 = \mathbf{x}_c^*$.

2. Make local perturbation, say, set other five points $\mathbf{x}_f^2, \mathbf{x}_f^3, \mathbf{x}_f^4, \mathbf{x}_f^5, \mathbf{x}_f^6$ around $\mathbf{x}_f^1$ to form 2 by 6 matrix $\mathbf{x}_f$ as listed in last two rows and first six columns in Table 5.1.

3. Find $\mathbf{x}_c^i$ by (5.2.3) to form 2 by 6 matrix $\mathbf{x}_c$ as listed in first two rows and six columns in the same table.

4. Find the transform operator $\mathbf{P}_1$ through

$$\mathbf{P}_1 = \mathbf{x}_c \mathbf{x}_f^{-1} = \mathbf{x}_c \mathbf{x}_f^T (\mathbf{x}_f^T)^{-1} \mathbf{x}_f^{-1} = \mathbf{x}_c \cdot \mathbf{x}_f^T \cdot (\mathbf{x}_f \cdot \mathbf{x}_f^T)^{-1} \qquad (5.2.6)$$

From (5.2.6) we have $\mathbf{P}_1 = \begin{bmatrix} 0.934 & 0.003 \\ 0.052 & 1.013 \end{bmatrix}$, and $\mathbf{x}_f^7 = \mathbf{P}_1^{-1}(\mathbf{x}_c^*)$.

5. Make PE according to $\mathbf{R}_f(\mathbf{x}_f^7) \approx \mathbf{R}_c(\mathbf{x}_c^7)$ and obtain $\mathbf{x}_c^7$. Both $\mathbf{x}_f^7$ and $\mathbf{x}_c^7$ are inserted into base-point family as listed in the seventh column of Table 5.1.

6. Repeat the above procedure and obtain

$$\mathbf{P}_2 = \begin{bmatrix} 1.015 & -0.003 \\ 0.107 & 1.009 \end{bmatrix}$$

consequently

$$\mathbf{P}_3 = \begin{bmatrix} 1.056 & -0.006 \\ 0.134 & 1.007 \end{bmatrix}, \mathbf{P}_4 = \begin{bmatrix} 1.083 & -0.008 \\ 0.154 & 1.005 \end{bmatrix}, \mathbf{P}_5 = \begin{bmatrix} 1.101 & -0.009 \\ 0.165 & 1.005 \end{bmatrix}$$

and $\mathbf{x}_c^8, \cdots, \mathbf{x}_c^{10}; \mathbf{x}_f^8, \cdots, \mathbf{x}_f^{11}$ which are listed in Table 5.1.

7. If (5.2.5) holds, the procedure stops. $\mathbf{x}_f^{11} = \mathbf{P}_5^{-1}(\mathbf{x}_c^*)$ is the final solution in the fine model.

It is observed that

$$\mathbf{P}_5 \mathbf{x}_f^{11} = \mathbf{P}_5 \begin{bmatrix} 16.27 \\ 209.34 \end{bmatrix} = \begin{bmatrix} 16.01 \\ 213.01 \end{bmatrix} \approx \begin{bmatrix} 16.00 \\ 213.00 \end{bmatrix}$$

Therefore $\parallel \mathbf{f}^{final} \parallel = \parallel \mathbf{x}_c^{final} - \mathbf{x}_c^* \parallel \leqslant \epsilon$ is satisfied simultaneously. However these two conditions are not always satisfied simultaneously and are not required to be satisfied simultaneously. The only condition to stop the procedure is (5.2.5).

5.2.2   Aggressive Space Mapping Algorithm (ASM)

The ASM process starts from optimizing the coarse model response $\mathbf{R}_c(\mathbf{x}_c^*)$ to satisfy all the required specifications first. The initial fine model solution $\mathbf{x}_f^1$ corresponding to $\mathbf{x}_c^*$ can be determined by one of the two options: if the fine model and the coarse model have the same contents and dimensions, for example, all are physical dimension parameters, we can simply set $\mathbf{x}_f^1 = \mathbf{x}_c^*$; if the two models have different contents and dimensions, for example, the coarse model is LC lumped circuit with values of capacitors and inductors as variables, while in fine model the variables are physical dimension parameters, to complete with the procedures in ASM, we are not able to set $\mathbf{x}_f^1 = \mathbf{x}_c^*$. However, in the latter case, it is not difficult to find appropriate formulas (knowledge) to calculate circuit elements. One may denote the knowledge relations as $\mathbf{K}$ or $\mathbf{x}_{cf} = \mathbf{K}^{-1}(\mathbf{x}_c)$. The variable space $\{\mathbf{x}_{cf}\}$ is usually the corresponding dimension variables of the circuit parameter space $\{\mathbf{x}_c\}$ and shares the same physical contents and dimensions as those of the variable space $\{\mathbf{x}_f\}$ in the fine model. The dimension of $\{\mathbf{x}_{cf}\}$ and that of $\{\mathbf{x}_c\}$ are not necessarily the same and $\mathbf{K}^{-1}$ is just a symbol for inverse relation[7].

Through satisfying $\parallel \mathbf{R}_f(\mathbf{x}_f^1) - \mathbf{R}_c(\mathbf{x}_c^1) \parallel < \epsilon$, the response $\mathbf{R}_f(\mathbf{x}_f^1)$ of the fine model and the response $\mathbf{R}_c(\mathbf{x}_c^1)$ of the coarse model are matched, where ϵ is a given error for a satisfactory matching. This is a mapping process: $\mathbf{x}_c^1 = \mathbf{P}(\mathbf{x}_f^1)$. One can measure the misalignment of the two models by defining an error function $\Delta\mathbf{x}_c^1 = \mathbf{x}_c^1 - \mathbf{x}_c^*$. If $\parallel \Delta\mathbf{x}_c^1 \parallel = \parallel \mathbf{x}_c^1 - \mathbf{x}_c^* \parallel < \eta$ is met, where η is a given error for a satisfactory solution, then $\mathbf{R}_f(\mathbf{x}_f^1)$ simultaneously satisfy the required specifications since $\mathbf{R}_c(\mathbf{x}_c^*)$ approaches to $\mathbf{R}_c(\mathbf{x}_c^1)$ and $\mathbf{R}_c(\mathbf{x}_c^1)$ approaches to $\mathbf{R}_f(\mathbf{x}_f^1)$. If the inequality condition is not fulfilled, the solution can be improved by solving the following non-linear equation:

$$\Delta\mathbf{x}_c^1 = \mathbf{x}_c^1 - \mathbf{x}_c^* = 0 \tag{5.2.7}$$

for the implicit solution of $\mathbf{x}_f$.

The non-linear equation can be solved by Newton's method such that

$$\mathbf{B}_j(\mathbf{x}_f^{j+1} - \mathbf{x}_f^j) = -\Delta\mathbf{x}_c^j \tag{5.2.8}$$

where $\mathbf{B}_j$ is a Jacobian matrix. The approximation to the Jacobian matrix is updated by the classic Broyden's formula[6]

$$\mathbf{B}_{j+1} = \mathbf{B}_j + \frac{\Delta \mathbf{x}_c^{j+1}(\Delta \mathbf{x}_f^j)^T}{(\Delta \mathbf{x}_f^j)^T \Delta \mathbf{x}_f^j} \qquad (5.2.9)$$

where $\Delta \mathbf{x}_f^j = \mathbf{x}_f^{j+1} - \mathbf{x}_f^j$, $\mathbf{B}_1 = [I]$. With $\mathbf{x}_f^j$, $\Delta \mathbf{x}_c^j$, $\mathbf{B}_j$, one may find $\mathbf{x}_f^{j+1}$ through (5.2.8). $\mathbf{x}_c^{j+1}$ could be found through the mapping $\mathbf{x}_c^{j+1} = \mathbf{P}(\mathbf{x}_f^{j+1})$. If (5.2.10) is satisfied, then stop.

$$\|\Delta \mathbf{x}_c^{j+1}\| = \|\mathbf{x}_c^{j+1} - \mathbf{x}_c^*\| < \eta \qquad (5.2.10)$$

Otherwise set $j = j+1$ to get $\mathbf{B}_{j+1}$ through (5.2.9) and repeat the process until satisfactory results are obtained.

The steps are summarized as follows:

1. Initialize $\mathbf{x}_f^1 = \mathbf{x}_c^*$, $\mathbf{B}_1 = [I]$, based on $\| \mathbf{R}_f(\mathbf{x}_f^1) - \mathbf{R}_c(\mathbf{x}_c^1) \| \leqslant \epsilon$ to find the mapping $\mathbf{x}_c^1 = \mathbf{P}(\mathbf{x}_f^1)$ and get $\Delta \mathbf{x}_c^1 = \mathbf{P}(\mathbf{x}_f^1) - \mathbf{x}_c^*$. Stop if $\| \Delta \mathbf{x}_c^1 \| \leqslant \eta$, otherwise go to the next step.

2. Solve $\mathbf{B}_j \Delta \mathbf{x}_f^j = -\Delta \mathbf{x}_c^j$ for $\Delta \mathbf{x}_f^j$, $j \geqslant 1$.

3. Set $\mathbf{x}_f^{j+1} = \mathbf{x}_f^j + \Delta \mathbf{x}_f^j$.

4. Find the mapping $\mathbf{x}_c^{j+1} = \mathbf{P}(\mathbf{x}_f^{j+1})$, based on $\| \mathbf{R}_f(\mathbf{x}_f^{j+1}) - \mathbf{R}_c(\mathbf{x}_c^{j+1}) \| \leqslant \epsilon$.

5. Compute $\Delta \mathbf{x}_c^{j+1} = \mathbf{P}(\mathbf{x}_f^{j+1}) - \mathbf{x}_c^*$. If $\| \Delta \mathbf{x}_c^{j+1} \| \leqslant \eta$, stop; otherwise go to the next step.

6. Update $\mathbf{B}_j$ to $\mathbf{B}_{j+1}$, $\mathbf{B}_{j+1} = \mathbf{B}_j + \dfrac{\Delta \mathbf{x}_c^{j+1}(\Delta \mathbf{x}_f^j)^T}{(\Delta \mathbf{x}_f^j)^T \Delta \mathbf{x}_f^j}$.

7. Set $j = j+1$; go to Step 2.

Using this approach, we can obtain a progressively improved design after each iteration step.

The main difference between SM and ASM is clearly seen. The criterion for stopping iteration is (5.2.5) in SM and is (5.2.10) in ASM. The updating of the iterations through $\mathbf{P}$ in SM and through $\mathbf{B}$ in ASM.

Instead of waiting for EM analysis at several base points in SM, ASM aggressively exploits every available EM analysis through updating iterations the by Broyden formula.

To make a comparison, we use the same example as that in Section 5.2.1 to do the optimization design of microstrip patch antenna. The results are given in Table 5.2.

Table 5.2   Model parameters (Unit: mils).

i	1	2	3
x_{c1}^i	16.10	16.00	16.00
x_{c2}^i	216.75	213.08	213.02
x_{f1}^i	16.00	15.90	15.90
x_{f2}^i	213.00	209.25	209.17

The results show that ASM converges faster than SM.

The ASM algorithm may also be summarized as follows:

- Starting from initial optimized parameters of coarse model

$$\mathbf{x}_c^* \to \mathbf{x}_f^1 \xrightarrow[\substack{\left\|\mathbf{R}_f(\mathbf{x}_f^1)-\mathbf{R}_c(\mathbf{x}_c^1)\right\|<\epsilon \\ \mathbf{x}_c^1=Map(\mathbf{x}_f^1)}]{} \mathbf{x}_c^1 \xrightarrow[\Delta\mathbf{x}_c^1=\mathbf{x}_c^1-\mathbf{x}_c^*]{} \Delta\mathbf{x}_c^1$$

$$\to \begin{cases} \left\|\Delta\mathbf{x}_c^1\right\| < \eta, & stop \\ \left\|\Delta\mathbf{x}_c^1\right\| > \eta, & go\ to\ iteration\ process\ (j=1,\ j=j+1) \end{cases}$$

- Go through with the iteration process

$$\Delta\mathbf{x}_c^j \xrightarrow[\substack{\mathbf{B}_j\Delta\mathbf{x}_f^j=-\Delta\mathbf{x}_c^j \\ \mathbf{B}_1=[I] \\ \mathbf{B}_{j+1}=\mathbf{B}_j+\frac{\Delta\mathbf{x}_c^{j+1}(\Delta\mathbf{x}_f^j)^T}{(\Delta\mathbf{x}_f^j)^T\Delta\mathbf{x}_f^j}}]{} \Delta\mathbf{x}_f^j$$

$$\xrightarrow[\mathbf{x}_f^{j+1}=\mathbf{x}_f^j+\Delta\mathbf{x}_f^j]{} \mathbf{x}_f^{j+1} \xrightarrow[\substack{\left\|\mathbf{R}_f(\mathbf{x}_f^{j+1})-\mathbf{R}_c(\mathbf{x}_c^{j+1})\right\|<\epsilon \\ \mathbf{x}_c^{j+1}=Map(\mathbf{x}_f^{j+1})}]{} \mathbf{x}_c^{j+1}$$

$$\xrightarrow[\Delta\mathbf{x}_c^{j+1}=\mathbf{x}_c^{j+1}-\mathbf{x}_c^*]{} \begin{cases} \left\|\Delta\mathbf{x}_c^{j+1}\right\| < \eta, & stop \\ \left\|\Delta\mathbf{x}_c^{j+1}\right\| > \eta, & back\ to\ \Delta\mathbf{x}_c^j\ (set\ \Delta\mathbf{x}_c^j=\Delta\mathbf{x}_c^{j+1}) \end{cases}$$

- Final outcome

$$\mathbf{x}_c^{j+1} \cong \mathbf{x}_c^*$$

$$\mathbf{R}_f(\mathbf{x}_f^{j+1}) \cong \mathbf{R}_c(\mathbf{x}_c^{j+1}) \cong \mathbf{R}_c(\mathbf{x}_c^*)$$

$$Notice\ that$$

$$\mathbf{R}_f(\mathbf{x}_f^1) \neq \mathbf{R}_c(\mathbf{x}_c^*)$$

5.2.3   Using the Closed Form Created by the Full Wave Solver as a Coarse Model in ASM

In performing the parameter extraction, the optimization algorithms such as genetic algorithm (GA) are available. It is worth noting that the step of the parameter extraction (PE) takes most of the computer time in SM technique and also is a crucial part in it. Better PE may lead to faster convergence and need less computer time, otherwise the result may even diverge. Moreover, although the calculation of the coarse model needs insignificant CPU time, it would be still very time consuming if hundreds of iterations have to be done in the GA algorithm in order to match the two responses from the coarse and the fine models. Therefore, an alternative coarse model was proposed in [8]. A quadratic function (response function) which is very simple to calculate was chosen as the coarse model for parameter extraction in the design of the patch antenna. Obviously, the calculation of the quadratic function is much faster than the full-wave analysis. The function used is described as follows:

$$\text{VSWR} = a(f + \alpha x_1 + \beta x_2)^2 + b(f + \alpha x_1 + \beta x_2) + c \qquad (5.2.11)$$

where f is the sampling frequency, and x_1, x_2 are structural parameters to be optimized. α and β are weighs of x_1 and x_2 respectively. a, b, c are coefficients to be determined.

It can be observed that as long as the initial values of x_1 and x_2 obtained from the optimal coarse model response and the VSWR values at three sampling frequency points are given by using the full wave solver, a, b and c can be determined by solving a set of linear equations.

In (5.2.11), both the frequency and the structure parameters are involved. The generation of (5.2.11) needs full wave analysis for three times only. Nevertheless this formula has no physical meaning. It serves as a good substitute to the coarse model.

We use the same example as that in Figure 5.1 to verify the idea. The objectives of the optimization are specified as:

$$\text{VSWR} \leqslant 1.35 \; at \; 15.9\text{GHz} \; and \; 16.1\text{GHz}; \quad \text{VSWR} \leqslant 1.1 \; at \; 16.0\text{GHz}.$$

The parameters to be optimized are x_1, x_2. W and ϵ_r are fixed ($W = 315$mil $= 8$mm, $\epsilon_r = 2.7$). In the coarse model, the discretization is 10 grids per wavelength, and the computer time for each frequency is 1 second. Whereas in the fine model, the discretization is 30 grids per wavelength, and the computer time for each frequency is 40 seconds.

According to the observation $\alpha = 0.1$, $\beta = 2.5$ are chosen. Based on the initial values $\mathbf{x}_c^* = [16.00 \; 213.00]^T$, performing the frequency-sampling through a full wave solver to the frequency response of VSWR at three frequencies respectively, we may obtain a, b and c by solving a set of linear equations.

The response of the function obtained from the optimal coarse model is shown in Figure 5.2 along with the fine model response at the start point. Figure 5.3 presents the response of the function obtained from the optimal coarse model response and the final response of the fine model. The algorithm terminated in five iterations, requiring five fine-model simulations only. Table 5.3 lists values of optimization parameters at each iteration step of the ASM technique using the response function and the fine model.

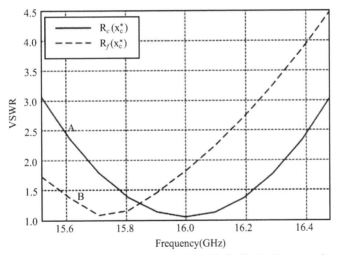

Figure 5.2   A: Optimized response from the coarse model ($\mathbf{R}_c(\mathbf{x}_c^*)$), B: Response from the fine model by using the initial optimized parameters of coarse model ($\mathbf{R}_f(\mathbf{x}_c^*)$).

Table 5.3   Model parameters (Unit: mils).

i	1	2	3	4	5
x_{c1}^i	15.76	15.54	16.35	16.28	16.12
x_{c2}^i	217.88	211.07	214.58	211.86	214.22
x_{f1}^i	16.00	16.24	16.52	16.37	16.11
x_{f2}^i	213.00	208.12	209.50	208.88	209.19

In the above, the function of the closed form is chosen to be the quadratic function. However, the choice of it is problem dependent. Other polynomial functions or rational functions may also be chosen if necessary.

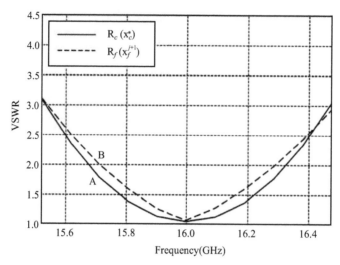

Figure 5.3   A: Optimized response from the coarse model ($\mathbf{R}_c(\mathbf{x}_c^*)$), B: Response from the fine model by using the fine parameters through SM after response matching ($\mathbf{R}_f(\mathbf{x}_f^{j+1})$).

5.2.4   Using the Closed Form Created by the Cavity Model as a Coarse Model in ASM

In Section 5.2.3, the analytical function is created with the full wave solver. This method is more general and may be used on problems other than microstrip patch antenna as long as the analytical function can be created. Specifically, the approximate input impedance formula in closed form for microstrip rectangular patch antennas has already been developed using the cavity model, and is given in Chapter 3. This closed form is a very suitable choice as the coarse model in the optimization design of microstrip patch antennas[9].

We use the same example as that in Figure 5.1 to verify the idea. Instead of using a coarse grid as the coarse model and the original SM, here the coarse model of the closed form from the cavity model and the ASM are used. Figure 5.4 and Figure 5.5 give the results.

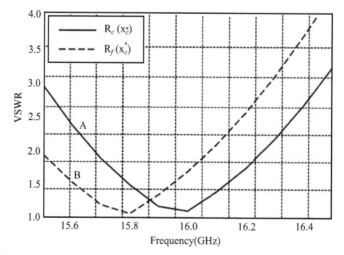

Figure 5.4   A: Optimized response from the coarse model ($\mathbf{R}_c(\mathbf{x}_c^*)$), B: Response from the fine model by using the initial optimized parameters of coarse model ($\mathbf{R}_f(\mathbf{x}_f^*)$).

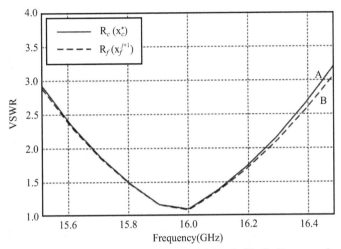

Figure 5.5 A: Optimized response from the coarse model ($\mathbf{R}_c(\mathbf{x}_c^*)$), B: Response from the fine model
by using the fine parameters with SM after response matching ($\mathbf{R}_f(\mathbf{x}_f^{j+1})$).

In SM technique, the choice of a coarse model is very important. Although the main concern is the computation time but not the accuracy, the value-range of the response between the coarse model and the fine model should have some overlap. If the overlap region is too small, the convergence in the iteration may never be achieved. The choice of a coarse model is problem dependent and there is still a lot of research to be done.

The SM only establishes a mapping between two parameter spaces at the optimal point after the optimization is completed based on the coarse model. In other words, it is a single-point mapping technique. In some cases of modeling and simulation, it is desired to find a multiple-point mapping that covers a range of parameters of interest. If so, we can use the mapped coarse model (enhanced coarse model) to replace the fine model without introducing too much error. This concern finally gave rise to a different concept, that is, the generalized space mapping (GSM)[10, 11]. The GSM attempts to establish a mapping between the parameters of the coarse model and the fine model in a certain range of interest through the optimization procedure. Consequently, the accuracy of this coarse model is significantly improved to such a level that it is almost as accurate as the fine model with almost the same computer time as that in the traditional coarse model.

The SM technique is still under development; there is a review paper summarizing this technique and introducing some recent developments[12].

5.3 Extrapolation and Interpolation Methods

5.3.1 One-Dimensional Asymptotic Waveform Evaluation (AWE)

A very powerful extrapolation method recently developed is the asymptotic waveform evaluation (AWE) method[1]. The concept of this method comes from the Taylor's expansion for a function. That is, if one knows the value and derivative information of a point, one can extract the information around this point. This concept is extended to a matrix equation in method of moments (MoM)[13]

$$Z(k)I(k) = V(k) \tag{5.3.1}$$

where Z is a square impedance matrix associated with the characteristics of the object to be analyzed, I is an unknown vector associated with the current distribution to be determined,

V is a known vector associated with the source or excitation and $k = 2\pi/\lambda = \omega\sqrt{\epsilon\mu}$. Since the matrix Z depends on k, it must be generated and solved repeatedly at each k in order to obtain a solution over an interested band of k, which is quite time consuming. The AWE method is a good candidate to solve this problem.

In the AWE method, $I(k)$ is expanded into a Taylor series

$$I(k) = \sum_{n=0}^{Q} M_n (k - k_0)^n \tag{5.3.2}$$

where k_0 is the expansion point, M_n denotes the unknown coefficients, and Q denotes the total number of such coefficients. Substituting (5.3.2) into (5.3.1), expanding $Z(k), V(k)$ into a Taylor series and matching the coefficients of the equal powers of $k - k_0$ on both sides yield the recursive relation for the moment vectors M_0 and M_n in terms of $Z^{-1}, Z^{(i)}, V^{(n)}$

$$M_0 = Z^{-1}(k_0)V(k_0) \tag{5.3.3}$$

$$M_n = Z^{-1}(k_0)\left[\frac{V^{(n)}(k_0)}{n!} - \sum_{i=1}^{n} \frac{Z^{(i)}(k_0)M_{n-i}}{i!}\right], \quad n \geqslant 1 \tag{5.3.4}$$

where the superscripts -1 and ( ) denote the inverse operator and ith or nth derivatives respectively.

The Taylor expansion has a limited bandwidth. To obtain a wider one, $I(k)$ is expanded into a better behaved rational Padé function. Matching the derivative information of the Taylor series and the Padé function by equaling the corresponding constant terms, the coefficients in the Padé function may be obtained. For the moments $m_n = [M_n]_{(r)}$ of an output r, we have

$$[I(k)]_{(r)} = \frac{\displaystyle\sum_{i=0}^{L} a_i(k - k_0)^i}{1 + \displaystyle\sum_{j=1}^{M} b_j(k - k_0)^j} = \frac{N(k - k_0)}{D(k - k_0)} \tag{5.3.5}$$

where $L + M = Q$. Substituting (5.3.2) into (5.3.5) and matching the coefficients of the same order of $k - k_0$, we have the matrix equation for solving b_j:

$$\begin{bmatrix} m_L & m_{L-1} & m_{L-2} & \cdots & m_{L-M+1} \\ m_{L+1} & m_L & m_{L-1} & \cdots & m_{L-M+2} \\ m_{L+2} & m_{L+1} & m_L & \cdots & m_{L-M+3} \\ \vdots & \vdots & \vdots & & \vdots \\ m_{L+M-1} & m_{L+M-2} & m_{L+M-3} & \cdots & m_L \end{bmatrix} \begin{bmatrix} b_1 \\ b_2 \\ b_3 \\ \vdots \\ b_M \end{bmatrix} = - \begin{bmatrix} m_{L+1} \\ m_{L+2} \\ m_{L+3} \\ \vdots \\ m_{L+M} \end{bmatrix} \tag{5.3.6}$$

and the coefficients a_i may be obtained by

$$a_i = \sum_{j=0}^{i} b_j m_{i-j} \quad 0 \leqslant i \leqslant L \tag{5.3.7}$$

It is seen that in AWE, the inverse operation for impedance matrix $Z(k)$ is needed only once, resulting in dramatic saving of the computer time with reasonable accuracy.

The adaptive algorithm for AWE is shown in Figure 5.6, where $E(f)$ is the desired quantity.

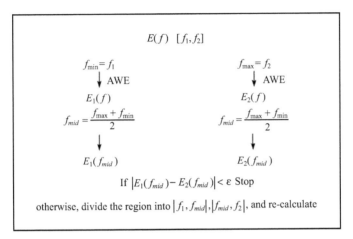

$$E(f) \quad [f_1, f_2]$$

$$f_{min} = f_1 \qquad\qquad\qquad f_{max} = f_2$$

$$\downarrow \text{AWE} \qquad\qquad\qquad \downarrow \text{AWE}$$

$$E_1(f) \qquad\qquad\qquad\qquad E_2(f)$$

$$f_{mid} = \frac{f_{max} + f_{min}}{2} \qquad\qquad f_{mid} = \frac{f_{max} + f_{min}}{2}$$

$$\downarrow \qquad\qquad\qquad\qquad\qquad \downarrow$$

$$E_1(f_{mid}) \qquad\qquad\qquad\qquad E_2(f_{mid})$$

$$\text{If } |E_1(f_{mid}) - E_2(f_{mid})| < \varepsilon \text{ Stop}$$

otherwise, divide the region into $|f_1, f_{mid}|, |f_{mid}, f_2|$, and re-calculate

Figure 5.6   Flowchart of the adaptive algorithm.

When single-point expansion does not satisfy the requirement over the whole frequency band, the multi-point expansion becomes necessary. The adaptive algorithm given in Figure 5.6 is available for this purpose.

An example of the application of one-dimensional AWE is given in Figure 5.7, where the dielectric constant of the substrate ϵ_r is 2.22, $L = 12.5$mm, $H = 8.79$mm, and the thickness of the substrate is 0.787mm. The frequency range is from 6.5GHz to 8.5GHz. The adaptive AWE is used to find the frequency response of the input impedance. Figure 5.8 gives a comparison between the results from MoM only and that from AWE.

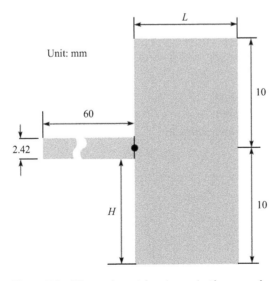

Figure 5.7   Microstrip patch antenna in the example.

Other than the frequency, the dielectric constant ϵ_r may also be taken as the variable in AWE[14]. In this case, the derivatives with respect to ϵ_r are needed. The approximate

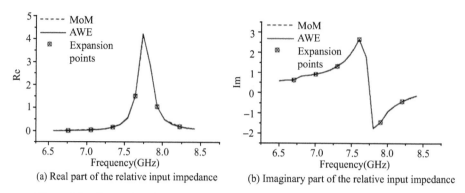

(a) Real part of the relative input impedance   (b) Imaginary part of the relative input impedance

Figure 5.8   Comparison between the results from the MoM only and from AWE.

images for the microstrip substrate given in (5.3.8)–(5.3.10) are suitable to be adopted[15].

$$G_e(\rho) \approx a\frac{\pi}{2}\left[L_0\left(\frac{\epsilon_r\rho}{h}\right) - L_0\left(\frac{\rho}{\mu_r h}\right)\right] - a^2\left[\frac{\exp(-jk_0p) - 1}{k_0\rho}\right] \qquad (5.3.8)$$

$$G_m(\rho) \approx 0 \qquad (5.3.9)$$

$$L_0(z) = H_0(z) - Y_0(z) \qquad (5.3.10)$$

where $G_e(\rho)$ and $G_m(\rho)$ are the expressions of the discrete full wave images for electric and magnetic fields, respectively. $H_0(z)$ is the Struve function of zero order, $Y_0(z)$ is the Neumann function of zero order, h denotes the thickness of the dielectric, ρ is the distance between the source point and the field point and $a = [(\epsilon_r\mu_r - 1)/\epsilon_r]k_0h$. In the use of (5.3.8)–(5.3.10), the following identities should be satisfied:

$$k_0\rho(k_0h/\epsilon_r)^2 \ll 1 \qquad (5.3.11)$$

$$k_0\rho(\mu_r k_0h)^2 \ll 1 \qquad (5.3.12)$$

$$(k_0h\sqrt{\epsilon_r\mu_r})^2 \ll 1 \qquad (5.3.13)$$

It means that (5.3.8)–(5.3.10) are only available for thin substrates. The example is the same as that in Figure 5.7. The frequency is 7.6GHz and the dielectric constant ϵ_r ranges from 1.5 to 2.9. The results for the real part of the relative impedance are shown in Figure 5.9.

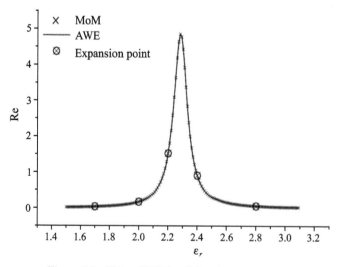

Figure 5.9   Using AWE for dielectric constant ϵ_r.

5.3.2 Two-Dimensional Asymptotic Waveform Evaluation (AWE)

Due to the differentiation rule $(df[u(x)]/dx = [df/du][du/dx])$, in addition to frequency, it is possible to take ϵ or μ as the variable. Moreover, with the multiple Taylor expansion, the AWE method may be used in the case of multi-dimension. In the microstrip structure, the sensitivity of the performance to the dielectric constant is also a factor to be concerned with. For this purpose, 2-D AWE has been developed to extrapolate the responses over frequency and permittivity simultaneously to characterize microstrip antennas; as a result, the response over certain frequency and dielectric constant ranges can be extrapolated from a single point (f, ϵ_r) accurately and quickly. The computation time is almost reduced by two orders compared to the conventional method[16].

Similar to (5.3.1), we start with the matrix equation in the following form:

$$Z(k, \epsilon_r)I(k, \epsilon_r) = V(k, \epsilon_r) \tag{5.3.14}$$

where Z, I and V are related to both the wave number k and the dielectric constant ϵ_r.

In accordance with the AWE method, $I(k, \epsilon_r)$ is expanded into a two-dimensional Taylor series to obtain the solutions of (5.3.14) over certain frequency and permittivity ranges. For simplicity, in the following we do not distinguish between the notations of the vectors and matrices. We have

$$I(k, \epsilon_{r0}) = \sum_{n=0}^{Q}\sum_{m=0}^{P} a_{nm}(k - k_o)^n(\epsilon_r - \epsilon_{r0})^m \tag{5.3.15}$$

$$
\begin{aligned}
a_{nm} = Z^{-1}\Bigg[& \frac{1}{(n+m)!}C_{m+n}^n\frac{\partial^{m+n}V}{\partial^n k \partial^m \epsilon_r} - \sum_{i=0}^{n-1}\sum_{j=0}^{m-1} a_{ij}\frac{1}{(n+m-i-j)!}C_{m+n-i-j}^{n-i} \\
& \cdot\frac{\partial^{n+m-i-j}Z}{\partial^{n-i}k\partial^{m-j}\epsilon_r} - \sum_{i=0}^{n-1} a_{im}\frac{1}{(n-i)!}\frac{\partial^{n-i}Z}{\partial^{n-i}k} - \sum_{j=0}^{m-1} a_{nj}\frac{1}{(m-j)!}\frac{\partial^{m-j}Z}{\partial^{m-j}\epsilon_r}\Bigg]
\end{aligned}
$$
$$\tag{5.3.16}$$

where k_o denotes the wave number at the expansion point, a_{nm} denotes the unknown coefficients and $P \times Q$ denotes the total number of such coefficients.

In order to get the coefficients a_{nm}, the derivatives of matrix I have to be generated. The closed forms in (5.3.8)–(5.3.10) are used to get all the derivatives.

The Taylor expansion has a limited bandwidth. To obtain a wider bandwidth, we represent $I(k, \epsilon_r)$ with a better rational Padé function:

$$I(k, \epsilon_r) = \frac{\displaystyle\sum_{i=0}^{X}\sum_{j=0}^{Y} b_{ij}(k - k_o)^i(\epsilon_r - \epsilon_{r0})^j}{\displaystyle\sum_{l=0}^{F}\sum_{m=0}^{G} c_{lm}(k - k_o)^l(\epsilon_r - \epsilon_{r0})^m} \tag{5.3.17}$$

where $C_{00} = 1$, $XF + YG + X + F + Y + G + 1 = PQ + P + Q$. If we make $Y = G$, the unknown coefficients b_{ij} and c_{ij} can be calculated by substituting (5.3.15) into (5.3.17); then multiplying (5.3.17) by the denominator of the Padé expansion, and matching the coefficients of the equal powers of $k - k_o$ and $\epsilon_r - \epsilon_{r0}$. This leads to the matrix equation

$$\begin{bmatrix} 1 & 0 & \cdots & 0 & 0 & \cdots & 0 & -a_{0,0} \\ 0 & 1 & \cdots & 0 & 0 & \cdots & -a_{0,0} & -a_{1,0} \\ \vdots & \vdots & & \vdots & \vdots & & \vdots & \vdots \\ 0 & 0 & \cdots & 1 & -a_{X-F,0} & \cdots & -a_{X-1,0} & -a_{X,0} \\ 0 & 0 & \cdots & 0 & -a_{X-F+1,0} & \cdots & -a_{X,0} & -a_{X+1,0} \\ \vdots & \vdots & & \vdots & \vdots & & \vdots & \vdots \\ 0 & 0 & \cdots & 0 & -a_{X+1,0} & \cdots & -a_{X+F,0} & -a_{X+F+1,0} \end{bmatrix} \begin{bmatrix} b_{0,n} \\ b_{1,n} \\ \vdots \\ b_{X,n} \\ c_{F,n} \\ \vdots \\ c_{0,n} \end{bmatrix}$$

$$= \begin{bmatrix} \sum_{i=0}^{n-1} c_{0,i} a_{0,n-i} \\ \sum_{i=0}^{n-1} c_{0,i} a_{1,n-i} + \sum_{i=0}^{n-1} c_{1,i} a_{0,n-i} \\ \vdots \\ \sum_{i=0}^{n-1}\sum_{j=0}^{X} c_{j,i} a_{X-j,n-i} \\ \sum_{i=0}^{n-1}\sum_{j=0}^{X+1} c_{j,i} a_{X-j+1,n-i} \\ \vdots \\ \sum_{i=0}^{n-1}\sum_{j=0}^{X+F+1} c_{j,i} a_{X+F-j+1,n-i} \end{bmatrix} \tag{5.3.18}$$

where n is from 1 to X. If we solve equation (5.3.18), b_{ij} and c_{ij} can be obtained, and the current vector $\mathbf{I}(k, \epsilon_r)$ can also be obtained by the calculated Padé model.

The example is the same as that in Figure 5.7. Figure 5.10 and Figure 5.11 show the real and imaginary parts of the input impedance as a function of frequency and permittivity, obtained by using the direct method and the AWE method, respectively. With a frequency increment of 0.01GHz and a permittivity increment of 0.01, the direct method requires 3,776,000 seconds to obtain the solution on a Personal Computer (1.2GHz AMD^{TM} K7 processor). With AWE ($Q = 5, P = 3, G = Y = 3, X = 2, F = 3$) to obtain the same accuracy, only 20,100 seconds are needed, which is 188.8 times faster than the direct method.

In the optimization of a design, people are interested in the size response of a structure as well. When the size is used as the variable, the AWE method requires the derivatives with respect to the size. Through fixing the mesh and introducing the rate of the size extension, the AWE method is extended into size dimension. Examples of calculating the capacitance of a microstrip patch and the input impedance of a dipole array confirm the validity of this concept[17, 18].

As one of the interpolation methods, the adaptive frequency sampling (AFS) technique is widely used. In traditional AFS, it is inevitable to invert an $N \times N$ matrix in order to solve for the coefficients of targeted rational interpolation functions, where N is the number of samples. The ill-conditioned matrix of a large N restricts traditional AFS techniques to an electromagnetic simulation accelerator. The general Stoer-Bulirsch (S-B) algorithm is employed in developing a new AFS scheme (S-B AFS). Since the S-B algorithm is a recursive tabular method and requires no matrix inversion, it can process a large number of sampling data for obtaining a rational interpolation function without suffering from the singularity problem[19]. This attribute virtually leads this AFS approach to an ultra broadband interpolation with a single rational function.

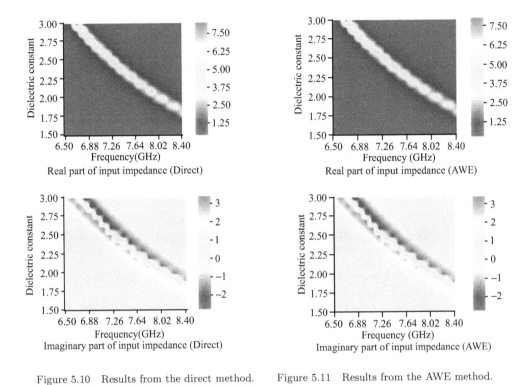

Figure 5.10   Results from the direct method.   Figure 5.11   Results from the AWE method.

5.4   Using the Model from Physical Insight to Create a Formula

The extrapolation and interpolation actually belong to the family of model-based parameter estimation (MBPE). A systematic introduction of MBPE in electromagnetics may be found in [20–22]. In this section, we are going to introduce the method of using a model from the physical insight to create a mutual impedance formula between two antenna elements. This method is essentially one kind of MBPE. The observation of physical insight comes from the concept of synthetic asymptote which was introduced in [23] and has found many applications such as those in [24, 25]. The formula obtained may be used to carry out the interpolation and even extrapolation in certain ranges with the variables of both the distance and the angle.

5.4.1   Mutual Impedance Formula Between Two Antenna Elements

Mutual coupling is very important in the design of antenna arrays, especially phased arrays. In the following, we focus on the elements of a microstrip patch antenna. However, the method is quite general. Many efforts during the last decades were devoted to finding the mutual coupling between two elements. The full wave analysis, of course, is the most accurate method to find the mutual couplings. However, it has been computed less tediously by some simplified models, such as the transmission line model, magnetic current approximation, etc. The simplified models of the patch are much faster but may be inaccurate, and frequently they may be restricted to a certain range of structures such as a thin substrate or regular patches.

Taking advantage of the knowledge of static variable separation in near field and the radiation pattern in far field, it is found that the mutual impedance between two arbitrary patches can be written in a generalized form of separated variables. The relatively small number of 12 coefficients can be determined by matching with full wave analysis through accurate commercial software such as the IE3D^{TM}.

The formula has a similar form as Bailey's formula[26] derived from the specific case of coupling between aperture antennas. Bailey's formula is constructed through observations on spectral domain and its integration[27]. The formula here is for the general case of coupling patches and aperture antennas, including the vertical source of probe excitation. It is constructed through the simpler, spatial domain with variable separations for both static and radiation fields. The spatial domain is simpler, and its variable separations ensure the completeness of the constituent functions in the formula, even for the complicated dielectric substrate of the patch antennas[28, 29].

1) The Division of Near, Far and Surface-Wave Regions

Like the fields produced by a point source, the mutual impedance between two antenna elements as extended sources can still be divided into three regions in terms of the center-to-center separation r between the patches in Figure 5.12. That is:

1. The near field for closely spaced adjacent patches with a static dependence of $1/r^{n+1}$ where $n \geqslant 1$, which forms the near asymptote of r;

2. The far field for widely separated patches, with a spherical wave dependence of $1/r$, this forms the far asymptote;

3. The surface wave zone for even wider separation with a cylindrical wave dependence of $1/\sqrt{r}$, this forms the "far-far" asymptote that may sometimes be neglected. The reason is that the surface wave may become significant only at $10\lambda_0$ or beyond; this distance could be outside the finite boundary of the antenna array.

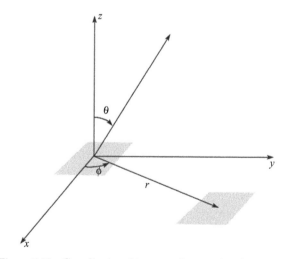

Figure 5.12   Coordinates of two coupling patches in an array.

2) The Static Field Region—Near Asymptote of Separation r

For coordinates defined in Figure 5.12, in the near field of static field region, according to the variable separation solution to the Laplace equation, the field of a point source must have the form:

$$V(r,\phi,\theta) = \sum_{n=0}^{\infty} \sum_{m=0}^{n} (a_{nm}r^n + \frac{b_{nm}}{r^{n+1}})P_n^m(\cos\theta)(c_m \cos m\phi + d_m \sin m\phi) \qquad (5.4.1)$$

where $P_n^m(\cos\theta)$ is the associated Legendre function. The coefficient a_{nm} is set to zero because V must be regular at infinite distance r in a coupling formula. The center-to-center separation r between adjacent patches in a practical array is usually a little larger than $\lambda_0/2$. Because r is not very small, and the general static potential terms of $1/r^{n+1}$ attenuate quickly, only the terms of the point source, dipole and quadrupole, i.e., $n = 0$ to 2, are left.

For a microstrip patch, each point source (charge or current) in a patch forms a dipole by ground plane reflection, and then can be moved to the center of the patch by an addition theorem as a series of multipoles[30]. This means that patch to patch coupling is a multipole to multipole coupling with a dependence of $1/r^{n+m+1}$, where n and m are the orders of the two coupling multipoles concerned.

Usually in a microstrip patch array, the substrate thickness is quite small compared to the free space wavelength λ_0, thus approximately the mutual coupling is on the plane of $\theta = \pi/2$. The associate Legendre polynomials in (5.4.1) are to be evaluated only at $\theta = \pi/2$ for the substrate surface and are constants. They then may be absorbed into the coefficients b_{nm}.

3) The Radiation Field Region—Far Asymptote of the Separation r

In the radiation zone, angle dependence of the field is the radiation pattern of the patch. With the distance dependence of $1/r$ and phase changing taken into account, the electric field E_z should take the form of:

$$E_z(r,\phi) = \frac{\exp(-jkr)}{r} \sum_{m=0}^{\infty} (c_m \cos m\phi + d_m \sin m\phi) \qquad (5.4.2)$$

The size of the patch is usually on the order of $\lambda_d/2$, where λ_d is the wavelength in the substrate; this means that in the far field the radiation pattern can be expressed simply by a few Fourier terms, say $m \leqslant 4$.

4) The Surface-Wave Field Region—the "Far-Far" Asymptote of Separation r

For the patch antenna using a substrate with a low dielectric constant, the surface wave may be negligible in comparison with the radiation wave[31]. If it is not negligible, the surface wave provides the "far-far-field." We see that surface wave is also a radiation from a small antenna, this means that surface-wave contribution may also be represented by the radiation pattern, but with distance dependence of $1/\sqrt{r}$ in (5.4.2). Similar to the radiation wave, the number m of Fourier terms in ϕ in surface wave is still less than 4.

We may add that the equation forms of (5.4.1) and (5.4.2) are always valid in the free space regions above the conductive patch and the dielectric substrate. The patch shape (e.g., rectangular or circular), the feed excitation and the grounded substrate can only change the coefficients b_{nm}, c_m and d_m in (5.4.1) and (5.4.2).

5) Reciprocity and Symmetry in Azimuth ϕ Between Two Coupled Patches

The mutual impedance between two patches a and b is given by

$$Z_{ab} = \frac{1}{I_a I_b} \iint_a \mathbf{E}_a \cdot \mathbf{J}_b dS_a dS_b \qquad (5.4.3)$$

where $\mathbf{E}_a$ and $\mathbf{J}_b$ are the electric field and current on the patch. Eqn. (5.4.3) is the formal link between the desired mutual impedance and what was discussed in 1) to 4). The reciprocity

$Z_{ba} = Z_{ab}$ of the mutual impedance in circuit theory, as well as in the formal link (5.4.3), means that we must have $Z_{ab}(\phi) = Z_{ab}(\phi + \pi)$ in all regions regardless of the shapes and feed-probe locations of the two patches. On the other hand, considering that the mutual coupling between the probe feeds of the two patches is equal for all azimuth ϕ, and the current distribution on the patch is symmetrical along x and y directions, we also have $Z_{ab}(\phi) = Z_{ab}(\pi - \phi) = Z_{ab}(-\phi)$. The symmetry and reciprocity require that only the $\cos 2m\phi$ terms remain (instead of the complete set of $\sin m\phi$ and $\cos m\phi$ in (5.4.1) and (5.4.2) which are non-zero in the mutual impedance of (5.4.3) between two patches). The reciprocity means that the mutual impedance between two patch antennas is always simpler in form than the field around a patch antenna, or the Green's function of a point source.

6) The Mutual Impedance Formula Between Two Patches

To create the formula, in principle, there are four significant terms in $1/r^{n+1}$ with $n = -1/2, 0, 1$ and 2 for the far asymptotic couplings of surface wave and radiation, and the near asymptotic couplings of static dipole and quadrupole. Three Fourier components of the azimuth angle ϕ satisfying the symmetry relations are used, that is: constant, $\cos 2\phi$ and $\cos 4\phi$. Thus, by the simple sum of synthetic asymptote, a total of $12(4 \times 3)$ terms with 12 unknown coefficients are to be determined by a numerical match. The mutual impedance between two patches can be written in a series with the unknown complex coefficients $c_{n,m}$ as

$$Z_{ab} = \eta_0 \frac{\exp(-jk_0 r)}{4\pi} \sum_{n=-1/2,0,1,2} \{[1/(k_0 r)^{n+1}][c_{n,0} + c_{n,2} \cos 2\phi + c_{n,4} \cos 4\phi]\} \quad (5.4.4)$$

where η_0 is the intrinsic impedance of free space. It is observed that (5.4.4) has a similar static term of a variable separated series from the Laplace differential equation, plus extra terms from radiation and surface waves.

The 12 unknown coefficients $c_{n,m}$ of (5.4.4) can be found by matching with a numerical solution (e.g. IE3D^{TM}), or measured values, between the center patch at the origin and the 12 coupling patches in a skeleton array as shown in Figure 5.13. When the surface wave is negligible for low dielectric constant substrate, only 9 coefficients need to be determined. In this case, the dotted patches 7, 8, 9 in Figure 5.13 are deleted.

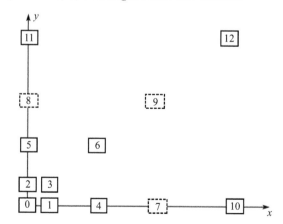

Figure 5.13   Layout of IE3D^{TM} computed mutual couplings between the zeroth patch and the nth patch ($n = 1$ to 12) for evaluation of the 12 $c_{n,m}$ coefficients.

The skeleton array gives the closest and farthest patch separations to sample the couplings of the near inductive, far radiation and surface wave fields at different angles ϕ. The boundary of the skeleton array may correspond to the boundary of an actual array to be analyzed. Evidently within the boundary, the intermediate patches may be selected arbitrarily without significantly increasing the small error in the resulting formula in (5.4.4).

5.4.2   Relationship Between Bailey's Formula and That in Formula (5.4.4)

In 1996, Bailey proposed a formula of mutual coupling for aperture coupling[26] similar in form to formula (5.4.4) for patch antenna coupling. His formula is for the mutual admittance between two apertures in the same ground plane:

$$Y_{ab} = \left\{ \left[A_1 \left(\frac{1}{k_0 r} \right)^2 + A_2 \left(\frac{1}{k_0 r} \right)^3 \right] \cos^2 \phi + \left[A_3 \left(\frac{1}{k_0 r} \right) + A_4 \left(\frac{1}{k_0 r} \right)^2 + A_5 \left(\frac{1}{k_0 r} \right)^3 \right] \right.$$

$$\left. \cdot \sin^2 \phi + \left[A_6 \left(\frac{1}{k_0 r} \right) + A_7 \left(\frac{1}{k_0 r} \right)^2 + A_8 \left(\frac{1}{k_0 r} \right)^3 \right] \sin^2 2\phi \right\} e^{-jkr}$$

$$(5.4.5)$$

His formula is postulated from observations on the spectral domain integration for mutual admittance between apertures on a plane. The unknown coefficients A_m are found by matching with numerical data, in the same way as that in finding $c_{n,m}$ in (5.4.4). The observation on the spectral domain, in terms of the patch separation r and the azimuth angle ϕ, in (5.4.5) applies equally well to mutual admittance and impedances between microstrip patches. This means that (5.4.5) can equally well be an impedance formula.

With identities of $\sin^2 \phi + \cos^2 \phi = 1$ and $\cos 2\phi = \cos^2 \phi - \sin^2 \phi$, one may convert the terms in the Bailey's formula (5.4.5) to resemble the formula (5.4.4), except for the lack of some radiation $1/k_0 r$ terms and all of the $1/\sqrt{k_0 r}$ surface wave terms specific to the patch antennas. The radiation and surface wave terms of $1/k_0 r$ and $1/\sqrt{k_0 r}$ with no ϕ dependence correspond to the radiations from the feed probe of the patch, as a vertical feed probe radiates omni-directionally around ϕ. The inclusion into (5.4.4) of the $1/k_0 r$ in all directions and the surface wave terms $1/\sqrt{k_0 r}$, therefore, makes formula (5.4.4) for mutual impedance more complete.

The completeness of formula (5.4.4) comes from its novel physical basis for finding the generalized form of separated variables. Based on the static variable separation in near field, radiation pattern in far field and the possible combination of these two by synthetic asymptote, all the possible terms, including the probe radiation term and surface wave terms, are presented in the formulation.

Being more complete with the probe radiation and surface wave terms, formula (5.4.4) should always be more accurate than Bailey's formula (5.4.5). With current distribution computed from IE3D^{TM} and not assumed beforehand, formula (5.4.4) is accurate for a wide range of practical problems such as for thick substrate and patches whose current distribution contains effectively many other modes in addition to the dominant (0,1) mode. Even for other kinds of elements this formula is still applicable because the information related to the element is involved in the full wave solver in sampling. However, Bailey's formula will fail for some problems due to its incompleteness in form, which can be observed from the following numerical comparisons.

5.4.3   Numerical Results

Formula (5.4.4) is used to compute the mutual coupling of several sets of microstrip patches. The results are compared with those from experiment, IE3D^{TM} simulation and

Bailey's formula. The coefficients are found from a 12-point matching with numerical data from IE3D^{TM}. Figure 5.13 shows the skeleton array of the patches for determining the coefficients in formula (5.4.4). For Bailey's formula, sampling points are fixed, as specified in his NASA report[26], i.e.,

$$x = 1.5W, 2.5W, 5W; \qquad y = 1.5L, 2.5L, 5L$$

Figure 5.14 and Figure 5.15 show the comparison between mutual coupling results from formula (5.4.4) and those from measurements by Jedlicka et al[32] for E-plane and H-plane mutual coupling $|S_{12}|$ versus normalized adjacent edge spacing d/λ_0 of rectangular patches and circular patches respectively. The sampling points for formula (5.4.4) are selected as

$$x = 1.25W, 2.4W, 4W, 7.5W; \quad y = 1.25L, 2.4L, 4L, 6.5L$$

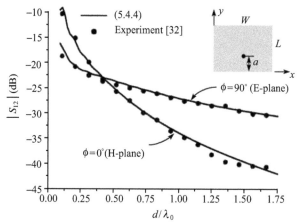

Figure 5.14   Comparison of mutual coupling from our method and experiment results from [32], for rectangular patches on substrate, $h = 3.05$mm, with patch size of $W = 50$mm, $L = 60$mm, at resonance $f_0 = 1.56$GHz. d is the distance between two adjacent edges.
(After Sun, Chow, and Fang [29],©2003 Wiley)

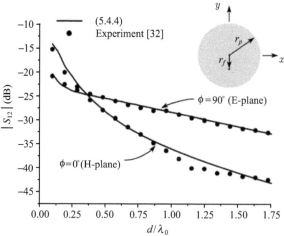

Figure 5.15   Comparison of mutual coupling from our method and experiment results from [32], for circular patches on substrate , $h = 1.575$mm, with patch radius of $r_p = 38.5$mm and a feed point location at radius $r_f = 11$mm, at resonance $f_0 = 1.44$GHz. d is the distance between two adjacent edges.
(After Sun, Chow, and Fang [29],©2003 Wiley)

for Figure 5.14 and

$$x = 2.5r_p, 5r_p, 8.2r_p, 11.3r_p; \quad y = 2.5r_p, 5r_p, 8.2r_p, 11.3r_p$$

for Figure 5.15. Good agreement between the two results is observed, despite the fact that the coefficients in our formula (5.4.4) are evaluated from IE3D^{TM} as stated above and not from the measurements of [32].

Figure 5.16 shows the mutual impedances versus center-to-center separation r, at different azimuth angles ϕ from Bailey's formula, our formula, and IE3D^{TM}. The elements are rectangular patches on substrate with parameters of $\epsilon_r = 2.55$, $h = 1.57$mm, with patch size of $W = 22.6$mm, $L = 17.6$mm and the probe feed is near the center of W with a distance $a = 5$mm. The operating frequency is at the resonance of $f = 5$GHz. The sampling points for formula (5.4.4) are selected as

$$x = 1.33W, 3W, 6W, 18.5W; \quad y = 1.3L, 3L, 8L, 23.8L$$

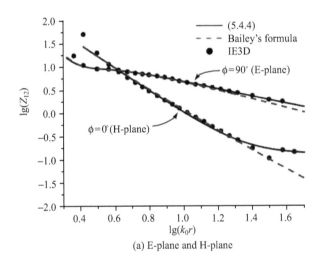

(a) E-plane and H-plane

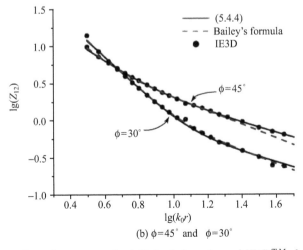

(b) $\phi=45°$ and $\phi=30°$

Figure 5.16   Mutual impedance from our method, Bailey's formula, and IE3D^{TM}, for patch on substrate , $h = 1.57$mm, with size of $W = 22.6$mm, $L = 17.6$mm, and probe feed near the center of W with distance $a = 5$mm, at resonance $f_0 = 5$GHz, r is the center-to-center distance between two patches.

(After Sun, Chow, and Fang [29],©2003 Wiley)

It is observed that both formula (5.4.4) and Bailey's formula give good results, except that there are errors from Bailey's formula when r is large in H-plane. This error comes from the neglect of the $1/k_0r$ radiation term and of the $1/\sqrt{k_0}r$ surface wave terms in Bailey's formula. When the substrate becomes thick, the radiation (as well as the surface wave) becomes stronger in comparison with the induction field. This means that this error of Bailey's appears at a smaller r. This early occurrence is observed in a later example in Figure 5.19.

At off-resonant frequency, say 5.4GHz, the patch current components J_x and J_y generally decrease except at the feed point. The magnitudes of currents at resonant frequency 5GHz and at off-resonant frequency 5.4GHz are shown in Figure 5.17 (a) and (b) respectively. This means relatively larger omni-directional ϕ terms in (5.4.4) from the feed probe, i.e. in the $1/\sqrt{k_0}r$ surface wave term and in the $1/k_0r$ radiation term. The relatively larger omni-directional terms at off-resonance are observable in the log-log plots in Figure 5.18, especially at the $\phi = 0°$ plot as an earlier appearance at $\lg(k_0r) = 0.9$ of surface wave at a slope of $1/2$, instead of the corresponding regular omni-directional term at resonance in Figure 5.16 at $\lg(k_0r) = 1.3$.

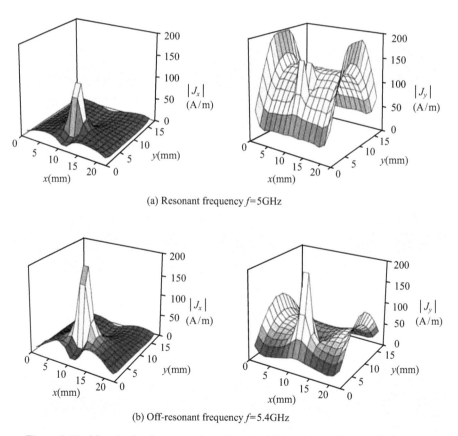

(a) Resonant frequency f=5GHz

(b) Off-resonant frequency f=5.4GHz

Figure 5.17   Magnitude of currents in x direction $|J_x|$ and in y direction $|J_y|$ for the patch shown in Figure 5.16. (After Sun, Chow, and Fang [29],©2003 Wiley)

For thick substrate of air with $h = 3$mm, Figure 5.19 compares the mutual impedance at resonance. The sampling points for our formula (5.4.4) are:

$$x = 1.3W, 3W, 6W, 15.7W; \quad y = 1.4L, 3L, 6L, 18.8L$$

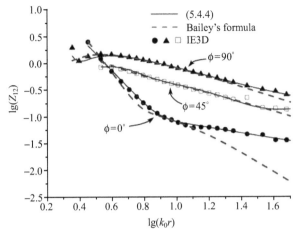

Figure 5.18   Mutual impedance for same patch set as in Figure 5.16 at $f = 5.4\text{GHz}$ which deviates from the resonance. r is the center-to-center distance between two patches.
(After Sun, Chow, and Fang [29],©2003 Wiley)

There is no surface wave in this case, however, the radiation from the longer probe now contributes more to the mutual impedance in the $1/r$ term without ϕ dependence. This means that Bailey's formula would fail earlier at $\lg(k_0r) = 0.8$ in the thick substrate case here, rather than at $\lg(k_0r) = 1.3$ in the thin substrate case in Figure 5.16.

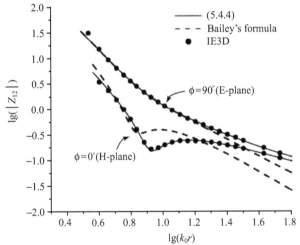

Figure 5.19   Mutual impedance for patches on substrate air, substrate thickness $h = 3\text{mm}$, with $W = 30\text{mm}$, $L = 25\text{mm}$, and $a = 2\text{mm}$, at resonance $f_0 = 5.45\text{GHz}$, r is the center-to-center distance between two patches. Three lines are from: (5.4.4), Bailey's formula and IE3D^{TM}.
(After Sun, Chow, and Fang [29],©2003 Wiley)

Formula (5.4.4) is also used to compute mutual couplings between slot coupled microstrip antennas. Figure 5.20 shows the comparison of mutual coupling from formula (5.4.4) IE3D^{TM} and the computation from [33]. Good agreements are observed as well. In this example, for formula (5.4.4) the sampling points are selected as

$$x = 1.2P_W, 1.5P_W, 2.4P_W, 5.8P_W; \quad y = 1.2P_L, 1.5P_L, 2.4P_L, 5.8P_L$$

Formula (5.4.4) being more complete with the probe radiation and surface wave terms, should be more accurate than the Bailey's formula (5.4.5). The above results show that our formula is indeed more accurate when compared with the IE3D^{TM} results and the measurement results.

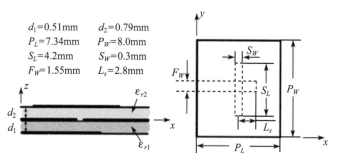

d_1=0.51mm d_2=0.79mm
P_L=7.34mm P_W=8.0mm
S_L=4.2mm S_W=0.3mm
F_W=1.55mm L_s=2.8mm

(a) Structure of the slot-coupled microstrip antenna

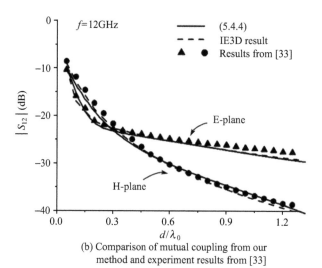

(b) Comparison of mutual coupling from our
method and experiment results from [33]

Figure 5.20   Comparison of mutual coupling from formula (5.4.4), IE3D^{TM} and the computation from [33]. The operation frequency is 12GHz. (After Sun, Chow, and Fang [29],©2003 Wiley)

The wideband closed-form counterpart of (5.4.4) has been developed and has been used to analyze the electrically large finite microstrip antenna arrays combined with a full wave analysis-based network method[41].

5.5   Using Models from the Artificial Neural Network (ANN) to Train Formula

5.5.1   Concept of the Artificial Neural Network (ANN)

Neural networks have recently gained attention as a fast, accurate and flexible tool for RF/microwave modeling, simulation and design. A tutorial has been published in [34] recently. From the viewpoint of application, ANN may be considered as using a model to train formula. This trained formula can be used to do the interpolation and extrapolation in certain ranges as well.

A neural network is a simplified mathematical model of a biological neural network. It consists of a collection of interconnected neurons. Here only a feedforward multilayer perceptron (MLP) neural network, pictorially represented in Figure 5.21 is introduced, which is suitable for device and circuit modeling[35].

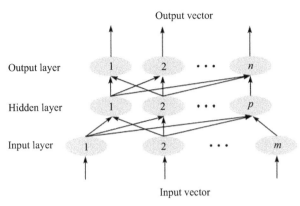

Figure 5.21   Three-layer neural network.

A feedforward multilayer neural network can be described as a mathematical tool which is capable of nonlinear mapping in higher dimension. It has been theoretically proven that a multilayer neural network with at least one hidden layer can model arbitrarily complex nonlinear input/output relationship $\mathbf{y} = F(\mathbf{x})$. Typically, a three-layer neural network is employed to model a nonlinear relationship.

The input space, $\mathbf{x}$ of dimension m, is mapped to the n-dimension output space represented as a layer of n neurons, through a hidden layer. This hidden layer has a fixed number of neurons, p, which can vary from problem to problem as will be discussed in the following. The outputs of any given neurons are the weighted linear combination of the outputs of all the neurons in the previous layer reflected off a nonlinear transfer function, the most commonly employed being the sigmoid.

1) Feedforward (Input Signal)

Mathematically, the neural network can be described as the mapping of a set of input vectors $\mathbf{x}$, whose kth sample is

$$\mathbf{x}_k = (x_{k1}, x_{k2}, \cdots, x_{km}) \tag{5.5.1}$$

to the corresponding output vector

$$\mathbf{y}_k = (y_{k1}, y_{k2}, \cdots, y_{kn}) \tag{5.5.2}$$

through a system of weighting factors and biases, which are defined as w_{ih}, b_h, for $i = 1, 2, \cdots, m$ and $h = 1, 2, \cdots, p$ and ν_{hj}, c_j, for $h = 1, 2, \cdots, p$ and $j = 1, 2, \cdots, n$, such that the n outputs are

$$y_{kj} = f(\zeta_{kj}) = \frac{1}{1 + e^{-\zeta_{kj}}} \tag{5.5.3}$$

where

$$\zeta_{kj} = \sum_{h=1}^{p} z_{kh}\nu_{hj} + c_j \tag{5.5.4}$$

Here, $f(\zeta)$ is the sigmoidal transfer function and z_{kh} is the output of the hth neuron in the hidden layer expressed as

$$z_{kh} = f(\gamma_{kh}) = \frac{1}{1 + e^{-\gamma_{kh}}} \tag{5.5.5}$$

where

$$\gamma_{kh} = \sum_{i=1}^{m} x_{ki} w_{ih} + b_h \tag{5.5.6}$$

It is seen that **x** is related to **y** by a set of sample data. If the set of samples $\mathbf{x}_k$, $k = 1, 2, \cdots, N$, is chosen such that it is representative of the entire Input/Output (I/O) space, then the objective mapping function F is learned by the neural network.

2) Back Propagation (Output Error)

The training of the neural network is the process during which the neural network learns the relationship F between the input and output samples presented to it. This relationship is learned over several training epochs, in which a large set of I/O data is repeatedly presented to the neural network. The weights and biases in (5.5.3)–(5.5.6) are adjusted automatically such that the error between the outputs as predicted by the neural network and the outputs of the training set is minimized.

Back propagation (BP) is probably the most common algorithm used today in training feed forward multilayer perceptron (MLP) neural networks. It is based on multilayer error-correction learning and is described as follows.

For a given set of input data, say $\mathbf{x}_k$, $k = 1, 2, \cdots, N$, whose corresponding output set is $\mathbf{d}_k$, if the neural network predicts the output to be $\mathbf{y}_k$, the batch-mode back propagation error E is defined as

$$E = \sum_{k=1}^{N} E_k = \frac{1}{N} \sum_{k=1}^{N} \left[\frac{1}{2} \sum_{j=1}^{n} (y_{kj} - d_{kj})^2 \right] \tag{5.5.7}$$

where E_k represents the individual mean-squared error of the kth sample. E is the error to be minimized during training. After each training epoch, during which the set of N data points is presented to the network, this error is determined, and the weights and biases are updated in the general direction of error minimization. The updating equations for the tth epoch, with momentum α and learning rate η are

$$\nu_{hj}^{t+1} = \nu_{hj}^{t} + \eta E^t \cdot \frac{\partial E^t}{\partial \nu_{hj}} + \alpha \left(\nu_{hj}^{t} - \nu_{hj}^{t-1} \right) \tag{5.5.8}$$

$$w_{ih}^{t+1} = w_{ih}^{t} + \eta E^t \cdot \frac{\partial E^t}{\partial w_{ih}} + \alpha \left(w_{ih}^{t} - w_{ih}^{t-1} \right) \tag{5.5.9}$$

$$b_{h}^{t+1} = b_{h}^{t} + \eta E^t \cdot \frac{\partial E^t}{\partial b_{h}} + \alpha \left(b_{h}^{t} - b_{h}^{t-1} \right) \tag{5.5.10}$$

$$c_{j}^{t+1} = c_{j}^{t} + \eta E^t \cdot \frac{\partial E^t}{\partial c_{j}} + \alpha \left(c_{j}^{t} - c_{j}^{t-1} \right) \tag{5.5.11}$$

The error sensitivities in the above equation are calculated using the following equations:

$$\frac{\partial E}{\partial \nu_{hj}} = \frac{\partial E}{\partial y_{kj}} \frac{\partial y_{kj}}{\partial \zeta_{kj}} \frac{\partial \zeta_{kj}}{\partial \nu_{hj}}$$

$$= \frac{1}{N} \sum_{k=1}^{N} (y_{kj} - d_{kj}) y_{kj} (1 - y_{kj}) z_{kh}$$

$$= \frac{1}{N} \sum_{k=1}^{N} \delta_{kj}^{(0)} z_{kh} \tag{5.5.12}$$

where the term $\delta_{kj}^{(0)}$, given by

$$\delta_{kj}^{(0)} = (y_{kj} - d_{kj})y_{kj}(1 - y_{kj}) \tag{5.5.13}$$

represents the local gradients at the jth neuron in the output layer for the kth sample

$$\frac{\partial E}{\partial w_{ih}} = \frac{\partial E}{\partial y_{kj}} \frac{\partial y_{kj}}{\partial \zeta_{kj}} \frac{\partial \zeta_{kj}}{\partial z_{kh}} \frac{\partial z_{kh}}{\partial \gamma_{kh}} \frac{\partial \gamma_{kh}}{\partial w_{ih}}$$

$$= \frac{1}{N} \sum_{k=1}^{N} z_{kh}(1 - z_{kh})x_{ki} \sum_{j=1}^{n} \nu_{hj}\delta_{kj}^{(0)} \tag{5.5.14}$$

$$\frac{\partial E}{\partial c_j} = \frac{\partial}{\partial c_j} \left[\frac{1}{2N} \sum_{k=1}^{N} \sum_{j=1}^{n} (y_{kj} - d_{kj})^2 \right]$$

$$= \frac{1}{N} \sum_{k=1}^{N} (y_{kj} - d_{kj})y_{kj}(1 - y_{kj})$$

$$= \frac{1}{N} \sum_{k=1}^{N} \delta_{kj}^{(0)} \tag{5.5.15}$$

$$\frac{\partial E}{\partial b_h} = \frac{\partial E}{\partial y_{kj}} \frac{\partial y_{kj}}{\partial \zeta_{kj}} \frac{\partial \zeta_{kj}}{\partial z_{kh}} \frac{\partial z_{kh}}{\partial \gamma_{kh}} \frac{\partial \gamma_{kh}}{\partial b_h}$$

$$= \frac{1}{N} \sum_{k=1}^{N} z_{kh}(1 - z_{kh}) \sum_{j=1}^{n} \nu_{hj}\delta_{kj}^{(0)} \tag{5.5.16}$$

The procedure is pictorially shown in Figure 5.22.

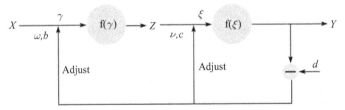

Figure 5.22   Feedforward and back propagation.

3) Training Algorithm

Using the above equations, a modification to the original back propagation training algorithm can be summarized in the following steps, which include a learning rate and momentum adaptation in order to improve the speed of convergence:

1. Initialization: Choose the number of hidden neurons q and initialize the weights and biases w_{ih}, b_h, ν_{hj}, and c_j with small random numbers. Choose initial values for η and α.

2. Input: Supply the training sample set $(\mathbf{x}_k, \mathbf{d}_k)$, $k = 1, 2, \cdots, N$ to the neural network.

3. Forward propagation: Compute the corresponding neural network output vector $\mathbf{y}_k$, $k = 1, 2, \cdots, N$ using (5.5.3)–(5.5.6).

4. Update (Back propagation of the error): Compute the batch back propagation error E from (5.5.7), the error sensitivities using (5.5.12)–(5.5.16), and adjust the parameters of the weights and biases using (5.5.8)–(5.5.11).

5. Termination condition check: If E is less than a specified tolerance value ε, end training; otherwise, go to next step.

6. Update learning rate and momentum: If E is larger than its previous value then decrease learning rate and momentum, i.e., $\eta = \gamma \times \eta$ and $\alpha = \gamma \times \alpha$ where the parameter γ is the learning rate adaptation, and go to Step 2. If E is smaller than its previous value then increase learning rate and momentum, i.e., $\eta = 1/\gamma \times \eta$ and $\alpha = 1/\gamma \times \alpha$, then go to Step 2.

4) Training Parameters

The efficiency of training depends on the following training parameters:

- Number of hidden neurons q :

  Deciding on the size of the hidden layer is a critical part of the design of a neural network model. Once the number of hidden layers is determined, the number of neurons in the hidden layer will determine the structure of the network. Unfortunately, there are no established methods to determine the appropriate number of hidden neurons required for a given problem. In general, a large number of hidden neurons are required to model complicated relationship. But too many may result in an overtrained network that tends to memorize rather than to generalize from data.

- Learning rate η :

  This parameter is an important training parameter representing the step size of the error convergence process. A small value of it affords stability but increases training time, while a large value of it decreases the stability of the training process. The step size can be a small fixed constant set by the user, for example, the user set $\eta = 0.1$ and step size remains 0.1 throughout. The step size can also be adaptive during training, that is, the user initially sets $\eta = 0.1$ and later η can be changed during training. For example, η may be set to $\eta = \eta/\gamma$, $\gamma = 0.8$, if E^t decrease steadily during the recent epochs and equals to $\eta = \eta\gamma$, $\gamma = 0.8$ otherwise.

- Momentum α :

  The momentum term is used to prevent the training algorithm from settling in local minimum. It also increases the speed of convergence. It is usually set to a positive value less than 1. Similarly α can also be adaptive during the training through the learning rate adaptation γ.

- Training tolerance ϵ :

  The critical learning parameter determines the accuracy of the neural network outputs. A small training tolerance usually increases learning accuracy but can result in less generalization capability as well as longer training time.

- Learning rate adaptation γ :

  An adaptive learning rate decreases training time by keeping the learning rate reasonably high while insuring stability.

5) Model Implementation and Data Generation

The m-input parameters of the neural network could be physical/geometrical parameters of a given device or circuit. The n-output parameters represent various responses of the device or the circuit under consideration. The neural network model is capable of mapping the relationship between the set of parameters defining the physical configuration of a system and its operational characteristics, and the set of parameters that can be used to analyze the signal integrity of the system.

To train and validate the neural network, two sets of data are required, the training set and test set. The neural network is first trained off-line using training samples of input-output data, and is then applied to simulation.

The training set data are obtained through repeated off-line simulation using an accurate simulation technique, such as an electrical CAD tool or an EM-field solver. The simulator is repeatedly called, each time with input variables randomly chosen from the input space. The training set can also be obtained as a collection of data from actual measurements, or from a look-up table if available. The number of data points needed in the training set is important to the training of the network. While training proceeds, if it is found that there is insufficient error convergence, it is indicative that the training set is probably not large enough, and should be made larger. Though a large training set would give a better representation of the Input/Output space to be modeled, too large a training set would result in a needless increase in the time invested in data generation and training.

The test set, obtained in an identical manner, should be large enough to be representative of the entire input space, and its contents should be different from those of the training set. The test set is used solely for the purpose of testing the accuracy of the model during and after training. If the test error is significantly higher than the training error, it means that the training set is not large enough, or the neural network is over-trained due to too many hidden neurons which lead to too much freedom in the I/O relationship represented by the neural network, where it tends to overfit the training data instead of generalizing.

5.5.2 Hybrid of AWE and ANN

The training is the most important step in the development of ANNs. The actual training process involves algorithms for finding values of weights associated with various neurons. This process can be viewed as one of optimization. Various well-known optimization techniques, such as genetic algorithm can be used for this purpose. This process is quite time consuming. For example, to train an ANN which is available over a wide frequency band, the computation should be carried out repeatedly at different frequencies. To overcome this difficulty, the asymptotic waveform evaluation (AWE) previously introduced may be used. This technique extrapolates the data from one point to a certain range based on the value and the high order derivatives at this point. This technique is computationally efficient due to involving the analytical relationships and is available to cases where the derivatives may be obtained. The 2-D AWE has been used to characterize the microstrip antenna in Section 5.3.2. such that the response over certain frequency and permittivity ranges can be extrapolated from single point simultaneously. In this case, the variables in the model are frequency, relative permittivity, position of feed line and the dimension of the patch. In the multilayer perceptrons (MLP) neural network shown in Figure 5.23, the top layer is the output layer and the input impedance and other scattering parameters can be obtained; the bottom layer is the input layer where frequency, relative permittivity, position of feed line and the dimension of patch are inputted. The other two layers are hidden layers, and they can be automatically treated in the software[36].

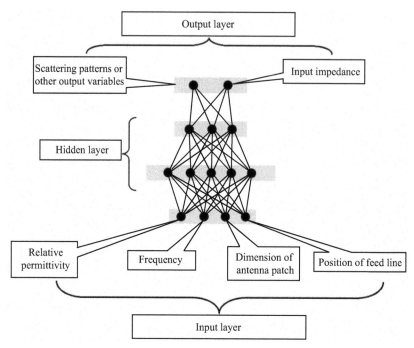

Figure 5.23   Multilayer perceptrons (MLP) structure.

In AWE, the differentiation operates on the Green's function, which does not involve the dimensions of the object to be analyzed. Therefore it is not able to obtain the response with respect to the dimensions through AWE. Although size dimensional AWE has been proposed[17, 18], the fixed mesh causes the variation of the density of the discretization and consequently limits its application to big changes of the size. Alternatively, in the following, the sampling data for training, varying with the dimension of the microstrip patch antenna and the position of the feed line, are calculated point by point. Even in this case, the speed of training is about one or two orders faster than that of direct training. With the two-dimensional AWE method and neural network technique in hand, we can accurately and efficiently construct the neural network model[37]. The flowchart is shown in Figure 5.24.

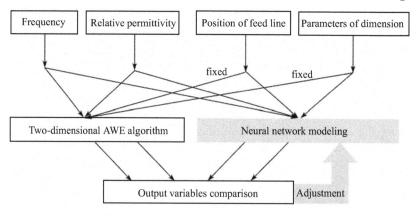

Figure 5.24   Process of the hybrid technique.

As the neural network model is constructed, the response of an object varying with each parameter can be immediately obtained. This trained model may be used in the optimization of microstrip structures other than microstrip antennas.

An example is a microstrip antenna consisting of a conducting patch residing on a dielectric substrate having thickness $h = 0.787$mm, as is shown in Figure 5.25. The increments of frequency, relative permittivity, position of feed line H and dimension of patch L are 0.01GHz, 0.01mm, 0.05mm and 0.1mm respectively. In order to get the response under the following specification: the frequency varies from 7.0GHz to 8.0GHz; the permittivity varies from 1.8 to 2.8; the dimension L varies from 12.1mm to 13.0mm; the feed line position H varies from 8.79mm to 3.79mm, the direct method requires 1,152,000,000 seconds to obtain the solution on a Personal Computer (1.2GHz AMD^{TM} K7 processor). To obtain the same accuracy, with a general neural network algorithm, including the training time, 576,070 seconds are needed. But with the hybrid method, only 34,630 seconds are needed, which is 33,266 times faster than the direct method and 16.6 times faster than the general neural network method (Table 5.4). The window of software "Neuralmodeler" for training is shown in Figure 5.26 and the final error is less than 0.01. Figure 5.27 and Figure 5.28 show the real and imaginary parts of the input impedance as a function of frequency, relative permittivity, for the given dimension L and position H by using the hybrid method of the two-dimensional

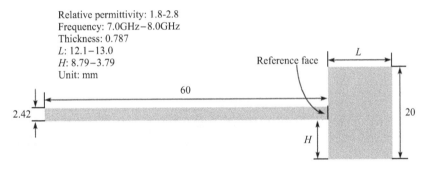

Figure 5.25   Antenna geometry.

Table 5.4   The comparison of neural network method and direct method.

Time		Neural network method		Direct method
		Hybrid method	General method	
Training	Sampling	34,560s	576,000s	No training
	Modeling	70s	70s	
Generating response		Almost zero	Almost zero	1,152,000,000s
Total time		34,630s	576,070s	1,152,000,000s

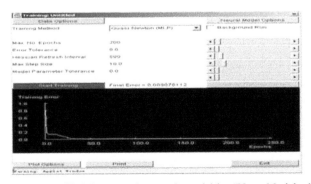

Figure 5.26   Training neural network model by "NeuroModeler."

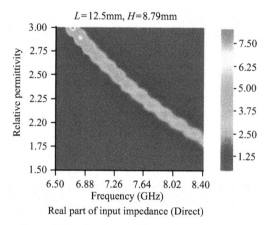

Figure 5.27   Real part of input impedance.

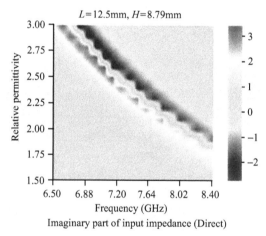

Figure 5.28   Imaginary part of input impedance.

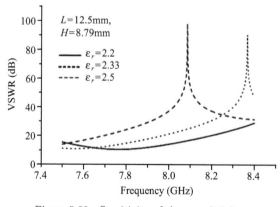

Figure 5.29   Sensitivity of the permittivity.

AWE method and neural network algorithm, respectively. Due to difficulty in presenting a four dimensional figure, the variables, dimension L and position H, are fixed. When the neural network model with four variables for this antenna patch is obtained, optimizing and designing will be an easy matter. We can use it to select the substrate of a different dielectric constant, and also to optimize the bandwidth by adjusting the dimension of the patch and position of the feed line. Because of its computational efficiency, the optimization and the observation of the sensitivity to the parameters such as the dielectric constant are realizable, as shown in Figure 5.29. In this example, only four variables are involved. The larger the number of the variables to be optimized, the larger the reduction of the computer time.

5.5.3   Hybrid of SM and ANN

In ANN, the learning data is usually obtained by either EM simulation or by measurement. Large amounts of learning data are typically needed to ensure model accuracy. Without sufficient learning samples, the neural models may not be reliable. However, as we have seen, this is very computation expensive. Hybridization of SM and ANN has been proposed as another choice to solve this problem. The fundamental idea is to construct a nonlinear multidimensional vector mapping function P from fine to coarse input space using an ANN. In so doing, the nonlinearity originally involved in ANN is partly transferred to the coarse model. The implicit knowledge in the coarse model not only allows to decrease the number of learning points needed, but also results in the reduction of the complexity and the improvement of the performance of the ANN. All of them make this approach more efficient with respect to the traditional neuro-modeling approach and it is called the space-mapped neuro-modeling (SMN)[38].

In the SMN approach, mapping from the fine to the coarse parameter space is implemented by an ANN. Figure 5.30 illustrates the concept of SMN. The key step in SMN is to find the optimal set of internal parameters of the ANN, in order that the coarse model response is as close as possible to the fine model response for all the learning points.

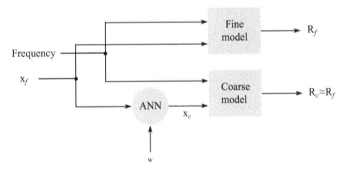

Figure 5.30   SMN concept.

The mapping can be found by solving the optimization problem

$$\min_{\mathbf{w}} \left\| \left[\mathbf{e}_1^T \mathbf{e}_2^T \cdots \mathbf{e}_l^T \right]^T \right\| \tag{5.5.17}$$

where vector $\mathbf{w}$ contains the internal parameters of the neural network (weights, bias, etc.) selected as optimization variables, l is the total number of learning samples, and $\mathbf{e}_k$ is the error vector given by

$$\mathbf{e}_k = \mathbf{R}_f(\mathbf{x}_{f_i}, freq_j) - \mathbf{R}_c(\mathbf{x}_c, freq_j) \tag{5.5.18}$$
$$\mathbf{x}_c = \mathbf{P}(\mathbf{x}_{fi}) \tag{5.5.19}$$

with

$$i = 1, \cdots, B_{p_{tr}} \tag{5.5.20-a}$$

$$j = 1, \cdots, F_p \qquad (5.5.20\text{-b})$$
$$k = j + F_p(i-1) \qquad (5.5.20\text{-c})$$

where $B_{p_{tr}}$ is the number of training base points for the input design parameters and F_p is the number of frequency points per frequency sweep. It is seen that the number of learning samples is $l = B_{p_{tr}} F_p$. Similarly, if the number of test base points is $B_{p_{te}}$, then the number of testing samples is $t = B_{p_{te}} F_p$. The specific characteristics of P depend on the ANN paradigm chosen whose internal parameters are in **w**.

Once the mapping is found, that is, once the ANN is trained, a space-mapped neuro-model for fast accurate evaluations is immediately available.

A rectangular microstrip patch antenna as shown in Figure 5.31 is taken as an example of using SMN. The height of the substrate is 0.5mm, and the dielectric constant is 2.7. The input parameters of the ANN include the operation frequency of the antenna, the width of the patch W and the location of the feedline X. The center frequency of the model is 16GHz. The frequency bandwidth is from 15.8GHz to 16.2GHz. The coarse grid ($\lambda_g/10$) and fine grid ($\lambda_g/30$) of IE3D^{TM} simulator are used as the coarse model and fine model respectively. The output of the model is S_{11}. The neuro-modeling can be obtained by the following steps.

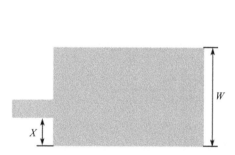

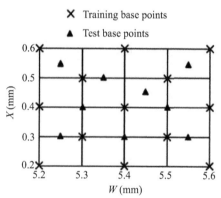

Figure 5.31   Microstrip patch antenna.   Figure 5.32   Location of the sampling points.

1. Determine the initial values of the neuro-modeling by using the coarse model optimizer. In this example, $W = 5.4101$mm and $X = 0.4064$mm.

2. Define the valid values in Table 5.5 of all parameters.

Table 5.5   Valid range of the input parameters.

	Initial value	Valid value
Frequency(f)(GHz)	16	15.8-16.2
Width(W)(mm)	5.41019	5.21019-5.61019
Feedline location(X)(mm)	0.4064	0.2064-0.6064

3. Select the sampling points and calculate the EM simulation at the sampling points. In this neuro-modeling, 13 training base points and 9 test base points are used and the frequency step is 0.02GHz. In other words, there are 13×21=273 learning points and 9×21=189 test points to be simulated using fine model. The training base points and test base points are selected as shown in Figure 5.32.

4. Find the coarse model parameters, for which the responses of the coarse model are closest to those of the fine model at the corresponding sampling points. First, assuming that the coarse model point has the same frequency as the fine model point, the optimal physical dimensions of the coarse model are obtained. At this point, the responses of all sampling frequencies have the least square error, and one-dimension scanning is then used to find the mapped frequency.

5. Train the neural network. Error of one input vector is used for updating the weights and biases. S_{11} and the error in S_{11} at the training points and test points are shown in Figures 5.33 and Figure 5.34. The average errors of training data and test data are correspondingly about 0.01 and 0.08. Assuming the maximal error vector of the ANN is $\mathbf{y}_m$, the maximal error in S_{11} caused by $\mathbf{y}_m$ is in the range of $[0.01, 0.1]$, so the error of the whole model would also be in this range. A better neural network algorithm, which has a smaller error vector $\mathbf{y}_m$, will surely have less error than the whole neuro-modeling. It can be seen that the accuracy of ANN and the sensitivity of the coarse model are important factors for the performance of the neuro-modeling. In theory, any coarse model can be used in this process. If the coarse model is the same as the fine model, the ANN mapping is a unit one; if the coarse model is a unit mapping, the neuro-modeling is just a traditional one.

6. Test the mapping $\mathbf{P}$ between parameters and their fine model responses.

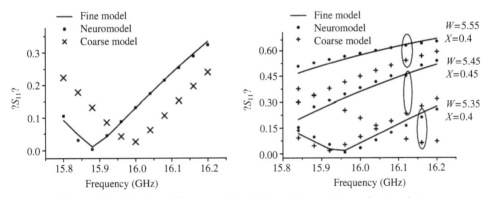

Figure 5.33 $|S_{11}|$ for different models at the training points and test points.

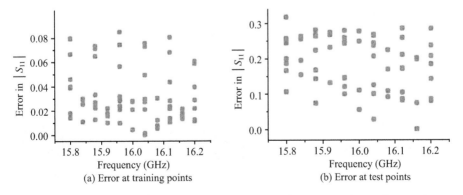

Figure 5.34 Error in $|S_{11}|$ of different models at several training points and test points.

5.5.4   Hybrid of SM/ANN and Adaptive Frequency Sampling (AFS)

In Section 5.5.3, a large quantity of frequency samples is always required in the SM technique to find the optimized solution in the coarse model. The S-B AFS technique introduced in [19] and combined with the SM-based neuro-modeling may be used to eliminate the computation cost for a large number of frequency sweeps[39]. Numerical results confirm the validity of this approach in both reducing the sampling time and in increasing the accuracy of the NN. In this scheme, a fine model is sampled at a set of base points $\mathbf{x}_f$ at frequency points $f \in [f_{\min}, f_{\max}]$ using S-B AFS. For the responses of the fine models $\mathbf{R}_f(\mathbf{x}_f, f_1, \cdots, f_n)$, curve fittings are carried out in the coarse model to satisfy $\mathbf{R}_c(\mathbf{x}_c, f_1', \cdots, f_n') \approx \mathbf{R}_f(\mathbf{x}_f, f_1, \cdots, f_n)$, where the frequency of the fine model may be projected to another band in order to get a better alliance between the responses of the two models. Enough frequency points in sampling ensure the consistency of the fitting solution, which is important in the later NN training. In NN, $\mathbf{x}_c$ and $\mathbf{x}_f$ are used as the input and output data, respectively. With a network trained, the response of the fine EM model can be well approximated by the responses of the coarse model at the mapped new points.

In order to verify this technique, an edge-fed microstrip antenna, shown in Figure 5.35, is investigated. In this example, we use a package Zeland IE3D^{TM} as a fine model, and the cavity model of a rectangular microstrip antenna as a coarse model. The parameters of the antenna are illustrated in Figure 5.35. The samples distributed in the fine model are equally spaced, as shown in Figure 5.36. The VSWRs of the antenna in the fine model are shown in Figure 5.37. The curve in Figure 5.37 is interpolated with 5 or 6 samples using S-B AFS method. If equal space sampling is used, in order to get the response with the same accuracy, 31 samples are required in each frequency sweep. That is, 83% of the sampling time is saved. The NN training is conducted by a package called "Neuromodeler." A back-propagation

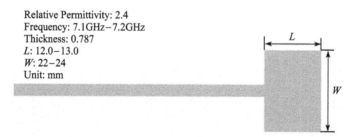

Relative Permittivity: 2.4
Frequency: 7.1GHz–7.2GHz
Thickness: 0.787
L: 12.0–13.0
W: 22–24
Unit: mm

Figure 5.35   Center edge-fed microstrip antenna.

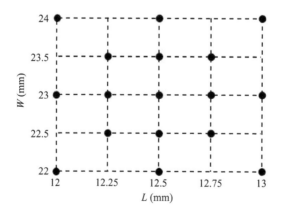

Figure 5.36   Location of the samples.

algorithm is used to find a neural network with three layers; one input layer, one output layer, and one hidden layer. The errors of the samples after the NN training are shown in Figure 5.38. The number of neurons on each layer is listed in Figure 5.39. The model comprising the NN and coarse model is tested at two test points. A good agreement can be observed in Figure 5.40. If the response of the circuit vibrates rapidly in the interested frequency band, the time saved will be more predominant. The S-B AFS may also be used in frequency space-mapped neuro-modeling (FSMN) to decrease the cost for frequency sweeps in sampling[38, 40].

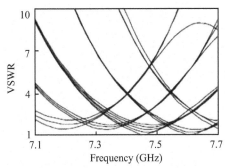

Figure 5.37   VSWR of the antenna
at all samples.

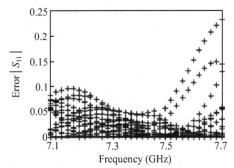

Figure 5.38   Error distribution
of coarse model.

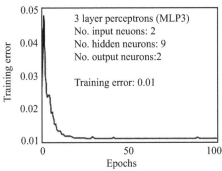

Figure 5.39   The curve of error convergence
using software Neuromodeler.

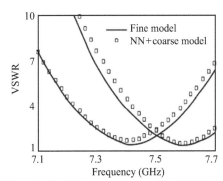

Figure 5.40   The comparison of the VSWR
of the fine model and NN+coarse model.

5.6   Summary

Some effective methods introduced in this chapter are quite general, although only examples of certain specific applications of microstrip antennas are presented. These methods may be used flexibly, alone or in hybridization, in order to make full use of the advantages of each. There is still a lot of research work to be done with these aspects, such as to investigate the capability of extrapolation of different models and information types; to exploit the research achievements from areas of Numerical Approximation, System Identification, Signal Processing and so on; and to apply the extra port method to the problems other than the linear dipole. In computational electromagnetics, accuracy and efficiency are always a trade-off. How to achieve a good balance is truly an engineering art. These applications in designing microstrip antenna arrays will be discussed in the following chapter.

Bibliography

[1] W. C. Chew, J. M. Jin, E. Michielssen, J. M. Song, Edited, *Fast and Efficient Algorithms in Computational Electromagnetics*, Artech House, 2001.

[2] Y. S. Chung, T. K. Sarkar, B. H. Jung and M. S-Palma, "An unconditionally stable scheme for the finite-difference time-domain method," *IEEE Trans. Microwave Theory Tech.*, vol. 51, no. 3, pp. 697–704, Mar., 2003.

[3] R. S. Chen, D. G. Fang, K. F. Tsang and Edward K. N. Yung, "Analysis of millimeter wave scattering by an electrically large metallic grating using wavelet-based algebraic multigrid preconditional CG method," *Inter. J. of Infrared and Millimeter Waves*, vol. 21, no. 9, pp. 1541–1559, Sept., 2000.

[4] Y. Xu, D. G. Fang, M. Y. Xia and C. H. Chan, "Speedy computation of the time-domain Green's function for microstrip structures," *Electronics Letters*, vol. 36, no. 22, pp. 1855–1857, Oct., 2000.

[5] J. W. Bandler, R. M. Biernacki, S. H. Chen, P. A. Grobelny and R. H. Hemmers, "Space mapping technique for electromagnetic optimization," *IEEE Trans. Microwave Theory Tech.*, vol. 42, no. 12, pp. 2536–2544, Dec., 1994.

[6] J. W. Bandler, R. M. Biernacki, S. H. Chen, R. H. Hemmers and K. Madsen, "Electromagnetic optimization exploiting aggressive space mapping," *IEEE Trans. Microwave Theory Tech.*, vol. 43, no. 12, pp. x22874–2882, Dec., 1995.

[7] K. L. Wu, R. Zhang, M. Ehlert and D. G. Fang, "An explicit knowledge-embedded space mapping technique and its application to optimization of LTCC RF passive circuits," *IEEE Trans. Components and Packaging Tech.*, vol. 26, pp. 399–406, 2003.

[8] R. Zhang, D. G. Fang and K. L. Wu, "Modeling of microstrip patch antenna in space mapping technique," *Cross Strait Tri-Regional Radio Science and Wireless Technology Conference*, Dec. 2000, Hong Kong, pp. 65–68.

[9] R. Zhang, *The Optimization Design of Microstrip Antennas*, Master Thesis, Nanjing University of Science and Technology, 2001.

[10] J. W. Bandler, N. Georgieva, M. A. Ismail, J. E. Rayas-Sanchez, and Q. J. Zhang, "A generalized space-mapping tableau approach to device modeling," *IEEE Trans. Microwave Theory Tech.*, vol. 49, no. 1, pp. 67–79, Jan., 2001.

[11] N. N. Feng and W. P. Huang, "Modeling and simulation of photonic devices by generalized space mapping technique," *J. of Lightwave Technology*, vol. 6, no. 36, pp. 1562–1567, Jun., 2003.

[12] J. W. Bandler et al, "Space mapping: the state of the art," *IEEE Trans. Microwave Theory Tech.*, vol. 52, no. 1, pp. 337–361, Jan., 2004.

[13] R. F. Harrington, *Field Computation by Moment Method*, The Macmillan Company, 1968.

[14] Y. Xiong, *The Interpolation and the Extrapolation in Computational Electromagnetics*, Master Thesis, Nanjing University of Science and Technology, 2003.

[15] A. Hoorfar, "Simple closed-form expressions for microstrip Green's functions in a magneto-dielectric substrate," *Mircrowave Opt. Tech. Letters*, vol. 8, no. 1, pp. 33–36, Jan., 1995.

[16] Y. Xiong, D. G. Fang and F. Ling, "Two-dimensional AWE technique in fast calculation of microstrip antennas," *Inter. Conference on Microwave and Millimeter Wave Technology Proceeding*, Aug. 2002, Beijing, pp. 393–396.

[17] G. B. Han, D. G. Fang, Y. Xiong and Y. Ding, "Size dimensional asymptotic waveform evaluation in electrostatic problems," *Chinese Journal of Radio Science*, vol. 19, no. 1, pp. 13–16, Jan., 2004.

[18] G. B. Han, D. G. Fang and W. M. Yu, "The application of size dimensional AWE in the dynamic electromagnetic problems," *Digest of National Conference on Microwave and Millimeter Wave Technology*, pp. 146–149, 2003.

[19] Y. Ding, K. L. Wu and D. G. Fang, "A broad-band adaptive-frequency-sampling approach for microwave-circuit EM simulation exploiting Stoer-Bulersch algorithm," *IEEE Trans. Microwave Theory Tech.*, vol. 51, no. 3, pp. 928–934, Mar., 2003.

[20] E. K. Miller, "Model-based parameter estimation in electromagnetics Pt.1," *IEEE Antennas & Propagation Magazine*, vol. 40, no. 1, pp. 42–52, Feb., 1998.

[21] E. K. Miller, "Model-based parameter estimation in electromagnetics Pt.2," *IEEE Antennas & Propagation Magazine*, vol. 40, no. 2, pp. 51–65, Apr., 1998.

[22] E. K. Miller, "Model-based parameter estimation in electromagnetics Pt.3," *IEEE Antennas & Propagation Magazine*, vol. 40, no. 3, pp. 49–66, Jun., 1998.

[23] Y. L. Chow and M. M. A. Salama, "A simplified method for calculation of the substation grounding resistance," *IEEE Trans. Power Delivery*, vol. 9, no. 4, pp. 736–742, Apr., 1994.

[24] Y. X. Sun, Y. L. Chow and D. G. Fang, "Impedance formulas of RF patch resonators and antennas of cavity model using fringe extensions of patches from DC capacitors," *Microwave and Optical Technology Letters*, vol. 35, no. 4, pp. 293–297, Nov., 2002.

[25] Y. L. Chow N. N. Feng and D. G. Fang, "A simple method for ohmic loss in conductors with cross-section dimensions on the order of skin depth," *Microwave and Optical Technology Letters*, vol. 20, no. 5, pp. 302–304, Mar., 1999.

[26] M. C. Bailey, "Technique for extension of small antenna mutual coupling data to larger antenna arrays," *NASA Technical Paper 3603*, Langley Center, Hampton VA., Aug., 1996.

[27] M. C. Bailey, "Closed-form evaluation of mutual coupling in a planar array of circular apertures," *NASA Technical Paper*, 3552, Langley Center, Hampton VA., Apr., 1996.

[28] Y. X. Sun, *CAD Formulas for Microstrip Antenna and Its Arrays*, Ph.D Dissertation, Nanjing University of Science and Technology, 2002.

[29] Y. X. Sun, Y. L. Chow, and D. G. Fang, "Mutual impedance formula between patch antennas based on synthetic asymptote and variable separation," *Microwave and Optical Technology Letters*, vol. 36, no. 1, pp. 48–53, Jan., 2003.

[30] W. R. Smythe, *Static and Dynamic Electricity* (third edition), McGraw-Hill Book Company, New York, 1968, p. 145, pp. 159–161, pp. 252–253.

[31] Y. L. Chow and W. C. Tang, "3D Green's functions of microstrip separated into simpler terms—behavior, mutual interaction and formulas of the terms," *IEEE Trans. Microwave Theory Tech.*, vol. 49, no. 8, pp. 1483–1491, Aug., 2001.

[32] R. P. Jedlicka, M. T. Poe and K. R. Carver, "Measured mutual coupling between microstrip antennas," *IEEE Trans. Antennas Propagat.*, vol. 29, no. 1, pp. 147–149, Jan., 1981.

[33] S. G. Pan and I. Wolff, "Computation of mutual coupling between slot-coupled microstrip patches in a finite array," *IEEE Trans. Antennas Propagat.*, vol. 40, no. 9, pp. 1047–1053, Sept., 1992.

[34] Q. J. Zhang, K. C. Gupta and V. K. Devabhaktuni, "Artificial neural networks for RF and microwave design-from theory to practice," *IEEE Trans. Microwave Theory Tech.*, vol. 51, no. 4, pp. 1339–1350, Apr., 2003.

[35] Q. J. Zhang and K. C. Gupta, *Neural Networks for RF and Microwave Design*, Artech House, 2000.

[36] Q. J. Zhang and his research team, software "NeuralModeler," Version 1.2.2 for Windows NT 4.0.

[37] Y. Xiong, D. G. Fang and R. S. Chen, "Application of 2-D AWE algorithm in training multidimensional neural network model," *Applied Computational Electromagnetics Society Journal*, vol. 18, no. 2, pp. 64–71, Jul., 2003.

[38] J. W. Bandler, A. Ismail, J. E. R. Sanchez, and Q. J. Zhang, "Neuromodeling of microwave circuits exploiting space-mapping technology," *IEEE Trans. Microwave Theory Tech.*, vol. 47, no. 12, pp. 2417–2427, Dec., 1999.

[39] Y. Ding and D. G. Fang, "Accelerated SM-based neuro-modeling exploiting S-B AFS technique," *Chinese J. of Microwaves*, vol. 21, no. 4, pp. 1–5, 2005.

[40] X. J. Zhang, D. G. Fang and Y. Ding, "Frequency space-mapped neuro-modeling technique exploiting S-B AFS for the design of microwave circuits," *Digest of APMC* 2005, pp. 2863–2865.

[41] H. Wang, D. G. Fang, B. Chen, X. K. Tang, Y. L. Chow, and Y. P. Xi, "An effective analysis method for electrically large finite microstrip antenna arrays," *IEEE Trans. Antennas Propagat.*, vol. 57, no. 1, pp. 94–101, 2009.

Problems

5.1 Check the data in Table 5.1.

5.2 Check the data in Table 5.2.

5.3 Check the data in Table 5.3.

5.4 Check the results in Figure 5.4 and Figure 5.5.

5.5 Prove (5.3.3)–(5.3.7).

5.6 Create a mutual coupling impedance formula for two microstrip patch antennas assuming that the term with $1/r^3$ distance dependence dominates, and then check with the result from a full wave analysis solver.

5.7 Repeat Prob. 5.6 assuming $1/r$ distance dependence.

5.8 Repeat Prob. 5.6 assuming $1/\sqrt{r}$ distance dependence.

5.9   1. Rewrite (5.3.5) into the following form

$$I(k)D(k) = N(k) \tag{5.6.1}$$

where

$$I(k) = m_0 + m_1(k - k_0) + m_2(k - k_0)^2 + \cdots$$
$$m_n = \frac{I^{(n)}|_{k=k_0}}{n!}$$
$$D(k) = 1 + b_1(k - k_0) + b_2(k - k_0)^2 + \cdots$$
$$N(k) = a_0 + a_1(k - k_0) + a_2(k - k_0)^2 + \cdots$$

Upon differentiating with respect to k, this leads to (omitting the explicit k dependence)

$$I'D + ID' = N'$$
$$I''D + 2I'D' + ID'' = N''$$
$$I'''D + 3I''D' + 3I'D'' + ID''' = N'''$$
$$\cdots\cdots\cdots\cdots\cdots \tag{5.6.2}$$

Matching the constant terms of both sides in (5.6.1) and (5.6.2) results in

$$m_0 = a_0$$
$$m_1 = a_1 - m_0 b_1$$
$$m_2 = a_2 - m_1 b_1 - m_0 b_2$$
$$\cdots\cdots\cdots\cdots\cdots \tag{5.6.3}$$

Prove that (5.6.3) may yield (5.3.6) and (5.3.7).

2. For given function $\ln(1 + k)$, $k_0 = 0$, prove that the rational Padé function is

$$\ln(1 + k) \approx \frac{k^3 + 21k^2 + 30k}{9k^2 + 36k + 30}$$

3. Compare the behavior of the Padé function and the Taylor expansion of $\ln(1 + k)$ (See reference [20]).

CHAPTER 6

Design of Conventional and DBF Microstrip
Antenna Arrays

6.1 Introduction

In certain applications, system requirements can be met with a single microstrip element. However, as in the case of conventional microwave antennas in communication and radar, the realization of some requirements such as high gain, beam scanning and difference pattern in mono-pulse radar is possible only when discrete radiators are arrayed to form a linear, planar or volume configuration depending on the intended application.

Due to several attractive advantages of a microstrip antenna array such as low weight, low profile with conformability and low manufacturing cost, in many military, space, and commercial applications microstrip arrays are going to replace the conventional high-gain antennas, for example the array of horns, helices, slotted waveguides, or parabolic reflectors. However, advantages of the microstrip arrays will still be offset by three inherent drawbacks: small bandwidth, relatively high feed line loss, and low power-handling capability. To minimize these effects, accurate analysis techniques, optimum design methods and innovative array concepts are imperative to the successful development of a microstrip array antenna.

All the analysis techniques available and the user-friendly software developed based upon them give engineers powerful tools in the design of antenna and arrays. However, any technique or software cannot, by itself, generate an array design. The basic design configuration has to originate from human experience, knowledge, and innovation. In this chapter, the design methodology of the normal microstrip antenna arrays is discussed.

For the normal microstrip antenna arrays, a module based full wave analysis approach is utilized in which a given antenna is used as a building block for a higher gain antenna[1]. In this case, the mutual coupling effects are partly involved for the consideration of both high analysis efficiency and accuracy. For the large phased arrays or ultra low side lobe level arrays, high analysis accuracy is required. To completely take into account the mutual coupling effect in a finite array environment as well as to maintain the relatively high efficiency, the element-by-element based full wave analysis approach may be used. In this method, a N-element array is represented by a multi-port network; each element corresponds to a port. The interaction between them is included through the entire domain basis function in method of moments (MoM)[2–4]. For the middle size arrays, the integrated based full wave analysis approach as is described in [5] and [6] can be used when it is necessary. Many commercial software packages such as IE3D^{TM}[7] are available for this purpose.

All the design methods described above are for the conventional microstrip antenna arrays and each element is weighted by the radio frequency (RF) feed network. As we have described in Chapter 2, in a digital beamforming (DBF) antenna system, the received signals are detected and digitized at the element level. Digital beamforming is based on capturing the RF signals at each of the antenna elements and converting them into baseband signals. The beamforming is carried out by weighting these digital signals, thereby adjusting their amplitude and phase in such a way that when added together they form the desired beam. The DBF antenna system is used to produce simultaneous independently steerable multi-beams. Each beam could also be adaptively nulled, to produce nulls at the direction of interferences. Therefore, in the design of a DBF antenna array, both the uniformity of the elements and the mutual couplings have to be carefully considered. The mutual coupling reduction and the prediction of the adaptive nulling performance are important. One way of reducing the mutual coupling is to put an electromagnetic band gap (EBG) structure or absorbing material between elements. The DBF microstrip antenna array takes advantage of doing that, because it is exempted from the RF feed network, providing plenty of space. These factors are the features needed in designing the DBF microstrip antenna array and will be discussed in this chapter.

6.2   Feeding Architecture

It is the function of the feed to provide correct element excitation and an impedance match at its input. A wide variety of feed architectures is available. It is necessary to review them to grasp the significance and limitations of the feeds adapted for microstrip arrays. The feed systems discussed can be either co-planar with the radiating elements, or situated in a separate transmission-line layer. In the present discussion, only the most important types that often challenge the skills of the antenna designers are covered. Series or parallel or their combinations are the basic forms for feed networks and will be introduced in detail.

6.2.1   Series Feed

In a series feed configuration, multiple elements are arranged linearly and fed serially by a single transmission line. Figure 6.1 (a) and Figure 6.1 (b) illustrate two different configurations of the series feed method: in-line feed (series-connected series feed)[8, 9] and out-of-line feed (shunt-connected series feed)[10]. The in-line feed array occupies the smallest space with the lowest insertion loss, but generally has the least polarization control and the narrowest bandwidth.

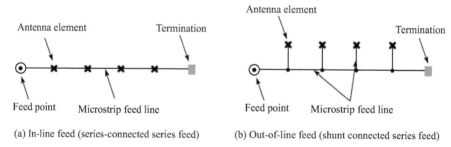

(a) In-line feed (series-connected series feed)   (b) Out-of-line feed (shunt connected series feed)

Figure 6.1   Series-fed linear arrays.

Two-dimensional series-fed arrays are also possible. A complete end-fed array is shown in Figure 6.2 (a) and a complete center-fed array is shown in Figure 6.2 (b). The hybrid of end-fed and center-fed may also be formed.

The series-fed array is classified into two types: a resonant array if the termination is an open- or short-circuit and a traveling-wave array if the feed line is terminated with a matched load[11]. In a resonant array, the spacing between two elements is one wavelength on the line, so any reflected power still produces a broadside beam. For the same reason, the bandwidth of a resonant array is very narrow. With a slight frequency shift, the one-wavelength spacing no longer holds, thereby causing the multiple bounced waves to travel back into the input port as mismatched energy. For the traveling-wave arrays, it is possible to radiate all of the power input to the antenna. The termination is a matched load, to absorb any power not radiated. The traveling-wave array has a wider impedance bandwidth. In addition, its main beam scans as frequency changes. Both in-line and out-of-line feed arrays can be designed to be of the resonant type or the traveling-wave type.

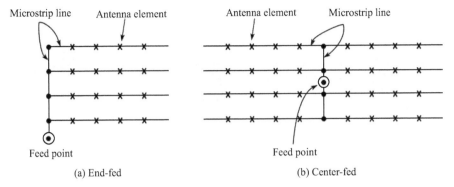

(a) End-fed   (b) Center-fed

Figure 6.2   Series-fed planar array.

6.2.2   Parallel Feed

The parallel feed is also called the corporate feed. Two basic forms of the parallel fed linear arrays are shown in Figure 6.3 (a), Figure 6.3(b), which shows that the power is equally split at each junction; however, different power divider ratios can be chosen to generate a tapered distribution across the array. If the elements are fed by a power divider with identical path lengths from the feed point to each element, the beam position is independent of the frequency and the feed is broadband. By incorporating proper progressive phase shifters or line extensions, the beam direction can be controlled.

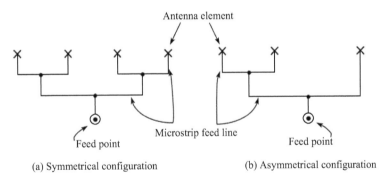

(a) Symmetrical configuration   (b) Asymmetrical configuration

Figure 6.3   Basic linear parallel feed network.

A one-dimensional parallel feed can be arranged to form a two-dimensional one as shown in Figure 6.4. The basic sub-array configuration can be extended to larger arrays with specifically 2^n elements per side.

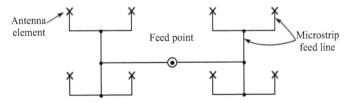

Figure 6.4   Basic planar parallel feed.

Corporate-fed arrays are fairly straightforward to design, especially for uniform distribution. By using the module based full wave analysis approach, different sized arrays can be assembled quite quickly. One of the drawbacks of corporate feeds is that they take up considerable room. As the array size increases, so does the line length to any element. The losses become larger, thereby reducing antenna efficiency. Also, feed line radiation can become a problem. Not only is there some spurious radiation from the straight sections, but also more importantly, every bend and T-junction are sources of spurious radiation. The main effects of the spurious radiation are increased side lobe level and reduced gain.

6.2.3   Hybrid Series/Parallel Feed

One- or two-dimensional arrays can have combinations of parallel and series feeds. These are sometimes referred to as hybrid feeds. Figure 6.5 presents some examples. The hybrid-fed array achieves a wider bandwidth than does a purely series-fed array having the same aperture size. Of course, having partial parallel feed, the insertion loss of a hybrid array is higher than that of a purely series-fed array. This hybrid technique gives the designer a chance to make design trade-offs between bandwidth and insertion loss.

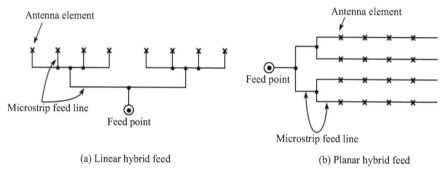

(a) Linear hybrid feed   (b) Planar hybrid feed

Figure 6.5   Hybrid-fed arrays with a combination paralled/series feed.

6.2.4   Single-Layer or Multilayer Design and Other Considerations[12]

A microstrip array can be designed in either a single-layer or multilayer configuration. The factors that determine the choice are complexity and cost, side lobe/cross-pol level, number of discrete components, polarization diversity, and bandwidth. When the given electrical requirements are more relaxed, a single-layer design will generally suffice. Because all the transmission lines and patch elements are etched on the same layer, single layer design has the advantage of lower manufacturing cost. However, when extremely low side lobe level or cross-pol radiation (e.g., less than −30dB) is required, the double-layer design seems to be a better choice. With all the transmission lines etched on the second layer behind the radiating patch layer, the double layer's ground plane will shield most of the spurious radiation of the lines.

It is often desirable to design a microstrip array with larger element spacing so that more space can be made available for transmission lines and discrete components. However, to avoid the formation of high grating lobes, element spacing is limited to less than one free-space wavelength for broadside beam design and less than 0.6 free-space wavelength for a wide-angle scanned beam. In designing a wide-angle scanned microstrip phased array, substrate thickness, dielectric constant, and element spacing are all important parameters that need to be considered for reducing mutual coupling effects and avoiding scan blindness[13].

6.3   Design of Power Divider and Transmission on the Transformer

To implement the parallel feed, the power dividers are indispensable. There is no general guideline for the optimized integrated design of a feed network based on bandwidth or other criterion. Usually a good design is considered as having well-matched lines at every stage, such that the feed network will suffer less from mismatch losses and radiation leakage losses. In this section, several typical power dividers or couplers are introduced and the transmission line design formulas are given[11, 12, 14] and are shown in Figures 6.6–6.13. Although the results from the formulas are not accurate enough, they may serve as an initial guess in the optimization by using a full wave solver or the coarse model in space mapping technique. The

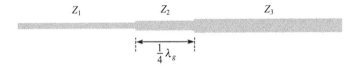

Figure 6.6   Microstrip transmission line quarter-wave impedance transformer.

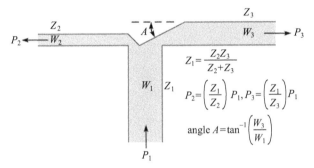

Figure 6.7   Microstrip two-way power divider.

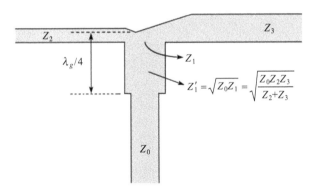

Figure 6.8   Microstrip two-way power divider with a quarter-wave transformer.

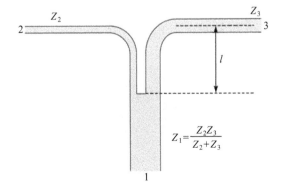

Figure 6.9   In-line unequal power divider.

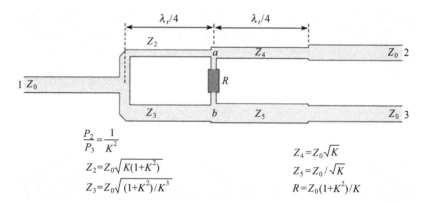

$$\frac{P_2}{P_3}=\frac{1}{K^2}$$

$$Z_2=Z_0\sqrt{K(1+K^2)}$$

$$Z_3=Z_0\sqrt{(1+K^2)/K^3}$$

$$Z_4=Z_0\sqrt{K}$$

$$Z_5=Z_0/\sqrt{K}$$

$$R=Z_0(1+K^2)/K$$

Figure 6.10   Branch type isolated in-line (Wilkinson) power divider.

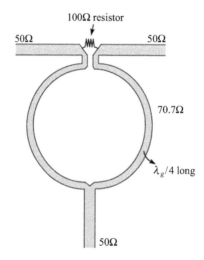

Figure 6.11   Ring type isolated in-line (Wilkinson) power divider.

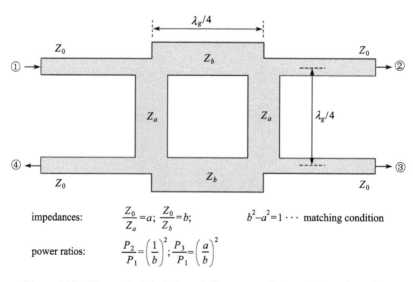

impedances: $\dfrac{Z_0}{Z_a}=a;\ \dfrac{Z_0}{Z_b}=b;$ $b^2-a^2=1$ ⋯ matching condition

power ratios: $\dfrac{P_2}{P_1}=\left(\dfrac{1}{b}\right)^2;\dfrac{P_3}{P_1}=\left(\dfrac{a}{b}\right)^2$

Figure 6.12   Microstrip hybrid branch-line power divider with two branches.

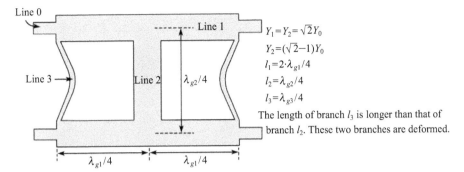

Figure 6.13   Microstrip hybrid branch-line power divider with three branches.

exact structure parameters including the distance l in Figure 6.9 will be finally determined from the full wave solver.

The estimation of spurious radiation from microstrip lines has been done by using the full wave analysis in [15]. The highest radiation occurs when the length of the line is near resonance, and the terminations are either open or short circuits.

It is also observed that a matched load termination at one of the terminals of the microstrip interconnection reduces the spurious radiation significantly, as compared to the radiation levels for other terminations that can cause resonances to occur. The distribution of power, voltage and current on the feed network may be obtained simply from the transmission line formulas. These results are helpful in observing the matching mechanism and the spurious radiation.

Consider the microstrip transmission line quarter-wave impedance transformer shown in Figure 6.14. The formulas for voltage, current and power are given in the following according to the derivation in Chapter 4. The results in Figure 6.14 may be obtained from (6.3.1)–(6.3.4).

$$\frac{V_x}{V_0} = \frac{Y_{c1}/\sinh\gamma_{c1}x}{Y_x + Y_{c1}\coth\gamma_{c1}x} \tag{6.3.1}$$

$$\frac{I_x}{I_0} = \frac{Y_x/\sinh\gamma_{c1}x}{Y_{c1} + Y_x\coth\gamma_{c1}x} \tag{6.3.2}$$

$$Y_x = Y_{c1}\frac{Y_{c1} + Y_L\coth\gamma_{c1}(L-x)}{Y_L + Y_{c1}\coth\gamma_{c1}(L-x)} \tag{6.3.3}$$

$$P_x = \frac{1}{2}V_x(I_x)^* \tag{6.3.4}$$

For example, when $x = 0$

$$V_{x=0} = V_0 = \frac{1}{Y_{0r} + Y_{0l}} = \frac{1}{(1/64) + (1/64)} = 32\text{V} = |V_0|$$

$$I_{x=0} = I_0 = V_0 Y_{x=0} = V_0\frac{(Y_{c1})^2}{Y_L} = 32 \times \left(\frac{1}{80}\right)^2 \times 100 = 0.5\text{A} = |I_0|$$

$$P_{x=0} = \frac{1}{2}V_0(I_0)^* = \frac{1}{2} \times 32 \times (0.5) = 8\text{W} = \text{Re}(P_{x=0})$$

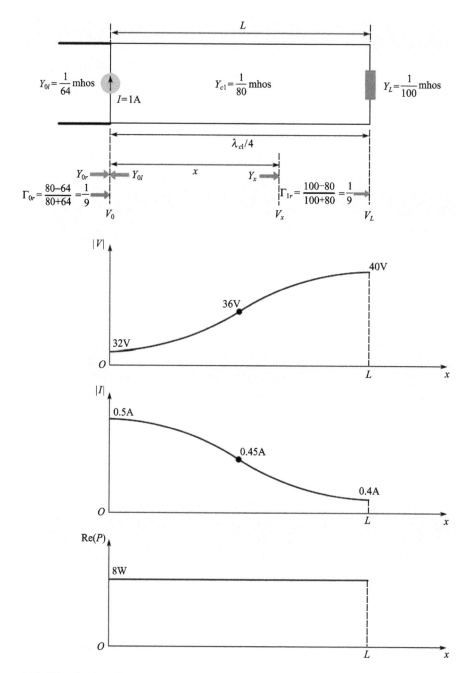

Figure 6.14   Distribution of power, voltage and current along the transmission line (subscript l denotes looking left; r, looking right).

when $x = 0.5L$

$$Y_{x=0.5L} = Y_{c1}\frac{Y_{c1} - jY_L}{Y_L - jY_{c1}} = \frac{1}{80}(0.98 + j0.22)$$

$$V_{x=0.5L} = V_0\frac{Y_{c1}/(j\sqrt{2}/2)}{Y_{x=0.5L} + Y_{c1}(-j)} = \frac{32(-j\sqrt{2})}{0.98 - j0.78} = -j28.9(0.98 + j0.78)$$

$$|V_{x=0.5L}| = \mathrm{Re}(V_{x=0.5L}) = 36\mathrm{V}$$

$$I_{x=0.5L} = Y_{x=0.5L}V_{x=0.5L} = \frac{-j28.9}{80}(0.98+j0.22)(0.98+j0.78)$$
$$= -j0.36(0.79+j0.98)\text{A} \tag{6.3.5}$$

$$|I_{x=0.5L}| = \text{Re}(I_{x=0.5L}) = 0.45\text{A}$$

$$P_{x=0.5L} = \frac{1}{2}V_{x=0.5L}(I_{x=0.5L})^* = \frac{1}{2}(-j28.9)(0.98+j0.78)(0.353+j0.28)$$
$$= (8-j1.85)\text{W} \tag{6.3.6}$$

(minus imaginary part means that the storage energy is capacitive)

$$|P_{x=0.5L}| = \text{Re}(P_{x=0.5L}) = 8\text{W}$$

when $x = L$,

$$V_{x=L} = V_L = -jV_0\frac{Y_{c1}}{Y_L} = -j32 \times \frac{100}{80} = -j40\text{V}$$
$$|V_{x=L}| = 40\text{V}$$
$$I_{x=L} = I_L = -jI_0\frac{Y_L}{Y_{c1}} = -j0.5 \times \frac{80}{100} = -j0.4\text{A}$$
$$|I_{x=L}| = 0.4\text{A}$$
$$P_{x=L} = \frac{1}{2}V_{x=L}(I_{x=L})^* = \frac{1}{2}(-j40)(j0.4) = 8\text{W} = \text{Re}(P_{x=L})$$

Formulas 6.3.1–6.3.4 are also useful in analyzing the frequency response of the transformer.

6.4   Design Examples of Microstrip Antenna Arrays

In practice, it is often necessary to design an antenna system that must meet the system requirements for a certain application. Accordingly, different design strategies will be taken to satisfy different system constraints. For example, to satisfy the size limitation, the element spacings in a certain direction must be compressed; to design a low side lobe level antenna, the current on the array aperture must follow some specified distribution, say Dolph-Tschebyscheff distribution or Taylor distribution, etc. Generally speaking, the primary goal for antenna array design is to acquire the required performance characteristics over a specified frequency band.

In this section, several types of microstrip antenna arrays are introduced including compact, wide bandwidth, low side lobe level, single layer monopulse and LTCC arrays.

6.4.1   Design of a 16GHz Compact Microstrip Antenna Array

The requirements to the array are:
- polarization: linear
- gain: > 22.5dB
- bandwidth: 80MHz
- beamwidth: $3.5° - 4.5°$ horizontal and $9° - 11°$ vertical
- side lobe level: < -11dB
- size: 230mm×100mm.

One of the first tasks in patch design is the selection of a suitable substrate material. Considering all the factors, including the price and availability of the material, a substrate with $\epsilon_r = 2.7$, $h = 0.5$mm, $\tan\delta = 10^{-4}$ was chosen in this case.

One of the most basic configurations for microstrip antennas is a resonant rectangular patch which has resonant length L and width W. The design procedure provided in Chapter 3 is applied.

After obtaining the preliminary parameters, a MoM based full wave solver IE3D^{TM}[17] is employed to determine the feeding position and fine tune the design to obtain the desired antenna characteristics. The dimension of the patch and its VSWR behavior are shown in Figure 6.15 and Figure 6.16 respectively.

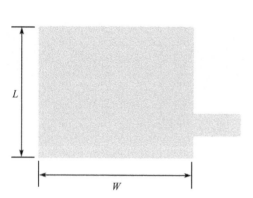

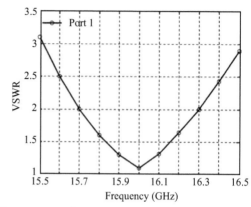

Figure 6.15   Microstrip patch antenna with   Figure 6.16   VSWR for the patch shown in Figure 6.15.
$W = 311$mils, $L = 207$mils.

The parallel feed to rectangular patches is realized through feeding to the non-radiating edge for its simplicity in feed line and therefore less feed network losses. It is shown in [16, 17] that when $W/L = 1.5$, the cross-polarization will be reduced to the minimum. The 2×2 sub-array is shown in Figure 6.17.

From the consideration of gain, bandwidth and mutual coupling, the spacing between two adjacent patches of the normal design is about 0.75 free-space wavelength. To satisfy the size limitation, spacing in the vertical direction is reduced to 0.63 free-space wavelength, resulting in strong mutual coupling between the feed line and one edge of the patch. To solve this problem, the normal T-junction was reshaped into a curved one to move it down the main feed line[18] as is shown in Figure 6.17.

The commercial software IE3D^{TM} based on the full wave analysis is applied to the design of this sub-array. The results are used as the basic building block in a loaded feed network shown in Figure 6.18. This network is designed by using IE3D^{TM} again. The layout of a

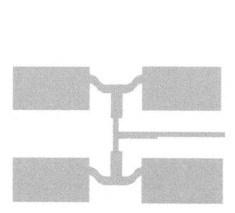

Figure 6.17   Geometry of 2×2 sub-array.   Figure 6.18   Geometry of a loaded feed network.

16×8 element array is configured repetitively in the above way, which is called the element-
or module-loaded based full wave analysis method. Through this method, the computation
time is reduced dramatically. The sacrifice is that only part of the mutual coupling effect,
i.e., only the mutual coupling in the 2×2 array, is considered in this method. As this array
is of uniform amplitude and phase distribution, this neglect of mutual coupling does not
introduce much difference to the performance of the array.

    The array is fed by a coaxial line. The center conductor of the cable is attached to the
patch, while the outer conductor line is attached to the back side of the printed circuit board.
It has been shown that this feed mechanism introduces an inductive reactance in series to
the antenna array[19]. In addition, when the probe diameter is larger than the width of the
feed line connected to it, a pad has been placed to insure good welding of the probe to the
microstrip line. The discontinuity at the junction of the probe and the feed line may also
introduce large reactance to the circuit. All these factors make the matching between the
antenna and feed line very difficult. In order to cancel this reactance introduced by the
probe feed, stub tuning illustrated in Figure 6.19 is applied. The resulting input impedance
of the array is shown in Figure 6.20.

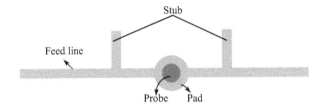

Figure 6.19   Wideband impedance matching using tuning stub.

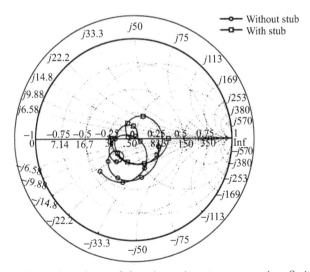

Figure 6.20   Input impedance of the microstrip antenna array in a Smith Chart,
with or without tuning stubs.

    It is observed that the unwanted reactance could be cancelled and thus the wideband
impedance matching is realized. Another important role of these two stubs is that they can
give a useful degree of freedom for tuning, when the experimental VSWR of the array is not
satisfying.

Figures 6.21 and Figure 6.22 show the compact microstrip antenna array and its characteristic VSWR behavior, respectively. It is observed that the bandwidth is 3.13% for VSWR< 1.5, and 6.25% for VSWR< 2 (center frequency is 16GHz).

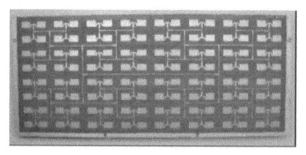

Figure 6.21   Compact microstrip antenna array.

Fifty antenna arrays were manufactured and measured. The typical characteristic of VSWR and the radiation pattern obtained by far-field measurement method are shown in Figures 6.23 and Figure 6.24 respectively. The mean values are: bandwidth 6.14% (VSWR< 2), gain 23.7dB, beamwidth 4.11° (H-plane) and 9.55° (E-plane), side lobe level −11.6dB. The data show the success of the design.

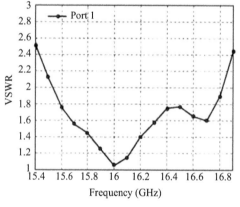

Figure 6.22   VSWR of the array.

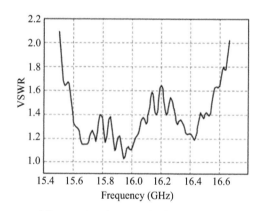

Figure 6.23   Measured VSWR behavior.

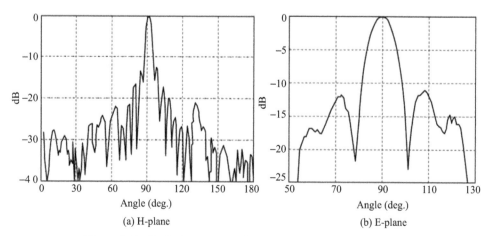

(a) H-plane

(b) E-plane

Figure 6.24   Radiation pattern obtained by far-field measurement.

6.4.2 Design of a Low Side Lobe Level Microstrip Antenna Array

In order to achieve low side lobe level of an array, a tapered distribution on the array aperture is needed. If a parallel feed network is used, various complicated power dividers must be designed, which will cause difficulty in impedance matching. Consequently, a simple structure of a series-fed array with shunt-connected square-shaped microstrip antennas is employed. The corner-fed square patches have been chosen because they provide a high input impedance well suited for series array. It is also very easy to feed each element on the corner; a tapered distribution is readily obtained using quarter-wavelength transformers along the line. In order to get a broadside pattern, the spacing between two elements should be one wavelength on the line or a half-wavelength with alternate elements to keep the in-phase condition.

The full wave IE3D^{TM} solver is again applied to the module of 1×8 linear array and the whole array is also analyzed using module-loaded based full wave analysis. Since in this case, the element spacings are not very close and the side lobe level is not very low, the mutual coupling effects will not be so serious as to significantly degrade the performances the antenna array. This has been validated by the experiment.

The parameters of the substrate chosen are $\epsilon_r = 2.7$, $h = 0.5$mm and $\tan \delta = 10^{-4}$. For the requirement of -18dB side lobe level, taking into account the manufacturing tolerance, a -25dB Taylor distribution is used for a design. Designed according to the normal design procedure, a 2×16 low side lobe level microstrip antenna array was constructed and is shown in Figure 6.25. The simulated results about the input impedance of the array are shown in Figure 6.26. It is observed that the impedance bandwidth (VSWR< 2) is 5%.

Figure 6.25   Geometry of a 2×16 low side lobe level microstrip antenna array.

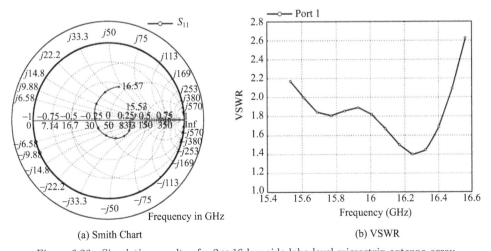

(a) Smith Chart

(b) VSWR

Figure 6.26   Simulation results of a 2×16 low side lobe level microstrip antenna array designed by the normal procedure.

Generally speaking, the quarter-wavelength impedance transformer is appropriate for the design of a resistive matching network. However, it is found that the input impedance seen at the point A in Figure 6.27 is not pure resistance. As a result, a more useful and flexible technique called stub matching is applied here, as shown in Figure 6.28[20].

Figure 6.27   Normal impedance matching using a quarter-wavelength transformer.

Figure 6.28   Advanced impedance matching using a single tuning stub.

Figure 6.30 gives the simulation results for an advanced 2×16 low side lobe level microstrip antenna array shown in Figure 6.29. The VSWR curve in Figure 6.30 indicates that good impedance matching results in a larger bandwidth, which is 11.5% (VSWR< 2), 5.7%(VSWR< 1.5).

Figure 6.29   Geometry of the advanced 2×16 low side lobe level microstrip antenna.

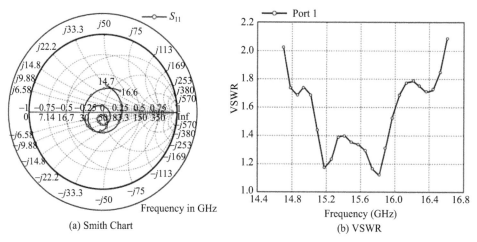

(a) Smith Chart   (b) VSWR

Figure 6.30   Simulation results of the 2×16 low side lobe level microstrip antenna array designed by the advanced procedure.

The comparison of the experimental VSWR behavior of the array using different matching techniques is shown in Figure 6.31. It is observed that the measured bandwidth of the normal scheme is 3.7% (VSWR< 2). By using the modified scheme with stub matching, the measured bandwidth is extended to 9.1% (VSWR< 2). The measured radiation pattern in the H-plane of the designed 2×16 low side lobe level microstrip antenna array is also shown in Figure 6.32. It is seen the achieved side lobe level is -18.5dB.

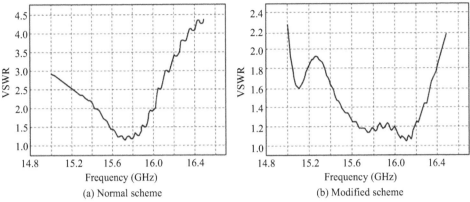

(a) Normal scheme   (b) Modified scheme

Figure 6.31   Comparison of the measured VSWR for different matching techniques.

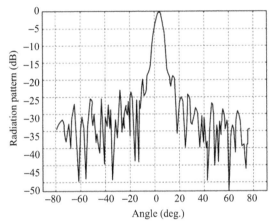

Figure 6.32   Measured radiation pattern in the H-plane of the advanced 2 × 16 low side lobe
level microstrip antenna array.

6.4.3   Design of a Compact Single Layer Monopulse Microstrip Antenna Array With Low Side Lobe Levels

Monopulse, also called a simultaneous lobe comparison, has been developed as a solution to overcoming the erroneous angle indication, slow searching speed of the lobe switching and conical scan techniques used in radar tracking systems.

In traditional monopulse radar systems, the common types of the antennas are Cassegrain parabolic antennas or lens antennas. The monopulse comparator in such systems is usually very complicated and heavy. To make full use of the advantages of microstrip antennas, some microstrip monopulse antennas have been presented[21–23]. In [21], a low cost K-band microstrip patch monopulse antenna has been proposed. This antenna array is formed by 2 × 2 elements and is located at the central part enclosed by the comparator network. This structure is impossible to use in the sub-array design. In [22], a microstrip monopulse antenna has been designed at millimeter wave frequency. In this array structure, the comparator network is situated at the center of the antenna. But the comparator makes a large blockage on the antenna aperture, making it difficult to achieve the low side lobe level. In [23], a low-cost and simple-structured bi-directionally-fed microstrip patch array has been introduced. However, only one dimensional monopulse performance is obtained, and the side lobe level is higher than –10dB. Recently a low-cost monopulse radial line slot antenna has been proposed [24]. The radiation elements and feed network of this antenna are placed on two layers to eliminate some negative effects, such as spurious radiation from the feed network and the blockage of the comparator. To alleviate these problems, recently a monopulse microstrip antenna array in a single layer was proposed[25]. This antenna can achieve two-dimensional monopulse performances, and the single layer structure is simple and cost effective.

A. Specification and Structure of the Antenna Array

The performance of the antenna array used in the monopulse radar system is listed in the following:
- Range of operating frequency: 13.85GHz − 15.1GHz
- Bandwidth: 5.6% (VSWR< 2, central frequency is 14.25GHz)
- Polarization: linear
- Side lobe levels (SLL) of the sum pattern: < -17dB
- Null depths of the difference patterns: < -30dB
- Half-power beamwidth in the E plane: 5.5°
- Half-power beamwidth in the H plane: 4.7°
- The maximum gain at the operating frequencies: 24.5dBi

The size of the antenna is 280mm×260mm.

The structure of the antenna as shown in Figure 6.33 includes (1) four sub-arrays connected to the four ports of the comparator; (2) the comparator with four outputs connected to the sum, H-plane difference, E-plane difference and the matching load respectively; (3) the snake lines which are used to provide the phase delay needed. Each sub-array is formed by 8 × 8 patches with the parallel feed. The comparator is formed by four 3dB hybrid junctions. As the figure shows, the structure in a single layer is very compact. But this compactness brings some challenging issues in design which will be discussed. The MoM based full wave commercial software IE3D^{TM} is used in the simulation.

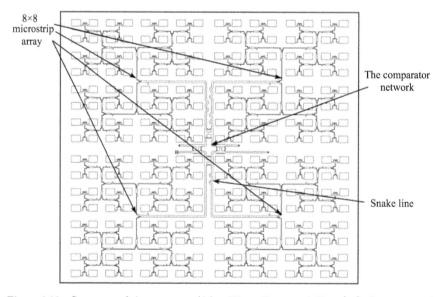

Figure 6.33   Structure of the antenna. (After Wang, Fang, and Chen [25], © 2006 IEEE)

B. The Limitation of Single Layer Structure to the Feed Architectures

The compact single layer structure also brings some limitations in choosing the feed architectures. If a series feed is used in this design as showed in Figure 6.34, the comparator network, placed in the middle of the antenna, would make the spacing L between feed line and comparator network very small. The strong mutual coupling affects the magnitude and phase of the feed network and makes it very difficult to achieve low side lobe levels in two dimensions. To overcome this problem, a parallel feed is adopted in the design.

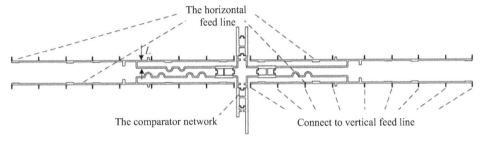

Figure 6.34   Partial structure of the series feed monopulse antenna.
(After Wang, Fang, and Chen [25], © 2006 IEEE)

C. The Effect of the Blockage on the Side Lobe Level

The comparator with the cross configuration as shown in Figure 6.33 limits the spacing between two sub-arrays. This blockage would deteriorate the side lobe level. Table 6.1 shows the deterioration of side lobe level caused by the blockage. The deterioration becomes more severe when the side lobe level is lower. This Table could serve as a guideline to design this kind of monopulse microstrip antenna.

Table 6.1   The deterioration of side lobe level caused by the blockage.
(After Wang, Fang, and Chen [25], © 2006 IEEE)

SLL(dB) without blockage	SLL(dB) with blockage
16	15.5
18	17.5
20	19
22	21
24	22
26	23
28	24
30	25

D. The Effect of Spurious Radiation to the Side Lobe Level

There are many discontinuities in the feed network and comparator, which produce the spurious radiation, causing limitation in achieving a lower side lobe level. In [26], Pozar has mentioned that when the feed network is printed on the same substrate as that of the antenna elements, the side lobe level will be limited in the range of -15dB$--25$dB. To have a good design, it is very important to evaluate the effect of the spurious radiation. Figure 6.35 and Figure 6.36 show the H and E plane radiation patterns. Table 6.2 shows the deterioration of the side lobe at different side lobe levels. For example, due to the spurious radiation, for a designed side lobe level of -25dB, the achievable side lobe level is only -22dB.

E. The Feed Network Losses

The losses due to the feed network are caused when the wave moves from input port to all the antenna elements. To evaluate the losses, the transmission coefficients between the input port and each element are calculated by using IE3D^{TM}. The total power fed to all the elements through the feed network is then determined. The difference between the power with the feed network and that without the feed network is due to the feed network losses. In this example, according to the method described above, the calculated value of the feed network losses is −2.93dB.

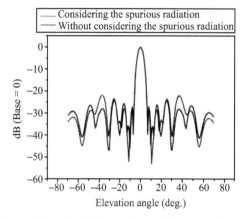

Figure 6.35   Radiation patterns with and without considering the spurious radiation in the H-plane. (After Wang, Fang, and Chen [25], © 2006 IEEE)

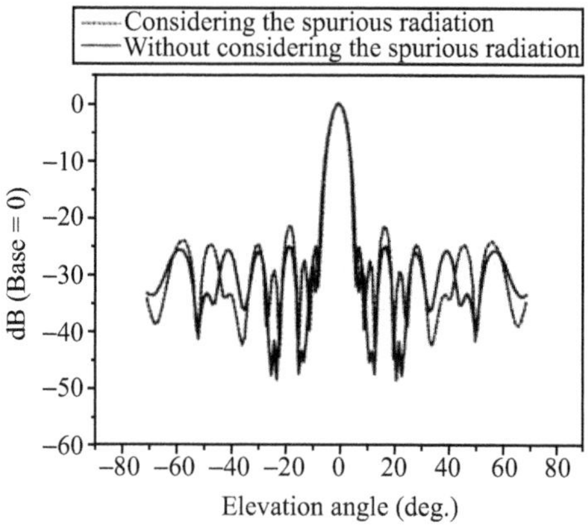

Figure 6.36   Radiation patterns with and without considering the spurious radiation in the E-plane. (After Wang, Fang, and Chen [25], © 2006 IEEE)

Table 6.2   The deterioration of the side lobe level (SLL) caused by spurious radiation (SR). (After Wang, Fang, and Chen [25], © 2006 IEEE)

SLL (dB) without SR	SLL (dB) with SR
16	16
18	18
20	20
22	22
24	22
25	22

F. Design of the Radiating Structure

One of the most basic configurations for a microstrip antenna is a resonant rectangular patch which has resonant length L and width W. The design procedure provided by Bahl and Bhartia[27] is applied to get the preliminary design. The parameters of the substrate are a dielectric constant 2.7, a thickness 5mm and a loss tangent 0.003. Figure 6.37 shows the dimension of microstrip patch.

The radiating structure is formed by 16×16 patches. The parallel feed to rectangular patches is realized through feeding to the non-radiating edge for its simplicity in feed line and therefore there are less feed network losses[17]. From consideration of gain, bandwidth and mutual coupling, the spacing between two adjacent patches of the conventional design is about 0.8 free-space wavelengths. Due to the tolerances and other uncertainties, in the conventional design of an array, the difference between the designed SLL and the required one is 1dB − 4dB. On both the E-plane and the H-plane, −24dB Taylor's distribution is chosen. Due to symmetry, only eight normalized excitation coefficients are given as follows: $1, 0.9595, 0.8758, 0.7557, 0.6223, 0.5067, 0.4316, 0.4062$. The nonuniform amplitude distribution is realized through many power dividers as shown in Figure 6.9. The geometry of the 2×2 sub-array is shown in Figure 6.38. As seen in the figure, the normal T-splitter is reshaped into a curved one to move down the main feed line and avoid the unwanted coupling between the feed line and the patch. $\mathrm{IE3D}^{TM}$ is also used. The computational accuracy is mainly dependent on the discretization. Usually 20 grids per wavelength combined with the automatic edge cell discretization would be sufficient[7] and we call it a fine model, which is accurate but time consuming. For example, in our case, to obtain frequency response, CPU time of 5870 seconds is required. To solve this problem, space mapping (SM)[63, 64] technique (Ref. Chapter 5) is used by taking 15 grids per wavelength without automatic edge cell discretization as the coarse model. Thus, the corresponding CPU time is reduced from 5870 seconds to 100 seconds. Parameters x_1, x_2 and x_3 shown in Figure 6.38 are taken as the variables to be optimized.

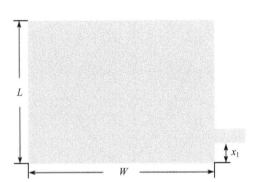

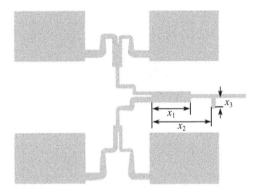

Figure 6.37   Dimension of a microstrip patch with $W = 8.4$mm, $L = 6$mm, $x_1 = 0.46$mm. (After Wang, Fang, and Chen [25], © 2006 IEEE)

Figure 6.38   Geometry of a 2 × 2 sub-array. (After Wang, Fang, and Chen [25], © 2006 IEEE)

The objective function is the VSWR at the input port of 2 × 2 sub-array, which is set to be less than 2 over the total bandwidth. The optimized result from coarse model is $x_c^* = [5.5, 7.59455, 0.71466]^T$ (all in mm). Figure 6.39 shows the responses of the coarse model and the fine model at the starting point. Figure 6.40 shows the final responses of the fine model and the coarse model, and $x_f = [5, 7.69455, 1.11466]^T$ (all in mm).

The total optimization time is several hours. However, if the SM method is not used, the time of design is dependent on experience. For example, the time needed for our first design without using the SM method was several days. Table 6.3 shows the values of the coarse model and the fine model at different stages of the iteration. The results of 2 × 2 sub-arrays are used as basic building block in a loaded feed network as shown in Figure 6.41. This network is designed by $\mathrm{IE3D}^{TM}$ again. The layout of a 16 × 16 element array is configured repetitively in the above way which is called the element-loaded based full wave analysis method. The sacrifice is that only part of the mutual coupling effect, i.e., only the mutual coupling in the 2 × 2 array is involved in this method. From our experience, it would not cause a problem except for the design of the ultra low side lobe level array. Through these two methods, computation time is reduced dramatically and the optimization design becomes feasible.

G. Design of a Monopulse Comparator

The diagram of a monopulse comparator is shown in Figure 6.42. This comparator is comprised of four 3dB hybrid couplers and several 90° delay lines. Figure 6.43 shows the geometry of an eight port monopulse comparator in the design. Compared with the structure proposed in [22], the structure in Figure 6.43 is more compact, which could decrease the blockage in the antenna, and achieve a lower side lobe level. In Figure 6.43, port 1,2,3,4 is the input port; port 5,6,7,8 is the output port. The output of port 5 is $(1+2)+(3+4)$; the output of port 6 is $(1+2)-(3+4)$; the output of port 7 is $(1+3)-(2+4)$; the output of port 8 is $(1+4)-(2+3)$. The performances of this comparator are analyzed by IE3D^{TM}. Table 6.4 shows the amplitude of S_{mn} (m denotes the number of the output port; n, the input port). Table 6.5 shows the phase of S_{mn} (m, n is defined as the same as that in the Table 6.4). As shown in Table 6.4 and Table 6.5, the output amplitudes by the same inputs at ports 1, 2, 3 and 4 could be calculated. The results are shown in Table 6.6. From Table 6.6, the null depth could be calculated at the central frequency, which is $20\lg(0.024/1.8376) \approx -38\text{dB}$.

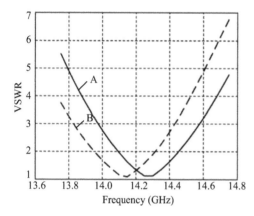

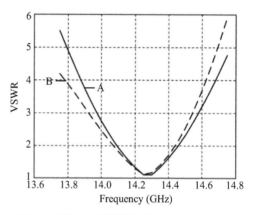

Figure 6.39   A: VSWR of the optimized coarse model; B: The fine model's VSWR at the start point. (After Wang, Fang, and Chen [25], © 2006 IEEE)

Figure 6.40   A: VSWR of the optimized coarse model; B: The final VSWR of the fine model. (After Wang, Fang, and Chen [25], © 2006 IEEE)

Table 6.3   **Values of the coarse model and the fine model at different stages of the iteration (1: first iteration, 2: second iteration). (After Wang, Fang, and Chen [25], © 2006 IEEE)**

	1	2
x_{c1}	6	5.5
x_{c2}	7.49455	7.59455
x_{c3}	0.31466	0.71466
x_{f1}	5.5	5
x_{f2}	7.59455	7.69455
x_{f3}	0.71466	1.11466

Figure 6.41   Geometry of a loaded feed. (After Wang, Fang, and Chen [25], © 2006 IEEE)

5: (1+2)+(3+4)   7: (1+3)−(2+4)

6: (1+2)−(3+4)   8: (1+4)−(2+3)

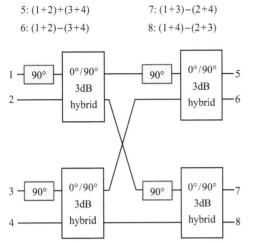

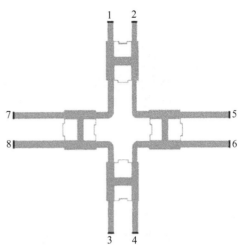

Figure 6.42   Block diagram of monopulse comparator. (After Wang, Fang, and Chen [25], © 2006 IEEE)

Figure 6.43   Geometry of the microstrip monopulse comparator. (After Wang, Fang, and Chen [25], © 2006 IEEE)

Table 6.4   The amplitude of S_{mn}, units: dB.
(After Wang, Fang, and Chen [25], © 2006 IEEE)

$\text{Amp}[S(5,1)]$	$\text{Amp}[S(6,1)]$	$\text{Amp}[S(7,1)]$	$\text{Amp}[S(8,1)]$
−6.750	−6.811	−6.646	−6.656
$\text{Amp}[S(5,2)]$	$\text{Amp}[S(6,2)]$	$\text{Amp}[S(7,2)]$	$\text{Amp}[S(8,2)]$
−6.796	−6.837	−6.788	−6.767
$\text{Amp}[S(5,3)]$	$\text{Amp}[S(6,3)]$	$\text{Amp}[S(7,3)]$	$\text{Amp}[S(8,3)]$
−6.717	−6.638	−6.630	−6.614
$\text{Amp}[S(5,4)]$	$\text{Amp}[S(6,4)]$	$\text{Amp}[S(7,4)]$	$\text{Amp}[S(8,4)]$
−6.773	−6.712	−6.684	−6.736

Table 6.5   The phase of S_{mn}, units: deg. (After Wang, Fang, and Chen [25], © 2006 IEEE)

$\text{Ang}[S(5,1)]$	$\text{Ang}[S(6,1)]$	$\text{Ang}[S(7,1)]$	$\text{Ang}[S(8,1)]$
100.80	102.00	101.20	103.7
$\text{Ang}[S(5,2)]$	$\text{Ang}[S(6,2)]$	$\text{Ang}[S(7,2)]$	$\text{Ang}[S(8,2)]$
11.07	13.02	−168.20	−167.30
$\text{Ang}[S(5,3)]$	$\text{Ang}[S(6,3)]$	$\text{Ang}[S(7,3)]$	$\text{Ang}[S(8,3)]$
12.27	−168.90	13.65	−168.10
$\text{Ang}[S(5,4)]$	$\text{Ang}[S(6,4)]$	$\text{Ang}[S(7,4)]$	$\text{Ang}[S(8,4)]$
−76.98	101.60	102.50	−77.64

Table 6.6   The output amplitudes when the same signals are input.
(After Wang, Fang, and Chen [25], © 2006 IEEE)

Freq(GHz)	Port5 Mag	Port6 Mag	Port7 Mag
14.25	1.8376	0.0015	0.0240

H. Measurements

The prototype antenna is shown in Figure 6.44. The measurements carried out include antenna pattern and gain at 14.25GHz and VSWR of three ports. Figures 6.45–6.48 show the E- and H-plane sum and difference patterns of the monopulse antenna, respectively. Figure

6.49 shows VSWR of three ports. From the measured data, the sum channel side lobe levels in the E- and H-plane are less than −17dB, respectively. The predicted deterioration of the side lobe level is caused by the tolerance, spurious radiation, comparator blockage and other uncertainties. The difference in side lobe level between the predicted and measured data is about 1dB. This shows that the prediction overestimates the side lobe level. In the measured H plane sum pattern, we are still not clear for the asymmetrical phenomenon. The null depths are less than –30dB in both planes. The bandwidth (VSWR< 2) of the antenna is 5.6%. However, the measured results show that the VSWR curve shifts slightly from the designed frequency to the higher frequency. This fact suggests that bandwidth has the potential to be improved through second round correction if necessary. The sum channel gain is measured as 24.5dBi with half power beamwidth in the E- and H-planes of 5.5 and 4.7 degrees, respectively. The method used to measure the gain of the antenna is the gain-comparison method. The efficiency of this array is 20%, which is lower than the conventional microstrip antenna array.

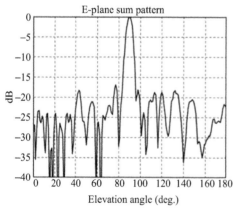

Figure 6.44   The prototype antenna. (After Wang, Fang, and Chen [25], © 2006 IEEE)

Figure 6.45   The measured E-plane sum pattern.   Figure 6.46   The measured H-plane sum pattern.
(After Wang, Fang, and Chen [25], © 2006 IEEE)   (After Wang, Fang, and Chen [25], © 2006 IEEE)

The prediction on the performance could be taken as the guideline for the design of this kind of antennas. Measured results show that the performance of this antenna is acceptable in most of the cases. This antenna is particularly useful in lightweight monopulse radar applications. However, in the proposed design, the compactness of the antenna is achieved at the price of efficiency.

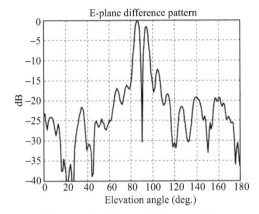

Figure 6.47   Measured E-plane difference.
(After Wang, Fang, and Chen [25], © 2006 IEEE)

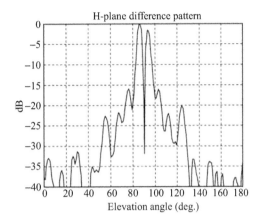

Figure 6.48   Measured H-plane difference.
(After Wang, Fang, and Chen [25], © 2006 IEEE)

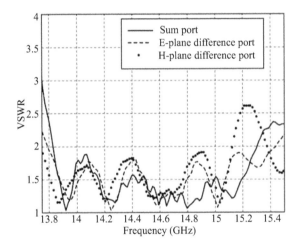

Figure 6.49   Measured VSWR. (After Wang, Fang, and Chen [25], © 2006 IEEE)

6.4.4   Design of an Integrated LTCC mm-Wave Planar Antenna Array

A. Outline of the Antenna Array

With the increasing demands of commercial mm-wave applications such as Collision Avoidance Radar and Local Multi-points Distribution System (LMDS), a multilayer Low Temperature Co-fired Ceramic (LTCC) large-scale antenna array has attracted some attention due to its flexibility in manufacturing, the capability of passive integration and the low production cost. One potential application is to build a microstrip antenna array in an LTCC substrate. A 256 element antenna array operating at 29GHz on a 12-layer $12.7 \times 12.7 \text{cm}^2$ LTCC tile was proposed[28]. Figure 6.50(a) shows a photo of the array antenna, which is directly fed by a piece of WR28 waveguide on the backside. A quasi-cavity-backed patch antenna is used as the radiating element as shown in Figure 6.50(b). This quasi-cavity-backed patch antenna can achieve a better radiation performance and higher efficiency than those of its counterparts without the cavity. To reduce the loss and unwanted radiation from the feeding network, a mixed feeding network configuration comprising a laminated waveguide (LWG), microstrip line and required transitions is used.

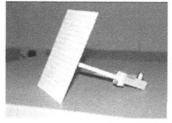

(a) Photograph of the Integrated LTCC Array
Antenna

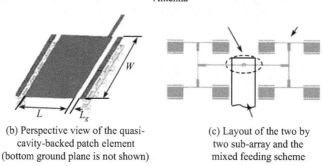

(b) Perspective view of the quasi-
cavity-backed patch element
(bottom ground plane is not shown)

(c) Layout of the two by
two sub-array and the
mixed feeding scheme

Figure 6.50   Integrated LTCC antenna array. (After Huang, Wu, Fang, and Ehlert [28], ©2005 IEEE)

B. Radiating Element

It is known that the bandwidth of a traditional patch antenna is proportional to the substrate thickness. To achieve a wider bandwidth, a thicker substrate can be used. However, working with the high dielectric constant substrate, a thicker substrate will lead to a higher surface wave loss and consequently degrade the radiation efficiency. For example, an antenna capable of achieving a 4% 2:1 VSWR bandwidth about 29GHz on Dupont® 943 LTCC substrate (with dielectric constant of 7.5, a loss tangent of 0.002, and a thickness of 0.447mm), the simulated radiation efficiency using $IE3D^{TM}$ is less than 78%. To improve the radiation efficiency, a quasi-cavity-backed patch (QCBP) antenna is introduced[29], as shown in Figure 6.50(a). The length of radiation edge and non-radiation edge are denoted as W and L, respectively. Two grounded grid-like conducting walls, comprising several metal strips and filled via-holes, are introduced to minimize the excitation of the surface wave and thus to improve the radiation efficiency. Using the $IE3D^{TM}$ EM simulation software, a numerical analysis has been conducted to study the radiation performance of a two by two sub-array, as depicted in Figure 6.50(c), with the proposed antenna element. As shown in Table 6.7, by using the grid-like conducting walls the radiation efficiency of the sub-array (η) can be as high as 94.6% in contrast to the efficiency of 77.9% for the case without the walls. The separation distance of the wall to the edge of patch antenna, L_g, should be kept close to the

Table 6.7   Simulated radiation efficient of the sub-array.
(After Huang, Wu, Fang, and Ehlert [28], ©2005 IEEE)

L_g(mm)	D(dB)	η(%)	W(mm)	L(mm)
0.127	13.1	94.6	2.49	1.49
0.152	13.1	92.3	2.49	1.52
0.178	13.1	90.4	2.49	1.55
Traditional Patch	13.0	77.9	2.54	1.57

extension length of the fringe field of the patch, in order to maximize the radiation efficiency. Due to the leaking effect of the meshed wall structure, as shown in Table 6.7, the simulated optimal L_g is found to be 0.127mm, which is less than the theoretic extension length of 0.178mm.

C. Mixed Feeding Network

Owing to the feature of no radiation loss and low insertion loss, a laminated waveguide (LWG) is considered to be one of the most effective transmission lines for LTCC mm-wave applications. A three-dimensional laminated waveguide is built by depositing metal planes on the top and bottom surfaces of a multilayered substrate and using a pair of grid-like conductive walls as sidewalls[30]. Assuming low loss and no leakage for the LWG, a mixed feeding network is proposed as shown in Figure 6.50(a). The main trunk of the feeding network is constructed by the LWGs. Since extending the LWG feeding network to each element in the array antenna significantly increases the complicity of the implementation with negligible loss reduction, as illustrated in Figure 6.51, the sub feeding network of all the 2×2 sub-arrays utilizes traditional microstrip lines. The laminated waveguide feeding network and the 2×2 sub-arrays are separated by an internal ground plane, which serves as the bottom ground of the array and the top metal wall of the LWG feeding network as shown in Figure 6.52. A laminated waveguide to microstrip line T-junction[31] can be used to connect the laminated waveguide feeding network to the microstrip line feeding network. To provide the LTCC array with an interface to an air waveguide system, a broadband transition between a laminated waveguide and a WR28 standard waveguide (LWG-to-WG transition)[32] has been developed and is integrated in the feeding network.

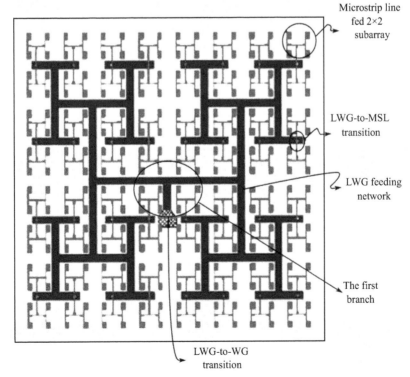

Figure 6.51 Proposed mixed feeding network configuration.
(After Huang, Wu, Fang, and Ehlert [28], © 2005 IEEE)

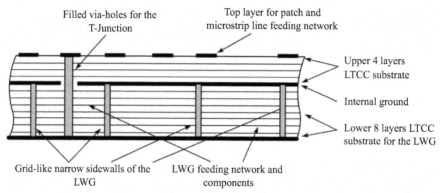

Figure 6.52   Vertical structure of the feeding network.
(After Huang, Wu, Fang, and Ehlert [28], © 2005 IEEE)

D. Loss Analysis and Experimental Results

A prototype of a patch antenna array with proposed quasi-cavity-backed elements and a prototype of the same patch antenna array without cavity-backing are fabricated using a 12-layer substrate of Dupont® 943 Green Tape^{TM}. An identical feeding network structure is used in the two prototypes. In the 12-layer substrate, the LWG feeding network is built in the lower eight layers and the antenna elements and microstrip line feeding network are built in the upper four layers, as shown in Figure 6.52. The thickness of each layer is 0.11mm. The 16×16 elements in the array antenna are excited equally. To verify the concept of the proposed mixed feeding network and also save the space for other loaded LWG components, only the first branch of the main trunk is implemented by LWG in the experimental array. Two types of required transitions, namely the transition from air waveguide to LWG and the T-junction from LWG to microstrip line, have been integrated in the experimental feeding network. Simulated results obtained from ANSOFT® HFSS^{TM} show that the insertion losses of the proposed mixed feed network, and a traditional microstrip edge feeding network are 3.7dB and 9.6dB respectively, where the cross-sectional dimension of LWG is 2.5mm by 0.22mm, and the microstrip trace width of 100 ohm microstrip line used in the microstrip line feeding network is 0.1mm. The simulated insertion loss of the experimental feeding network is 6.63dB. Although the experimental feeding network is just a portion of the proposed mixed feeding network, the improvement over the microstrip line feeding network is significant enough to verify the concept of the proposed mixed feeding network. Based on the calculated radiation efficiency presented in Table 6.7, we can conclude from the simulation, that the gain of a QCBP array with mixed feeding network and a conventional element array with a microstrip line feeding network is about 26.46dB and 20.42dB, respectively. Even for the experimental array, in which LWG is used only for the first branch of the feeding network and the quasi-cavity-backed elements are used, about 24.23dB gain can be achieved. Figure 6.53 illustrates the measured E-plane radiation pattern of both fabricated array prototypes. It can be observed that the improvement of the measured gain of the one with quasi-cavity-backed elements over the one without the cavity-backed elements is about 0.62dB, which is slightly less than the theoretic gain of 0.84dB as revealed in Table 6.7. The measured gain to the experimental QCBP array and patch array are 23.53dB and 22.91dB, respectively. The measured gain is about 0.7dB less than the simulated result. This difference is possibly caused by the mismatch of the junctions in the feeding the network, which is not accounted for in the loss analysis.

Based on the application demands, there are other design examples of microstrip antenna arrays considering simple structure, easy feeding, wide bandwidth, high gain, and low cross-polarization level[70],[71].

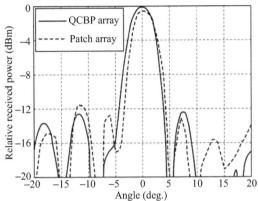

Figure 6.53   Measured E-plane radiation patterns of the array using patch elements and the QCBP elements. (After Huang, Wu, Fang, and Ehlert [28], © 2005 IEEE)

6.5   Mutual Coupling in Finite Microstrip Antenna Arrays

6.5.1   Mutual Coupling Effects and Analysis

Mutual coupling between the antenna elements in an antenna array is a potential source of performance degradation, particularly in a highly congested environment and in the applications of phased arrays, ultra-low side lobe level arrays, and adaptive nulling arrays. A lack of knowledge of the correct in-array impedance value may result in impedance mismatch, errors in the desired pattern and increased side lobe level. The usual element-by-element approach does not include the effects of scattering from the neighboring elements, because, in obtaining the mutual impedance between two elements of an array, the physical presence of other open-circuited elements is ignored. Accurate evaluation of mutual coupling effects could be achieved by full wave analysis through numerical methods, e.g., the method of moments (MoM). However, it is time consuming and needs large computer resources, especially for large arrays. When it is used to calculate moderate sized arrays, some approximations are made to alleviate the burden of computation. One approximation technique is to use only a few complex shaped basis functions for every antenna element[33]. This technique is based on the fact that the currents on the radiator are caused by the feeding source and the field originating from surrounding elements. Another technique is to reduce the time consuming Sommerfeld form of the Green's function to a relatively simple and accurate closed-form Green's function[34]. In [35, 36], approximate analytical expressions for spatial Green's functions taking into account the terms proportional to $1/\sqrt{R}$, $1/R$, and $1/R^2$ are developed. A new expansion wave concept is then introduced to solve the integral equations describing mutual coupling in the microstrip arrays. This approach yields a much smaller number of unknowns than with conventional rigorous technique without losing the accuracy. Actually, for large arrays, this still results in a large number of unknowns. The formulas presented in [33–36] include the effects of scattering from neighboring elements on the mutual impedance of an array as well as edge effects. However, these methods still require an intensive computation effort even if some approximations have been made, because the whole array problem still must be handled.

There is also much research work devoted to rapid numerical techniques to reduce the computation effort in full wave analysis. One powerful numerical technique is CG-FFT[37–39], which combines the conjugate gradient (CG) method with the fast Fourier transform (FFT).

Another more efficient method was proposed in [40], where the integral equation was discretized in the spatial domain by means of a full wave discrete image technique[41, 42]. The resulting system was solved using BCG algorithm in conjunction with FFT. Also, a fast array decomposition method (ADM) for accurately modeling finite arrays of complex three-dimensional structure was developed[43]. However, the analysis of a large finite array, requiring a massive number of unknowns to model, still remains a challenging problem. An alternative way to deal with the large finite array is to approximate it as an infinite one, thus the problem is reduced to the analysis of one single unit using a Floquet-type representation of fields[44, 45]. To account for the edge effects existing in a practical finite array, a windowing approach was proposed to obtain the characteristics of the finite array through convolving the infinite-array characteristics with the amplitude distribution function of the array[46]. This concept was extended to analyze the finite array by means of a finite array periodic Green's function, with an iterative approach to estimate current distribution functions for the array elements[47]. The final solutions for current distributions are used to construct the window function for the finite array. The method of infinite array combined with convolution technique was derived for large array analysis but has given good results also for small arrays. This approach is attractive for its simplicity; however, it may still be unsatisfactory in predicting input impedance of elements close to the edge of the array, especially when studying aperture arrays on ground planes[48]. To understand and describe the edge physical mechanism in a full-wave scheme, a hybrid asymptotic MoM method has been developed for analyzing the large periodic arrays[48, 49]. This work is primarily concerned with the efficient evaluation of currents on the array elements, not only for the assessment of truncation effects on the radiation pattern, but especially for the determination of the antenna input parameters (impedance and/or scattering parameters). However, the applicability of the present formulation is restricted to the case where the surface wave excitation is not significant and the array must be of periodic structure. Another Floquet-model-based analysis is presented in [50], where the self and mutual impedance between elements in an array environment are obtained from Floquet impedances, i.e., the active impedances of an element in an infinite array, by assuming that the array elements are excited with uniform amplitude and linear phase distributions in both directions. The technique is appropriate for estimating the performance of finite-array antennas with arbitrary amplitude and phase distributions and is simple for handling complex-element structures. However, it does not take into consideration the edge effects.

One method that can dramatically reduce the time consumption is the element-by-element analysis which is based on circuit theory[2]. Each antenna element in the array is taken as a basis function in the moment method, and elements of the impedance matrix are self and mutual impedance in isolated environment[51, 52]. However, although this procedure is much better than that of others where mutual interactions are not considered, it is not always successful. The reason is that the influence of the array environment is not taken into account. This influence comes from the surface current on parasitic elements even when they are open-circuited and can be involved in both the finite linear dipole array and finite microstrip patch array[53–55].

6.5.2   Mutual Coupling in a Linear Dipole Array of Finite Size

A. Element-by-Element Method

In the conventional element-by-element method, each element in the array is taken as a base function. For an N-element array, the equivalent generator voltage and current at each element terminal is related by the following matrix equation:

$$([Z] + [Z_T])[I] = [V_g] \qquad (6.5.1)$$

Where $[V_g]$ is the equivalent generator voltage and $[I]$ is the terminal current at the port of each patch, respectively. $[Z_T]$ is the generator terminating impedance matrix. The diagonal terms of the moment matrix $[Z]$ are self-impedances, and the off-diagonal terms are the mutual impedances. $[Z]$ is the impedance matrix where the diagonal elements are self-impedances and the off-diagonal ones are mutual impedances.

The general voltage excitation can be written as

$$V_{gn} = A_n e^{jk_0(ux_n + vy_n)} \qquad (6.5.2)$$

where, $u = \sin\theta\cos\phi$, $v = \sin\theta\sin\phi$, (x_n, y_n) is the nth element coordinate, (θ, ϕ) is the scan angle, and A_n is the amplitude of the voltage excitation at the nth element. Given the generator voltage $[V_g]$, the induced current $[I]$ can be solved from equation (6.5.1). This current $[I]$ is not of the same distribution of excitation $[V_g]$ due to the influence of mutual coupling.

Note that the formula is capable of treating the forced excitation case ($Z_T = 0$) as well as the free-excitation case ($Z_T \neq 0$). The forced excitation may be used to find the radiation impedances of the arrays without a feed network. When the array is terminated with a feed network, the free-excitation model should be used.

The input impedance looking into the nth element, which is a function of scan angle θ and ϕ, can be written as

$$Z_{in}^n(\theta, \phi) = \frac{V_n}{I_n} \qquad (6.5.3)$$

Form (6.5.3) is the input impedance of the element with mutual coupling effect taken into consideration. The active reflection coefficient at the nth port can then be calculated as [56]

$$|R^n(\theta, \phi)| = \left| \frac{Z_{in}^n(\theta, \phi) - Z_{in}^n(0,0)}{Z_{in}^n(\theta, \phi) + Z_{in}^{n*}(0,0)} \right| \qquad (6.5.4)$$

The conventional way of finding the impedance matrix $[Z]$ in (6.5.1) is to assume that the parasitic open-circuited elements have no influence on the self and mutual impedances. Based on this assumption, the diagonal elements of the matrix $[Z]$ are equal and each is the self-impedance of an element, and the off-diagonal ones are the mutual-impedances between two related elements. The properties of single radiators as well as the mutual coupling in two element arrays have been investigated by a number of authors, and could be found through accurate and rigorous full wave analysis or some simplified methods.

B. Extra Port Method

In the past, the finite array of printed dipoles has been studied by the element-by-element analysis[56], in which it is emphasized that in formulating the impedance matrix, the only approximation is in limiting the number of expansion modes per element, and that the presence of all elements is accounted for in the complete solution. In fact, if this single-mode approximation is assumed, the resulting self and mutual impedances calculated by a spectral domain moment method in an impedance matrix are still obtained from an isolated environment.

It can be observed from (6.5.1) that mutual interactions between elements are accounted for through currents at each port. However, due to electromagnetic interactions, surface current will be induced on parasitic elements when an element in the array is excited. If isolated self and mutual impedances are used in (6.5.1), the influence of this current can not be found, and there will be errors between full wave analysis and the conventional element-by-element method. To solve this problem, a method of defining extra ports on parasitic elements was proposed to take account of induced surface current[53], thus to get accurate impedance matrix of dipole arrays.

Figure 6.54 is the illustration of finding the in-array self impedance of dipole 1 with a parasitic dipole 2 presented. When dipole 1 exited with a 1A current source, it will induce surface current on itself as well as on the parasitic element. Due to the open circuit boundary condition, the distribution of surface current should take the form as shown in Figure 6.54(b). It is observed that the current distribution on the parasitic dipole is like two separate excited dipoles. Then, we can divide the parasitic dipole into two sub dipoles, and define an extra port at the center of each sub dipole to identify the induced surface currents as shown in Figure 6.54(c). Through this arrangement, the parasitic dipole could be considered as two sub dipoles with the ports short-circuited. Therefore, their influences could be found through the conventional element-by-element method, where only isolated impedance values are needed.

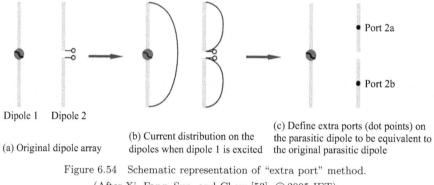

Dipole 1   Dipole 2

(a) Original dipole array

(b) Current distribution on the dipoles when dipole 1 is excited

(c) Define extra ports (dot points) on the parasitic dipole to be equivalent to the original parasitic dipole

Figure 6.54   Schematic representation of "extra port" method.

(After Xi, Fang, Sun, and Chow [53], © 2005 IET)

Similarly, in finding the in-array mutual impedance $Z_{i,j}$ of an N-element dipole array, all the parasitic elements are divided into two sub-elements. The voltage and current at each element terminal could be related by

$$
\begin{bmatrix} V_{1a} \\ V_{1b} \\ \vdots \\ V_i \\ \vdots \\ V_j \\ \vdots \\ V_{Na} \\ V_{Nb} \end{bmatrix} = \begin{bmatrix} Z_{1a,1a}^{iso} & Z_{1a,1b}^{iso} & \cdots & Z_{1a,i}^{iso} & \cdots & Z_{1a,j}^{iso} & \cdots & Z_{1a,Na}^{iso} & Z_{1a,Nb}^{iso} \\ Z_{1b,1a}^{iso} & Z_{1b,1b}^{iso} & \cdots & Z_{1b,i}^{iso} & \cdots & Z_{1b,j}^{iso} & \cdots & Z_{1b,Na}^{iso} & Z_{1b,Nb}^{iso} \\ \vdots & \vdots & & \vdots & & \vdots & & \vdots & \vdots \\ Z_{i,1a}^{iso} & Z_{i,1b}^{iso} & \cdots & Z_{i,i}^{iso} & \cdots & Z_{i,j}^{iso} & \cdots & Z_{i,Na}^{iso} & Z_{i,Nb}^{iso} \\ \vdots & \vdots & & \vdots & & \vdots & & \vdots & \vdots \\ Z_{j,1a}^{iso} & Z_{j,1b}^{iso} & \cdots & Z_{j,i}^{iso} & \cdots & Z_{j,j}^{iso} & \cdots & Z_{j,Na}^{iso} & Z_{j,Nb}^{iso} \\ \vdots & \vdots & & \vdots & & \vdots & & \vdots & \vdots \\ Z_{Na,1a}^{iso} & Z_{Na,1b}^{iso} & \cdots & Z_{Na,i}^{iso} & \cdots & Z_{Na,j}^{iso} & \cdots & Z_{Na,Na}^{iso} & Z_{Na,Nb}^{iso} \\ Z_{Nb,1a}^{iso} & Z_{Nb,1b}^{iso} & \cdots & Z_{Nb,i}^{iso} & \cdots & Z_{Nb,j}^{iso} & \cdots & Z_{Nb,Na}^{iso} & Z_{Nb,Nb}^{iso} \end{bmatrix} \begin{bmatrix} I_{1a} \\ I_{1b} \\ \vdots \\ I_i \\ \vdots \\ I_j \\ \vdots \\ I_{Na} \\ I_{Nb} \end{bmatrix} \quad (6.5.5)
$$

The superscript *iso* denotes Z values in an isolated environment. $V_i(V_j)$ and $I_i(I_j)$ stand for the voltage and current, on the un-split dipole. The voltages on all the sub dipoles $V_{1a}, V_{1b}, \cdots, V_{Na}, V_{Nb}$ are set to be zero for the short-circuit condition, which corresponds to the state that all the parasitic elements are open-circuit.

It should be noted that in (6.5.5), all elements in the Z impedance matrix are isolated values and can be found through available methods.

For the calculation of self-impedance in the array environment ($i = j$), we assume that V_i is the total voltage on port i when the ith feed line is driven with a 1A source ($I_i = 1A$). Through solving the above matrix equation (6.5.5), the self-impedance in an array environment is easily found as:

$$Z_{i,i}^{inar} = \frac{V_i}{I_i}, \qquad i = 1, 2, \cdots, N \tag{6.5.6}$$

For the calculation of mutual impedance in the array environment ($i \neq j$), we assume that V_i is the total voltage on port i when the jth feed line is driven with a 1A source ($I_j = 1A$). Then, the mutual impedance in an array environment $Z_{i,j}^{inar}$ between element i and j is

$$Z_{i,j}^{inar} = \frac{V_i}{I_j}, \qquad i, j = 1, 2, \cdots, N \tag{6.5.7}$$

Replacing the original matrix $[Z]$ in formula (6.5.1) with the resulting in-array impedance matrix $[Z^{inar}]$, with the influence of parasitic elements taken into consideration, we have

$$([Z^{inar}] + [Z_T])[I] = [V_g] \tag{6.5.8}$$

With the generator voltage $[V_g]$ given, the accurate current $[I]$ could be solved from equation (6.5.8) by matrix inversion.

C. Examples

Several examples are given to validate the method. The correction procedure described above is first applied to achieve the corrected impedance matrix in an array environment. The results obtained from this impedance matrix are noted as "corrected result." Then these results are compared with the so-called "uncorrected result" and "full wave result."

The exact Z_{14} value is first compared with the result by using our method for a four uniformly spaced dipole array in Figure 6.55. Also shown is the isolated Z_{14} value with the inner two dipoles absent. Each dipole is assumed to have a length l, and a width w. They are printed on a substrate of thickness h, having a relative dielectric constant ϵ_r.

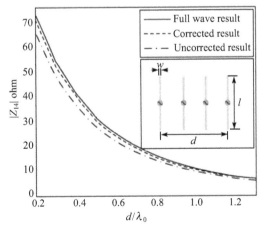

Figure 6.55   Comparison of exact, uncorrected and corrected values of Z_{14} in a four uniformly spaced dipole array. $f = 5$GHz, $\epsilon_r = 2.55$, $l = 0.478\lambda_0$, $w = 0.001\lambda_0$, $h = 0.15\lambda_0$.
(After Xi, Fang, Sun, and Chow [53], © 2005 IET)

From Figure 6.55, it is seen that although there is only a small difference between the full-wave result and the uncorrected one, we still find that the difference from results of the extra port method is even smaller.

However, the difference becomes larger in a scanning array. An example of a 3×3 array of center driven dipoles is shown in Figure 6.56. Each dipole is uniformly spaced by a distance d in x direction as well as in y direction. The magnitude of the active reflection coefficient for the center and edge element is plotted versus the E-plane scan angle in Figure 6.57(a) and (b), respectively. It is observed that the corrected solution converges with the exact solution in both cases.

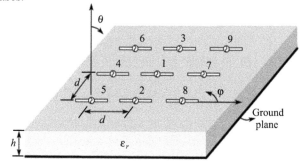

Figure 6.56   Geometry of uniformly spaced 3×3 dipole array.
$f = 5\text{GHz}$, $d = 0.37\lambda_0$, $\epsilon_r = 2.55$, $l = 0.478\lambda_0$, $w = 0.001\lambda_0$, $h = 0.15\lambda_0$.
(After Xi, Fang, Sun, and Chow [53], © 2005 IET)

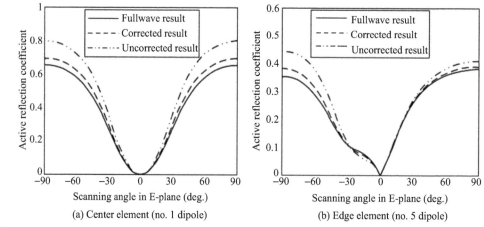

(a) Center element (no. 1 dipole)   (b) Edge element (no. 5 dipole)

Figure 6.57   Active reflection coefficient versus the E-plane scan angle for the 3×3 printed dipole array.
(After Xi, Fang, Sun, and Chow [53], © 2005 IET)

Assume next that we have a 1×5 linear scanning array with a spacing $d = 0.37\lambda_0$ and a dipole length of $l = 0.924\lambda$, as is shown in Figure 6.58. The parameters of this example are deliberately chosen to be more critical, in order to clearly illustrate the effectiveness of our method. Figure 6.59(a),(b) shows the active reflection coefficient of the center and edge element when the array scans in the H-plane.

The resulting H-plane radiation pattern is shown in Figure 6.60. Assume that a –40dB Tschebyscheff endfire pattern is desired. Without considering the mutual coupling effect, the array factor is plotted as the solid line. Now the mutual coupling effect is added by the element-by-element method using the corrected impedance matrix obtained from this method (broken line). Compared with full wave results from IE3D^{TM} (dashed line), the result of this method achieves a satisfying accuracy. For further comparison, also presented is the result from an isolated impedance matrix which is far from the full wave solution.

It is observed from the above simulated results, that compared with the rigorous MoM using small rooftop basis functions, the extra port method gives a more accurate result than the conventional element-by-element method does.

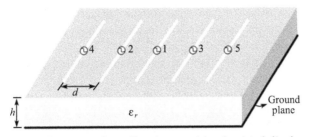

Figure 6.58 Geometry of a uniformly spaced 1×5 printed dipole array.
(After Xi, Fang, Sun, and Chow [53], © 2005 IET)

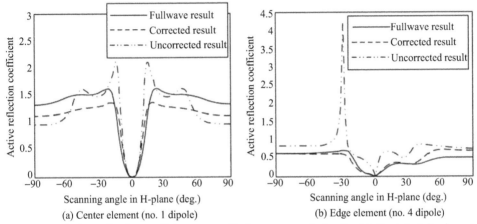

(a) Center element (no. 1 dipole)

(b) Edge element (no. 4 dipole)

Figure 6.59 Active reflection coefficient versus H-plane scan angle for the 1×5 printed dipole array.
(After Xi, Fang, Sun, and Chow [53], © 2005 IET)

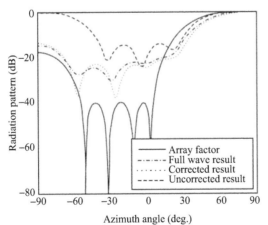

Figure 6.60 The H-plane radiation pattern of a 1×5 Tschebyscheff endfire.
(After Xi, Fang, Sun, and Chow [53], © 2005 IET)

6.5.3 Mutual Coupling in Finite Microstrip Patch Arrays

A. The Application of the Mutual Coupling Formula

The formula (5.4.4) for mutual coupling is very accurate compared with full wave analysis[57], based on the assumption that the patches concerned are isolated, without any nearby patches. This provides a first approximation of the array impedance matrix which we called an isolated impedance matrix $[Z^{iso}]$ because all the parasitic elements are ignored. A correcting procedure has been proposed in [54] which only needs the knowledge of the isolated element input impedance and mutual impedance of two isolated elements to finally lead to a corrected matrix $[Z^{inar}]$, that is, the in-array impedance matrix, which is a good approximation of the real one. Because this procedure makes use of the surface mode currents on the patches to calculate the mutual coupling effects, the current induced on the parasitic elements would naturally be involved. The primary relations are presented in the following.

Consider patch m in an N patch array, let v_m be the dominant mode voltage produced by a unit amplitude current mode. Then the port voltage V_m is

$$V_m = a_m v_m + j I_m X_{fm} \qquad (6.5.9)$$

where I_m is feed current, X_{fm} is the feed reactance of patch m; the dominant mode current amplitude on the mth patch may be written as

$$a_m = e_m I_m + \sum_{n \neq m} a_n M_{mn} \qquad (6.5.10)$$

where M_{mn} is the mutual coupling coefficient between element m and n, provided the assumption is made that only the dominant modes contribute to the coupling. e_m is the excitation coefficient. According to the definition of impedance matrix, the corrected matrix $[Z^{inar}]$ can be obtained through a series of matrix manipulations

$$[Z^{inar}] = [v]([U] - [M])^{-1}[e] + j[X_f] \qquad (6.5.11)$$

where $[U]$ is the identity matrix and $[v], [e], [X_f]$ are the diagonal matrices of dimension $N \times N$. In the case of identical patches and feeds

$$[e] = e[U], [v] = v[U], [X_f] = X_f[U],$$

then (6.5.11) reduces to

$$[Z^{inar}] = Z_S^{iso}([U] - [M])^{-1} + j X_f[U] \qquad (6.5.12)$$

where Z_S^{iso} is the "subtracted" input impedance of an isolated patch — the total input impedance minus the feed reactance; $[M]$ is the mutual coupling coefficient matrix. The terms M_{mn} can be determined by relating to the isolated mutual impedance Z_{mn}^{iso}, i.e., Z_{ab} of (5.4.4) in the following form

$$Z_{mn}^{iso} = \frac{v_m e_n M_{mn}}{1 - M_{mn} M_{nm}} \qquad (6.5.13)$$

Assuming identically shaped patches ($M_{mn} = M_{nm}$) and identical feeds ($e_m = e_n$), $[M]$ can be obtained by

$$M_{mn} = \frac{1}{2 Z_{mn}^{iso}}[-Z_S^{iso} + [(Z_S^{iso})^2 + 4(Z_{mn}^{iso})^2]^{1/2}], \qquad m \neq n \qquad (6.5.14)$$

$$M_{mn} = 0, \qquad\qquad\qquad m = n \qquad (6.5.15)$$

If the row and column norms of $[M]$ are less than one in magnitude, meaning the coupling is sufficiently weak, the following expansion for terms in (6.5.12) may be used to avoid the matrix inversion,

$$([U] - [M])^{-1} = [U] + [M] + [M]^2 + \cdots \qquad (6.5.16)$$

Finally, the resulting in-array matrix $[Z^{inar}]$ replaces the original matrix $[Z]$ of the conventional element-by-element method, that is

$$([Z^{inar}] + [Z_T])[I] = [V_g] \qquad (6.5.17)$$

With the generator voltage $[V_g]$ given, the accurate current $[I]$ could be solved from equation (6.5.17) by matrix inversion.

Fortunately, a formula for the self-impedance of an isolated patch is available[58, 59]. And instead of using the moment method, the mutual impedances terms $Z_{mn}^{iso}(m \neq n)$ between two isolated patches can be calculated by the closed form formula (5.4.4), which can significantly reduce the computer time, thus making the correction procedure more efficient in analyzing the finite array. The reason is that the time consumed in calculating the formula number to be determined can almost be neglected, compared with carrying the MoM for each pair of $Z_{mn}^{iso}(m \neq n)$. As a result, the larger the array, the more time can be saved. The only limitation comes from the assumption of the Jackson's procedure that the elements are characterized by a single radiation mode. However, this assumption will not lead to unacceptable errors, because a microstrip patch is a highly resonant structure, and near the resonance its current distribution can be well approximated by a single mode. This distribution is not greatly affected by the proximity of nearby elements in an array assuming that the elements are not too closely spaced[58].

B. Numerical Examples

The mutual coupling between elements will change the active radiation pattern of a finite array. It is also the cause of a blind angle in large phased arrays with a wide scanning range. These effects of mutual coupling are investigated with some examples in the following.

Consider a 5×5 microstrip patch array, as shown in Figure 6.61. The coordinates of two coupling patches in an array are shown in Figure 6.62. The array is taken as having uniformly spaced elements of identical shape, but this is not a necessary restriction.

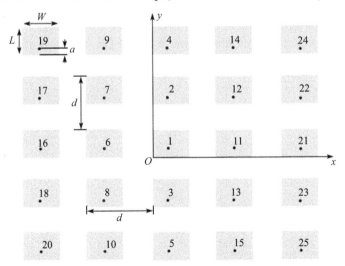

Figure 6.61   Geometry of a 5×5 microstrip patch array, $W = 22$mm, $L = 17.4$mm,
a probe feed is near the center of W with distance $a = 5$mm, $\epsilon_r = 2.55$,
substrate thickness $h = 1.57$mm, resonant frequency $f_0 = 5$GHz.

The mutual impedance formula (5.4.4) combined with the correction procedure in [54] is first applied to obtain the corrected impedance matrix in an array environment. The results obtained from this impedance matrix are noted as a "corrected result." Then these results are compared with the so-called "uncorrected result" (solution obtained from element-by-element method using an isolated impedance matrix) and the "full wave result" (solution from the exact impedance matrix calculated by the full wave analyzer IE3D^{TM}).

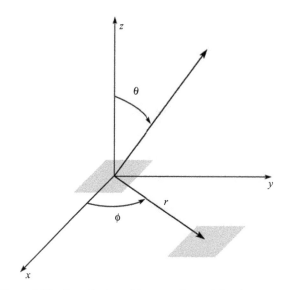

Figure 6.62   Coordinates of two coupling patches in an array.

As is well known, the input impedance of a certain element in an array generally differs from that of an isolated element because of mutual coupling. This effect is caused by the difference in coupling environments presented to each element. The active element input impedance also depends on the array excitation and therefore varies with the scan angle. In Figure 6.63(a), (b), the active reflection coefficient of the center element in the 5×5 array is plotted against beam scan angle θ in the H-plane and the E-plane, respectively. It is assumed that the array is conjugate matched to its broadside scan impedance. It is observed that the mutual coupling causes impedance mismatch of the patch. Compared with full wave results of the whole array, the element-by-element method using the corrected impedance matrix is shown to be more accurate than using an uncorrected one.

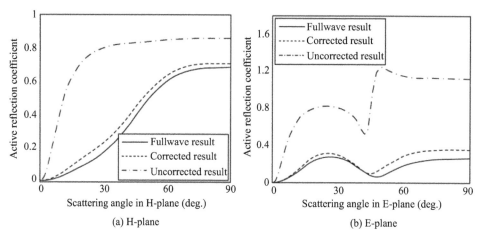

Figure 6.63   Active reflection coefficient versus scan angle for center patch of the 5×5 array.
(After Xi, Fang, Sun, and Chow [55], © 2005 Wiley)

Next let us assume that a −40dB Tschebyscheff pattern with a scanning angle $\theta = 20°$ is desired. Shown in Table 6.8 is the current distribution at the feed terminals of the elements in the array. The "uncorrected" currents are calculated by using the isolated impedance matrix. The "corrected" currents are obtained by corrected procedure and the "exact" currents are from full wave analysis. Although the former two methods both can incorporate the effect of mutual coupling, it is seen that the corrected results will be more accurate in predicting the current distortion, and is also validated by the E-plane pattern shown in Figure 6.64. It is observed that the side lobe level is far from the desired one of -40dB due to the influence of mutual coupling. Compared with full wave results from IE3D^{TM}, the corrected results achieve a more satisfying accuracy than uncorrected ones. For further comparison, also presented is the array factor which is obtained from standard Tschebyscheff synthesis where isotropic radiators (without mutual coupling) are assumed.

Table 6.8 Comparison between uncorrected, corrected and exact currents at element terminals. (After Xi, Fang, Sun, and Chow [55], © 2005 Wiley)

Element Number	Uncorrected Mag. A	Phase deg.	Corrected Mag. A	Phase deg.	Exact Mag. A	Phase deg.
1	0.200	−35.8	0.214	−21.7	0.215	−21.4
2	0.147	37.3	0.145	42.8	0.148	41.3
3	0.146	−95.0	0.162	−84.0	0.166	−81.4
4	0.025	132.0	0.035	102.1	0.035	98.5
5	0.072	−166.3	0.064	−148.0	0.068	−146.0
6	0.210	−34.6	0.234	−23.8	0.237	−24.2
7	0.149	36.0	0.154	40.9	0.158	39.5
8	0.158	−95.9	0.177	−87.7	0.181	−86.7
9	0.027	120.0	0.036	96.6	0.036	92.7
10	0.069	−169.7	0.063	−153.1	0.066	−152.4
16	0.181	−22.1	0.194	−16.5	0.197	−17.0
17	0.127	43.8	0.130	47.45	0.132	46.3
18	0.139	−84.9	0.145	−79.8	0.147	−79.0
19	0.031	115.9	0.032	101.8	0.032	97.4
20	0.052	−158.8	0.051	−141.6	0.053	−139.5

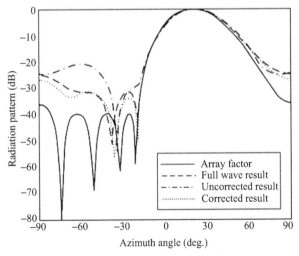

Figure 6.64 E-plane radiation pattern of a 5 × 5 Tschebyscheff array; the scanning angle is 20°. Spacing between elements is $d = 0.5\lambda_0$. (After Xi, Fang, Sun, and Chow [55], © 2005 Wiley)

For demonstration purposes, the computer time consumed in getting the exact impedance matrix at a single frequency point is given in Table 6.9 to validate the efficiency of the presented approach. All the results are obtained on a PC P4/2.0G. The rigorous moment method used for comparison is conducted with a mesh size of 20 cells per wavelength.

Table 6.9   Comparison of the computer time in obtaining the exact impedance matrix using different methods. (After Xi, Fang, Sun, and Chow [55], © 2005 Wiley)

	Method I	Method II	Method III
Time	1918	$0.25^{\#}$	168.25

- Method I: Full wave analysis;
- Method II: Jackson's approach (sampling by formula (1));
- Method III: Jackson's approach (sampling by MoM);
- #: This value does not include the time (approximately 84 seconds) for constructing the formula described in Section 2. Actually, this part can be regarded as a preprocessor. Once the formula is constructed, the time consumed in computing this analytical expression can be neglected.

6.6   Introduction to a Digital Beamforming Receiving Microstrip Antenna Array

In the design of a DBF microstrip antenna array, the challenging problems include the uniformity, cost, and mutual coupling. There are some possible solutions to these problems, such as spatial multiplexing technique[65], the techniques of using phase weighting and angle weighting that was introduced in Chapter 2[66, 67]. The basic concept of these techniques is to realize the DBF function through one channel. The amplitude and phase distribution on the aperture is obtained by time sequence. However, the strategy of "using time to buy space" is not always feasible. In this section, we will focus on the design of the full DBF microstrip antenna array. The main topics involved are the element uniformity, the reduction of the mutual couplings and the prediction of adaptive nulling performance in a DBF microstrip antenna array.

6.6.1   Description of the Antenna Array

In order to satisfy both the wide bandwidth and the simplicity requirements, the U Slot patch antenna was chosen. The optimization design was determined using the software IE3D^{TM}. The prototype of this antenna and its characteristics are shown in Figure 6.65 (a)–(d).

An 8×8 array was formed as shown in Figure 6.66. The VSWR characteristics of the 64-elements measured in the array environment with the elements other than the measured ones terminated by open circuit loads and matching loads are shown in Figure 6.67(a), (b). In the figures, the curves of eight elements, for example, elements no.1–no.8, are of the same type. It is seen that the VSWR characteristics satisfy the requirements very well.

6.6.2   Mutual Coupling Reduction of the Microstrip Antenna Array

We have seen in Section 6.5 that the mutual coupling effects may deteriorate the performance of the antenna array. In some critical cases, it is desired to reduce the mutual coupling. The reduction of the mutual coupling may be realized through the sieve test of the antenna elements. Four kinds of antenna elements with 10% bandwidth (VSWR< 2) are examined. They are coupled rectangular microstrip patches with an air gap (EMC patch), the conventional rectangular patch antenna (Patch), the patch antenna with a U-shape slot (U-slot), and the wideband stacked dipole antenna (Dipole). The simulated results for mutual couplings of the E-plane and the H-plane are shown in Figures 6.68 and Figure 6.69. It is seen that the mutual coupling of the EMC patch decays monotonically and rapidly as spacing increases. These results are useful in the application of adaptive nulling and will be discussed later.

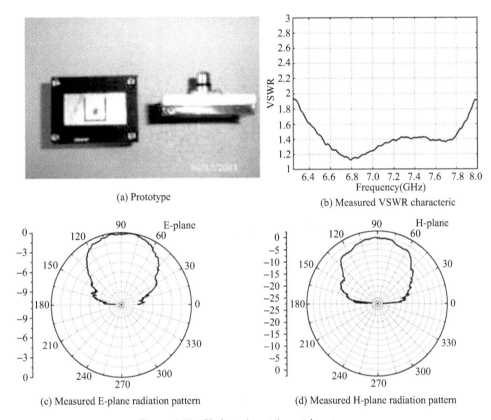

(a) Prototype

(b) Measured VSWR characteric

(c) Measured E-plane radiation pattern

(d) Measured H-plane radiation pattern

Figure 6.65   U-slot microstrip patch antenna.

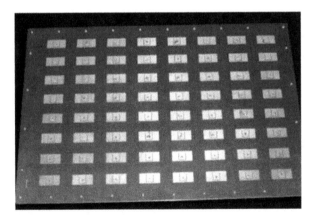

Figure 6.66   Picture of an 8 × 8 U-slot microstrip patch antenna array.

Another effective way for reducing the mutual coupling effects is utilizing the EBG structure[60, 61]. Based on the given operating frequency, the prototype of two patches with uniplanar compact electromagnetic bandgap (UC-EBG) structure was designed. The central spacing between two patches is $0.75\lambda_0$. The investigation is for closer spacing.

Figure 6.70 shows that the mutual coupling has at least 6dB reduction within the bandwidth from $6.56 - 7.04$GHz (VSWR< 2) by using the UC-EBG structure. Moreover, this UC-EBG structure is very easy to manufacture making it very attractive in high performance antenna array design. Especially in the digital beam forming antenna, there is plenty of space to locate the EBG structure because there is no RF feed network.

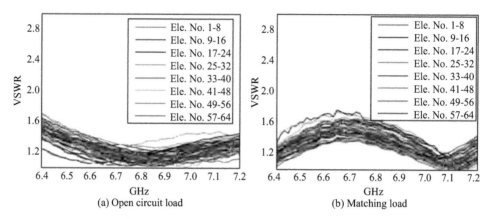

(a) Open circuit load   (b) Matching load

Figure 6.67   VSWR characteristics of an U-slot microstrip patch antenna array.

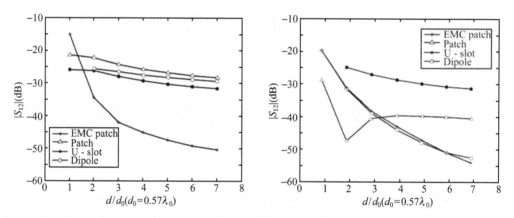

Figure 6.68   Normalized central spacing in E-plane.   Figure 6.69   Normalized central spacing in H-plane.

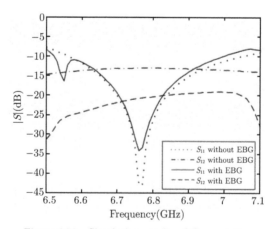

Figure 6.70   Simulation results of S_{11} and S_{12}.

6.6.3   Adaptive Nulling

An important function of an adaptive array is to suppress interferences. This is achieved by steering the nulls of the radiation pattern toward the interferences. However, the depths and the accuracy of the positions of the nulls will be significantly affected by the existence of the mutual coupling between the antenna elements. To take into account the mutual coupling, a full wave solver should be used. Four kinds of antenna elements given in Section 6.6.1 were examined. It was found that the EMC patch antenna element is the best one. The mutual coupling and the null depth were calculated for a 1×8 linear array at three different frequencies (6.5GHz, 6.8GHz, and 7.1GHz). The geometry of the array and the configuration of the interferences are shown in Figure 6.71. Angles θ_0, θ_1, θ_2, $\cdots$ are all smaller than $45°$, and the minimum difference allowed between them is $\Delta\theta$. The null depth was calculated at three different frequencies when the number of the interference sources is one, two and three. The results of in-array mutual coupling are given in Figure 6.72 and those of the null depth are given in Table 6.10.

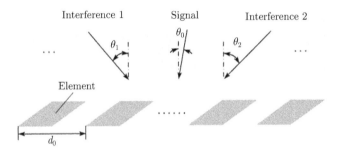

Figure 6.71   Geometry of the array.

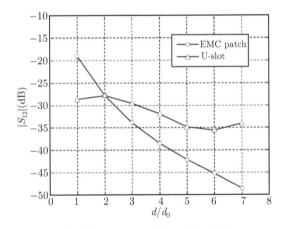

Figure 6.72   In-array mutual coupling of the arrays.

Table 6.10   **Null depth for two kinds of the antenna arrays** *.

Case	Num. of interf-erence sources	Type of Antennas	$\Delta\theta$ (deg.)	Average (dB)	Mean square Error(dB)	Worse (dB)	Best (dB)
1	1	U-slot	$10°$	-32.1	6.62	-16.0	-47.1
2	1	EMC Patch	$10°$	-38.7	6.92	-28.0	-59.0
3	2	U-Slot	$8°$	-32.3	6.67	-18.2	-56.1
4	2	EMC Patch	$8°$	-36.8	7.72	-17.6	-59.6
5	3	U-Slot	$5°$	-27.8	8.02	-8.3	-45.4
6	3	EMC Patch	$5°$	-30.5	9.01	-2.4	-57.6

* For each case, the number of the statistic tests related to θ is 36.

The ideal weights and IE3D software were used when calculating the null depth in Table 6.10. The results may be somewhat different with different software. Once the mutual coupling is reduced below a certain level, the simple compensation measure will no longer be effective. Moreover, for a microstrip patch antenna, the perfect compensation of the distortion of the current distribution is almost impossible. The reason is that the compensation can only be carried out on port level rather than on the base function level. This fact suggests that the mutual coupling reduction is more effective and important than the mutual coupling compensation in adaptive nulling[62].

In addition to the analysis in the frequency domain, the full-wave integrated time-domain analysis of the microstrip array antenna introduced in [68] is also a good candidate. In the analysis, a Gaussian pulse excitation is used, and the frequency response of the array antenna is effectively obtained over a broad frequency range via Fourier transform of the time-domain field response.

The reduction of mutual coupling between two finite microstrip antenna arrays is also very important in applications. A full wave analysis-based network method is proposed in [69] to perform the optimization design of mutual coupling reduction in certain bandwidths. The experiment result of −10dB mutual coupling reduction in 8% bandwidth has been achieved and is in good agreement with that from the simulation.

Bibliography

[1] J. Ashkenazy, P. Perknutter, and D. Treves, "A modular approach for the design of microstrip array antennas," *IEEE Trans. Antennas Propagat.*, vol. 31, no. 1, pp. 190–193, Jan., 1983.

[2] A. A. Oliner, R. G. Malech, "Mutual coupling in infinite scanning arrays," in *Microwave Scanning Antennas*, vol. 2, Chapter 3, R. C. Hansen, (Ed.) Peninsula Publishing, 1985.

[3] K. M. Lee and R. S. Chu, "Analysis of mutual coupling between a finite phased array of dipoles and its feed network," *IEEE Trans. Antennas Propagat.*, vol. AP-36, no. 12, pp. 1681–1699, Dec., 1988.

[4] K. C. Lee and T. H. Chu, "A circuit model for antenna array mutual coupling effects," *IEEE Antennas Propagation Symp. Dig.*, pp. 946–949, 1995.

[5] J. R. Mosig and F. E. Gardiol, "General integral equation formulation for microstrip antennas and scatterers." *IEE. Proc.*, Pt.H, vol. 132, no. 7, pp. 424–432, 1985.

[6] K. A. Michalski and D. Zheng, "Electromagnetic scattering and radiation by surface of arbitrary shape in layered media, part I: theory," *IEEE Trans. Antennas Propagat.*, vol. 38, no. 3, pp. 335–344, Mar., 1990.

[7] Zeland Software Inc. IE3D^{TM} *User's Manual*, Release 9.

[8] A. G. Derneyd, "Linearly polarized microstrip antennas," *IEEE Trans. Antennas Propagat.*, vol. 29, no. 6, pp. 174–178, Nov., 1981.

[9] T. Metzler, "Microstrip series arrays," *IEEE Trans. Antennas Propagat.*, vol.24, no. 1, pp. 846–851, Jan., 1976.

[10] J. Huang, "A parallel-series-fed microstrip array with high efficiency and low cross-polarization," *Microwave and Optical Tech. Lett.*, vol. 5, no. 5, pp. 230–233, May, 1992.

[11] J. R. James, P. S. Hall, and C. Wood, *Microstrip Antennas Theory and Design*, Peter Peregrinus, London, 1981.

[12] J. Huang and D. M. Pozar, "Microstrip arrays: analysis, design, and application," in *Advances in Microstrip and Printed Antennas* edited by K. F. Lee and W. Chen, John Wiley and Sons, Inc., 1997.

[13] D. M. Pozar, "Scanning characteristics of infinite arrays of printed antenna subarrays," *IEEE Trans. Antennas Propagat.*, vol. 40, no. 6, pp. 666–674, Jun., 1992.

[14] Editorial Group of Microstrip Circuits of Tsinghua University, *Microstrip Circuits*, People's Posts and Telecommunications Press, 1975.

[15] M. I. Aksun and R. Mittra, "Estimation of spurious radiation from microstrip etches using closed-form Green's functions," *IEEE Trans. Microwave Theory Tech.*, vol. 40, no. 11, pp. 2063–2069, Nov., 1992.

[16] J. R. James and P. S. Hall, *Handbook of Microstrip Antennas*, Peter Peregrinus Ltd., 1989.

[17] M. L. Oberhart, Y. L. To and R. Q. H. Lee, "New simple feed network for an array module of four microstrip elements," *Electronics Letters*, vol. 23, no. 9, pp. 436–437, 1987.

[18] R. Zhang, *Optimization and Design of Microstrip Antennas*, Master Thesis, Nanjing University of Science and Technology, 2001.

[19] K. R. Carver and J. M. Mink, "Microstrip antenna technology," *IEEE Trans. Antennas Propagat.*, vol. 29, no. 1, pp. 2–24, 1981.

[20] W. M. Yu, D. G. Fang, Y. P. Xi and W. X.Sheng, "The design of low side lobe microstrip antenna array," *Proceedings of National Conference on Electromagnetic Compatibility*, Oct., 2002, Tianjin, pp. 142–145.

[21] M. Jackson, "Low cost K-band microstrip patch monopulse antenna," *Microwave Journal*, vol. 30, no. 7, pp. 125–126, 1987.

[22] B. J. Andrews, T. S. Moore, A. Y. Niazi, "Millimeter wave microstrip antenna for dual polar and monopulse applications," *Third International Conference on Antennas and Propagation*, ICAP 83, 12–15 April 1983.

[23] S. G. Kim and K. Chang, "Low-cost monopulse antenna using bi-directionally-fed microstrip patch array," *Electronics Letters*, vol. 39, no. 20, pp. 1428–1429, 2003.

[24] S. C. Manuel, S. P. Manuel, V. I. Maria and F. J. J. Jose, "Low-cost monopulse radial line slot antenna," *IEEE Trans. Antennas Propagat.*, vol. 51, no. 2, pp. 256–262, 2003.

[25] H. Wang, D. G. Fang, and X. G. Chen, "A compact single layer monopulse microstrip antenna array," *IEEE Trans. Antennas Propagat.*, vol. 54, no. 2, pp. 503–509, 2006.

[26] D. M. Pozar and K. Barry, "Design considerations for low sidelobe microstrip arrays," *IEEE Trans. Antennas Propagat.*, vol. 38, no. 8, pp. 1176–1185, 1990.

[27] I. J. Bahl and P. Bhartia, "*Microstrip Antennas*," Artech House, Dedham, MA, 1980.

[28] Y. Huang, K. L. Wu, D. G. Fang and M. Ehlert, "An integrated LTCC mm-wave planar array antenna with low loss feeding network," *IEEE Trans. Antennas Propagat.*, vol. 53, no. 3, pp. 1232–1234, 2005.

[29] R. Zhang, D. G. Fang, K. L. Wu and W. X. Sheng, "Study on the elimination of surface wave by metal fences," *Proceedings Asia-Pacific Conference on Environmental Electromagnetics*, Shanghai, pp. 174–178, 2000.

[30] H. Uchimura, T. Takenoshita, and M. Fujii, "Development of a laminated waveguide," *IEEE Trans. Microwave Theory Tech.*, vol. 46, no. 12, pp. 2438–2443, Dec., 1998.

[31] Y. Huang, K. L. Wu and M. Ehlert, "An integrated LTCC laminated waveguide-to-microstrip line T-junction," *IEEE Microwave and Wireless Comp. Lett.*, vol. 13, no. 8, pp. 338–339, Aug., 2003.

[32] Y. Huang, K. L. Wu, "A broadband LTCC integrated transition of laminated waveguide to air-filled waveguide for millimeter wave applications," *IEEE Trans. Microwave Theory Tech.*, vol. 51, pp. 1613–1617, May, 2003.

[33] M. Kuipers, "Mutual coupling computation and effects in phased array microstrip antennas," *IEEE 48th Conference on Vehicular Technology*, pp. 1181–1185, 1998.

[34] S. Barkeshili, P. H. Pathak and M. Marin, "An asymptotic closed-form microstrip surface Green's function for the efficient moment method analysis of mutual coupling in microstrip antennas," *IEEE Trans. Antennas Propagat.*, vol. 38, no. 9, pp. 1374–1383, 1990.

[35] F. J. Demuynck, G. A. E. Vandenbosch and A. R. Van De Capelle, "The expansion wave concept-Part I: Efficient calculation of spatial Green's functions in a stratified dielectric medium," *IEEE Trans. Antennas Propagat.*, vol. 46, no. 3, pp. 397–406, 1998.

[36] G. A. E. Vandenbosch and F. J. Demuynck, "The expansion wave concept-Part II: A new way to model mutual coupling in microstrip arrays," *IEEE Trans. Antennas Propagat.*, vol. 46, no. 3, pp. 407–413, 1998.

[37] T. J. Peters and J. L. Volakis "Application of a conjugate gradient FFT method to scattering from thin planar material plates," *IEEE Trans. on Antennas Propagat.*, vol. 36, no. 4, pp. 518–526, 1988.

[38] Y. Zhuang, K. L. Wu, C. Wu and J. Litva, "Full-wave analysis of finite large printed dipole arrays using the conjugate gradient-FFT method," *Microwave Opt. Technol. Lett.*, vol. 6, no. 4, pp. 235–238, 1993.

[39] Y. Zhuang, K. L. Wu, C. Wu and J. Litva, "A combined full-wave CG-FFT method for rigorous analysis of large microstrip antenna arrays," *IEEE Trans. Antennas Propagat.*, vol. 44, no. 1, pp. 102–109, 1996.

[40] C. F. Wang, F. Ling and J. M. Jin, "A fast full-wave analysis of scattering and radiation from large finite arrays of microstrip antennas," *IEEE Trans. Antennas Propagat.*, vol. 46, no. 10, pp. 1467–1474, 1998.

[41] D. G. Fang, J. J. Yang and G. Y. Delisle, "Discrete image theory for horizontal electric dipole in a multiayer medium," *IEE Proc.-Microw. Antennas Propagat.*, vol. 135, pp. 297–303, 1988.

[42] Y. L. Chow, J. J. Yang, D. G. Fang and G. E. Howard, "Closed-form spatial Green's function for the thick substrate," *IEEE Trans. Microwave Theory Tech.*, vol. 39, no. 3, pp. 588–592, 1991.

[43] R. W. Kindt, K. Sertel, E. Topsakal and J. L. Volakis, "Array decomposition method for the accurate analysis of finite arrays," *IEEE Trans. Antennas Propagat.*, vol. 51, no. 6, pp. 1364–1372, 2003.

[44] D. M. Pozar and D. H. Schaubert, "Scan blindness in infinite phased arrays of printed dipoles," *IEEE Trans. Antennas Propagat.*, vol. 32, no. 6, pp. 602–610, 1984.

[45] D. M. Pozar and D. H. Schaubert, "Analysis of an infinite array of rectangular microstrip patches with idealized probe feeds," *IEEE Trans. Antennas Propagat.*, vol. 32, no. 10, pp. 1101–1107, 1984.

[46] A. Ishimaru, R. J. Coe, G. E. Miller and W. P. Geren, "Finite periodic structure approach to large scanning array problem," *IEEE Trans. Antennas Propagat.*, vol. 33, no. 11, pp. 1213–1220, 1985.

[47] S. K. N. Yeo and A. J. Parfiff, "Finite array analysis using iterative spatial Fourier windowing of the generalized periodic Green's function," *IEEE AP-S Symp. Dig.*, pp. 392–395, 1996.

[48] A. Neto, S. Maci, G. Vecchi and M. Sabbadini, "A truncated Floquet wave diffraction method for the full-wave analysis of large phased arrays. Part I: basic principles and 2-D cases," *IEEE Trans. Antennas Propagat.*, vol. 48, no. 4, pp. 594–600, 2000.

[49] A. Neto, S. Maci, G. Vecchi and M. Sabbadini, "A truncated Floquet wave diffraction method for the full-wave analysis of large phased arrays. Part II: Generalization to 3-D cases," *IEEE Trans. Antennas Propagat.*, vol. 48, no. 4, pp. 601–610, 2000.

[50] A. K. Bhattacharyya, "Floquet-modal-based analysis for mutual coupling between elements in an array environment," *IEE Proc.-Microw. Antennas Propag.*, vol. 144, no. 6, pp. 491–497, 1997.

[51] K. M. Lee and R. S. Chu, "Analysis of mutual coupling between a finite phased array of dipoles and its feed network," *IEEE Trans. Antennas Propagat.*, vol. 36, no. 12, pp. 1681–1699, 1988.

[52] K. C. Lee and T. H. Chu, "A circuit model for antenna array mutual coupling effects," *IEEE AP-S Symp. Dig.*, pp. 946–949, 1995.

[53] Y. P. Xi, D. G. Fang, Y. X. Sun and Y. L.Chow, "Mutual coupling in a linear dipole array of finite size," *IEE Proc.-Microw. Antennas Propagat.*, vol. 152, no. 5, pp. 324–330, 2005.

[54] D. R. Jackson, W. F. Richards and A. Ali-khan, "Series expansion for the mutual coupling in microstrip patch arrays," *IEEE Trans. Antennas Propagat.*, vol. 37, no. 3, pp. 269–274, 1989.

[55] Y. P. Xi, D. G. Fang, Y. X. Sun and Y. L. Chow, "Mutual coupling in finite microstrip patch arrays," *Microwave Opt. Technol. Lett.*, vol. 44, no. 6, pp. 577–581, 2005.

[56] D. M. Pozar, "Analysis of finite phased arrays of printed dipoles," *IEEE Trans. Antennas Propagat.*, vol. 33, no. 10, pp. 1045–1053, 1985.

[57] D. M. Pozar, "Input impedance and mutual coupling of rectangular microstrip antennas," *IEEE Trans. Antennas Propagat.*, vol. 30, no. 6, pp. 1191–1196, Nov., 1998.

[58] Y. X. Sun, Y. L. Chow, and D. G. Fang, "Impedance formulas of RF patch resonators and antennas of cavity model using fringe extensions of patches from DC capacitors, " *Microwave Opt Technol Lett*, vol. 35, no. 4, pp. 293–297, 2002.

[59] Y. X. Sun, Y. L. Chow, and D. G. Fang, "Mutual impedance formula between patch antennas based on synthetic asymptote and variable separation," *Microwave Opt. Technol. Lett.*, vol. 35, no. 6, pp. 466–470, 2002.

[60] F. Yang and Yahya Rahmat-Samii,"Microstrip antennas integrated with electromagnetic bandgap structures: a low mutual coupling design for array applications," *IEEE Trans. Antennas Propagat.*, vol. 51, no. 10, pp. 2936–2946, Oct., 2003.

[61] Z. Iluz, R. Shavit, and R. Bauer,"Microstrip antenna phased array with electromagnetic bandgap substrate," *IEEE Trans. Antennas Propagat.*, vol. 52, no. 6, pp. 1446–1453, Jun., 2004.

[62] D. G. Fang, C. Z. Luan and Y. P. Xi, "Mutual coupling in microstrip antenna array: evaluation, reduction, correction or compensation," (invited paper) *Proceedings of IEEE International Workshop on Antenna Technology*, 2005, Singapore, pp. 37–40.

[63] W. Bandler, R. M. Biernacki et al, "Space mapping technique for electromagnetic optimization," *IEEE Trans. Microwave Theory Tech.*, vol. 42, pp. 2536–2543, Dec., 1994.

[64] W. Bandler, R. M. Biernacki et al, "Electromagnetic optimization exploiting aggressive space mapping," *IEEE Trans. Microwave Theory Tech.*, vol. 43, pp. 2874–2881, Dec., 1995.

[65] J. D. Fredrick, Y. Wang and T. Itoh, "Smart antenna based on spatial multiplexing of local elements for mutual coupling reduction," *IEEE Trans. on Antennas and Propagat.*, vol. 52, no. 1, pp. 106–114, 2004.

[66] W. X. Sheng, D. G. Fang, "Angular superresolution for phased antenna array by phase weighting," *IEEE Trans. on Aerospace and Electronic Systems*, vol. 37, no. 4, pp. 32–40, 2001.

[67] D. G. Fang, W. X. Sheng, C. Zhang, and Z. Li, "Comparative study of two approaches in improving cross range resolution," *IEEE Int. Symp. Antennas and Propagation*, pp. 2438–2442, 1997.

[68] B. Chen, D. G. Fang, D. Zhou, and C. Gao, "A full-wave integrated time domain analysis of a microstrip array antenna," *Acta Electronica Sinica*, vol. 26, no. 3, pp. 5–9, Mar., 1998.

[69] H. Wang, D. G. Fang, B. Chen, X. K. Tang, Y. L. Chow, and Y. P. Xi, "An effective analysis method for electrically large finite microstrip antenna arrays," *IEEE Trans. Antennas Propagat.*, vol. 57, no. 1, pp. 94–101, 2009.

[70] H. Wang, X. B. Huang, D. G. Fang, and G. B. Han, "A microstrip antenna array formed by microstrip line fed tooth-like-slot patches," *IEEE Trans. on Antennas and Propagat.*, vol. 57, no. 4, pp. 1210–1214, 2007.

[71] H. Wang, X. B. Huang, and D. G. Fang, "A single layer wideband U-slot microstrip patch antenna array," *IEEE Antennas and Wireless Propagat. Letts.*, vol. 7, pp. 9–12, 2008.

Problems

6.1 In designing the feed network of a microstrip antenna array, the double mitered bend is often used as shown in the figure below. For the given parameters: strip width $w = 3.13$mm, substrate thickness $h = 1$mm, relative permittivity $\epsilon_r = 2.2$, $\tan\delta = 0.0011$, $l = 15$mm, the given frequency $f = 10$GHz, use the full wave solver to find the optimized parameters d_1 and d_2 which would produce the maximum transmission, $|S_{12}|$. (Answer: $d_1 = 4.83$mm, $d_2 = 3.42$mm for full wave solver IE3D^{TM}.)

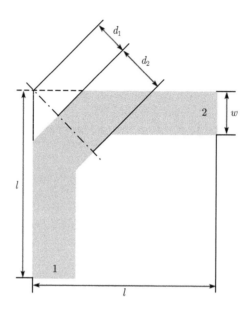

6.2   Design a 8×4 (8 in the x direction and 4 in the y direction) element uniform planar array. The array is formed by single layer microstrip rectangular patches. The central operating frequency $f = 10$GHz, bandwidth $\Delta f = 100$MHz (VSWR$\leqslant 1.5$), the parameters of the microstrip substrate are the same as in the above problem. The polarization is linear (vertical or horizontal). The requirement of the gain is $\geqslant 21.5$dBi.

CHAPTER 7

High Frequency Methods and Their Applications to Antennas

7.1 Introduction

The well-known methods of optics have found increasing use in the treatment of many electromagnetic problems when the wavelength is small compared to the size of the scatterer or antenna. In this chapter we will first examine the principles of geometrical optics followed by a brief discussion of the more general concept of physical optics. We will see that in many cases geometrical optics is inadequate to completely describe the behavior of the electromagnetic field and that it is necessary to include another field called the diffracted field. The diffracted field, when added to the geometrical field, permits us to solve many practical radiation and scattering problems in a more straightforward manner than any other way.

7.2 Geometrical Optics

The variation of the amplitude of the geometrical optics (GO) field within a ray tube is determined by the law of energy conservation since the rays are lines of energy flow[1]. Consider two wavefronts, L_0 and $L_0 + \Delta L$, as shown in Figure 7.1. The energy through cross section $d\sigma_0$ at P_0 must be equal to the energy flux through cross section $d\sigma$ at P. If S is the rate of energy flow per unit area, the condition of constant energy flow through the flux tube is thus

$$S_0 d\sigma_0 = S d\sigma \qquad (7.2.1)$$

In the case of electromagnetic waves the quantity S is the real part of the complex Poynting vector and we can assume that

$$S = \frac{1}{2}\sqrt{\frac{\epsilon}{\mu}}|\mathbf{E}|^2 \qquad (7.2.2)$$

From (7.2.1) and (7.2.2), we have

$$|\mathbf{E}_0|^2 d\sigma_0 = |\mathbf{E}|^2 d\sigma \qquad (7.2.3)$$

Solving for $|\mathbf{E}|$

$$|\mathbf{E}| = |\mathbf{E}_0|\sqrt{\frac{d\sigma_0}{d\sigma}} \qquad (7.2.4)$$

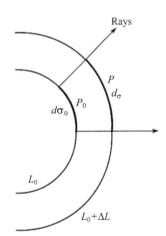

Figure 7.1   Energy flow through the flux tube between two wavefronts.

The next step is to express $d\sigma_0$, $d\sigma$ in terms of the radii of curvature. Consider the astigmatic ray tube picture in Figure 7.2. The principal radii of curvature of $d\sigma_0$ are ρ_1 and ρ_2 while the principal radii of curvature of $d\sigma$ are $(\rho_1 + l)$ and $(\rho_2 + l)$. We can write out the ratios

$$\frac{d\sigma_0}{\rho_1\rho_2} = \frac{d\sigma}{(\rho_1 + l)(\rho_2 + l)} \quad \text{or} \quad \frac{d\sigma_0}{d\sigma} = \frac{\rho_1\rho_2}{(\rho_1 + l)(\rho_2 + l)}$$

Thus

$$|\mathbf{E}| = |\mathbf{E}_0| \sqrt{\frac{\rho_1 \rho_2}{(\rho_1 + l)(\rho_2 + l)}} \tag{7.2.5}$$

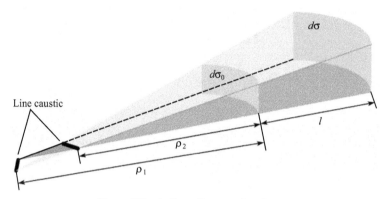

Figure 7.2   Astigmatic ray tube picture.

Note that the tube of rays converges to a line at $\rho_1 = 0$ and $\rho_2 = 0$ where the cross section of the ray tube goes to zero. Therefore, the amplitude of the geometrical optics field description becomes infinite there, although the actual field does not. The locus of points where the ray tube cross section exhibits such behavior is called a caustic. Caustics may be a point, a line or a surface. For example, considering a point source as shown in Figure 7.3, we can construct a ray tube from four rays and write

$$\frac{d\sigma_0}{\rho^2} = \frac{d\sigma}{(\rho + l)^2} \tag{7.2.6}$$

Thus

$$|\mathbf{E}| = |\mathbf{E}_0| \sqrt{\frac{\rho^2}{(\rho + l)^2}} = |\mathbf{E}_0| \frac{\rho}{\rho + l} \tag{7.2.7}$$

The caustic would be located at the point source in this case[1].

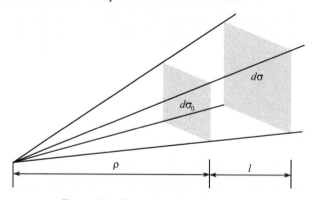

Figure 7.3   Ray tube of a point source.

In both (7.2.5) and (7.2.7), we note that as l becomes large, we have the usual inverse distance type field dependence found in the far zone of a three-dimensional source. Often however, one is concerned with two-dimensional problems where one of the radii of the curvature, say ρ_2, becomes infinite. In such problems

$$|\mathbf{E}| = |\mathbf{E}_0| \sqrt{\frac{\rho_1}{\rho_1 + l}} \tag{7.2.8}$$

This is a cylindrical wave and as l approaches to infinity, we have an amplitude dependence of the field at large distances of the form $1/\sqrt{l}$. Obviously, if both ρ_1 and ρ_2 are infinite, $|\mathbf{E}|$ is a constant for all values of l resulting in a plane wave.

The electric phase of the ray tube is given by $\exp(-j\beta l)$ and we may write for the amplitude and phase of the field in the ray tube of Figure 7.2

$$|\mathbf{E}| = |\mathbf{E}_0|\, e^{j\phi_0}\sqrt{\frac{\rho_1\rho_2}{(\rho_1 + l)(\rho_2 + l)}}\, e^{-j\beta l} = |\mathbf{E}_0|\, e^{j\phi_0} A(\rho_1, \rho_2, l) e^{-j\beta l} \qquad (7.2.9)$$

where $|\mathbf{E}_0|$ is the reference amplitude at $l = 0$, ϕ_0 is the reference phase at $l = 0$, $A(\rho_1, \rho_2, l)$ is the general spatial attenuation factor and $e^{-j\beta l}$ is the spatial phase delay factor.

Note that when l becomes less than $-\rho_2$, the quantity under the radical sign at $A(\rho_1, \rho_2, l)$ becomes negative and a phase jump of $\pi/2$ occurs when the observer passes through the caustic. While we can neither predict the amplitude nor the phase of the geometrical optics field at the caustic, we can determine the fields on either side of the caustic[1].

As an example of the use of the geometrical optics, consider the parabolic cylinder reflector with a central line source at its focus, as depicted in Figure 7.4[2]. Assume that the reflector extends vertically to $y = \pm D/2$ and thus that the secondary aperture field in the plane $x = a$ will also have a value in the range $-D/2 \leqslant y \leqslant D/2$. It is desired to find $P(y)$ in watts per square meter in this range, with $P(y)$ the power density in the secondary field at $x = a$ under geometrical optics assumptions.

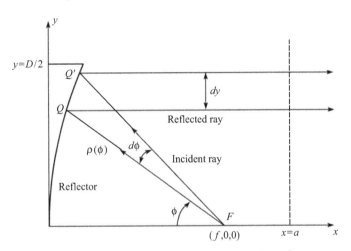

Figure 7.4   Ray geometry for a parabolic cylinder reflector.

The equation of the parabolic cylinder in polar coordinates is

$$\rho(\phi) = \frac{2f}{1 + \cos\phi} \qquad (7.2.10)$$

where f is the focal distance. From the condition of constant energy flow through the flux tube, the primary and secondary power flows are equal so that

$$P(y)dy = I(\phi)d\phi \qquad (7.2.11)$$

where $I(\phi)$ is the primary pattern in watts per radian-meter. From (7.2.10), we have

$$\frac{d\rho(\phi)}{d\phi} = \rho(\phi)\tan\frac{\phi}{2} \qquad (7.2.12)$$

Notice that

$$y(\phi) = \rho(\phi)\sin\phi \qquad (7.2.13)$$

The combination of (7.2.12) and (7.2.13) results in

$$\frac{dy(\phi)}{d\phi} = \frac{d\rho(\phi)}{d\phi}\sin\phi + \rho(\phi)\cos\phi = \rho(\phi) \qquad (7.2.14)$$

According to (7.2.11) and (7.2.14), the connection between primary and secondary power distributions is found to be given by the simple expression

$$P(y(\phi)) = \frac{I(\phi)}{\rho(\phi)} \qquad (7.2.15)$$

After having the field distribution on the aperture, the far-field pattern is easy to find by using the Huygens principle.

Another example is the microstrip reflectarray which is a fairly new antenna concept[3]. It consists of a very thin, flat reflecting surface and an illuminating feed. On the reflecting surface, there is an array of isolated microstrip patch elements with no feed network. The feed antenna illuminates these patch elements, which are individually designed to scatter the incident field with the phase needed to form a constant aperture phase. This operation is similar in concept to the use of a parabolic reflector shown in Figure 7.5. The surface of a paraboloidal reflector is formed by rotating a parabola around its axis. Its surface must be a paraboloid of revolution so that rays emanating from the focus of the reflector are transformed into plane waves. The design is based on geometrical optics as well. It does not take into account any diffraction from the rim of the reflector. Referring to Figure 7.5 and choosing a plane perpendicular to the axis and located at O', from the property of a paraboloid it follows that

$$FM + MM' = 2FO + FO' \qquad (7.2.16)$$

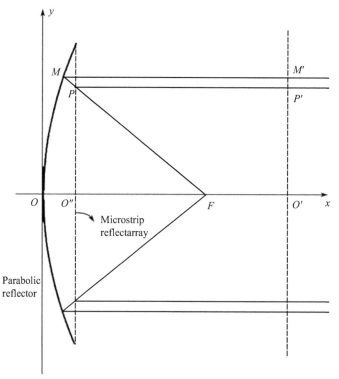

Figure 7.5 Two-dimensional configuration of a paraboloidal reflector and microstrip reflectarray.

and a flat reflecting surface perpendicular to the axis and located at O'', by geometrical optics

$$FP + PP' \neq 2FO'' + FO' \qquad (7.2.17)$$

This surface is formed by a microstrip patch array which is called a microstrip reflectarray. The phase ΔP for forming a constant aperture phase is provided by the individual reflected patch element with a different length of open-circuit-terminated transmission line stub, such that

$$FP + PP' + \Delta P = 2FO'' + FO' \qquad (7.2.18)$$

The element layout of a microstrip reflectarray is shown in Figure 7.6. To obtain the phase needed, patches with different sizes can also be used, since this changes the resonant frequency of the element, and hence its reflection phase[4].

This antenna concept combines some of the best features of the traditional parabolic reflector and the microstrip array technology. The major portion of the antenna, the reflecting surface, is a flat structure with a low profile. Without any power divider, the resistive insertion loss of this large array antenna is very small and is comparable to that of a parabolic reflector. The antenna, being a printed microstrip array, can be fabricated with a simple, low-cost etching process, especially when it is produced in large quantities.

Figure 7.6   Element layout of a microstrip reflectarray.
(Source: Seminar notes of Prof. J. Litva of McMaster University)

7.3   Physical Optics

The concept of physical optics (PO) can be considered to be somewhat more general than GO, since the results obtained from PO may often be reduced to those of GO in the high frequency limit. Consider a perfectly conducting body, the assumed PO surface current is

$$\mathbf{J}_{PO} = \hat{\mathbf{n}} \times \mathbf{H}_{total} \quad \text{in the illuminated region}$$
$$= 0 \quad \text{in the shadowed region} \qquad (7.3.1)$$

where $\hat{\mathbf{n}}$ is a unit normal vector outward from the surface of interest as shown in Figure 7.7.

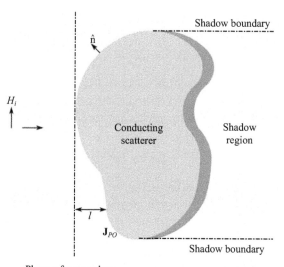

Figure 7.7   Geometry of a perfectly conducting scatterer illuminated by plane wave.

Let us derive an expression for the scattered electric field from such a body. Starting with the vector potential

$$\mathbf{A} = \iint\limits_{S} \frac{\mathbf{J}_{PO}e^{-j\beta R}}{4\pi R}dS \tag{7.3.2}$$

where R is the distance between the source point and observation point.

From the image theory, the tangential components of $\mathbf{H}$ at a perfect conducting plane are just twice those from the same source when the conducting scatterer is replaced by equivalent currents in free space. Thus the PO current is given by

$$\mathbf{J}_{PO} = 2(\hat{\mathbf{n}} \times \mathbf{H}_i) \tag{7.3.3}$$

if we assume the incident field phase to be zero at the phase reference planes. Making the customary far-field assumptions we can write

$$\mathbf{A} = \frac{e^{-j\beta l_0}}{4\pi l_0} \iint\limits_{S} 2(\hat{\mathbf{n}} \times \mathbf{H}_i)e^{-j\beta l}dS \tag{7.3.4}$$

from which the far-zone scattered field is given by $\mathbf{E}_s = -j\omega\mu\mathbf{A}$, or

$$\mathbf{E}_s = -\frac{j\omega\mu e^{-j\beta l_0}}{4\pi l_0} \iint\limits_{S} 2(\hat{\mathbf{n}} \times \mathbf{H}_i)e^{-j\beta l}dS \tag{7.3.5}$$

where l is the distance from the phase reference plane in Figure 7.7 to the scatterer and l_0 is the distance from the phase reference plane to the far-field observation point. It should be noted that this expression for the scattered field is frequency dependent in contrast to the GO expression which is frequency independent. Therefore, it might be intuitively inferred that PO provides a more accurate approximation to the scattered field. Let us make a PO calculation of the radar cross section (RCS) of the sphere, and then compare the results with those obtained via GO. From (7.3.5) we can write for the magnitude of $\mathbf{E}_s$

$$|\mathbf{E}_s| = \left| \frac{\omega\mu}{2\pi l_0} \iint\limits_{S} \hat{\mathbf{n}} \times \mathbf{H}_i e^{-j\beta l}dS \right| \tag{7.3.6}$$

According to the definition of RCS denoted by σ, we obtain

$$\sigma = \lim_{l_0 \to \infty} 4\pi l_0^2 \frac{|\mathbf{E}_s|^2}{|\mathbf{E}_i|^2} = \lim_{l_0 \to \infty} 4\pi l_0^2 \frac{|\mathbf{E}_s|^2}{\eta^2 |\mathbf{H}_i|^2}$$

$$= \frac{4\pi}{\lambda^2} \left| \frac{1}{\mathbf{H}_i} \iint_S (\hat{\mathbf{n}} \times \mathbf{H}_i) e^{-j\beta l} dS \right|^2 \tag{7.3.7}$$

For the case of the sphere shown in Figure 7.8, we note that the only component of current that will have a net contribution to the backscattered field is given by $2\hat{\mathbf{a}}_z \times (\hat{\mathbf{n}} \times \hat{\mathbf{a}}_x) H_i$. Due to the vector identity $\hat{\mathbf{a}}_z \times (\hat{\mathbf{n}} \times \hat{\mathbf{a}}_x) H_i = -(\hat{\mathbf{a}}_z \cdot \hat{\mathbf{n}}) \hat{\mathbf{a}}_x H_i$ we have

$$\sigma = \frac{4\pi}{\lambda^2} \left| \iint_S -(\hat{\mathbf{a}}_z \cdot \hat{\mathbf{n}}) e^{-j2\beta z} dS \right|^2 \tag{7.3.8}$$

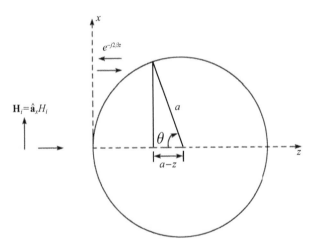

Figure 7.8   Scattering from a perfectly conducting sphere.

since $l = 2z$ due to the reference plane being the $z = 0$ plane. From Figure 7.8, we note that

$$-\hat{\mathbf{a}}_z \cdot \hat{\mathbf{n}} = \cos\theta = \frac{a - z}{a} \tag{7.3.9}$$

and that a surface area element is $dS = a^2 \sin\theta d\theta d\phi$. Since $z = a - a\cos\theta$ and $dz = a\sin\theta d\theta$, finally we have[1]

$$\sigma = \frac{4\pi}{\lambda^2} \left| \int_0^{2\pi} \int_0^a e^{-j2\beta z} \left(\frac{a - z}{a} \right) a \, dz \, d\phi \right|^2$$

$$= \frac{4\pi}{\lambda^2} \left| 2\pi \frac{1}{j2\beta} \left(a - \frac{1 - e^{-j2\beta a}}{j2\beta} \right) \right|^2 \tag{7.3.10}$$

The exponential term arises from the artificially imposed discontinuity in the current at the location of $\theta = \pi/2$ of the sphere. Since this discontinuity is nonphysical, so too is the exponential term in (7.3.10) and we must disregard it. Thus

$$\sigma = \lim_{\beta a \to \infty} \pi a^2 \left| \frac{1}{j} \left(1 - \frac{1}{2j\beta a} \right) \right|^2 = \pi a^2 \tag{7.3.11}$$

We see that in the high-frequency limit, the RCS of the sphere obtained via PO reduces to that from GO and is the geometric cross section of the sphere.

PO is an approximate method of considerable usefulness that can be expected to provide an accurate representation of the scattered field arising from a surface where the postulated PO current is reasonably close to the true current distribution. The PO current will be a reasonable representation of the true current if the field at the scatterer surface is correctly given by the GO surface field. Therefore, we can view PO as a high-frequency method that is an extension of GO[1].

7.4   Diffraction by a Conducting Half Plane With Normal Incidence

Since PO postulates a current only on the lit side and zero current on the shadowed side, the PO current alone is incapable of correctly predicting a nonzero field in the deep shadow region. By simple ray tracing it is quite apparent that GO is also incapable of correctly predicting a nonzero field in the shadow region. However, GO may be extended to include a class of rays, called diffracted rays, which permits the calculation of fields in the shadow region of a scatterer. Diffracted rays are produced, for example, as shown in Figure 7.9. It is these rays that account for a nonzero field in the shadow region. In addition, they also modify the GO field in the illuminated region.

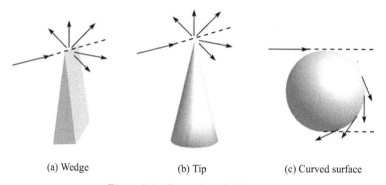

(a) Wedge   (b) Tip   (c) Curved surface

Figure 7.9   Examples of diffraction.

Because diffraction is a local phenomenon at high frequencies, the value of the field of a diffracted ray is proportional to the field value of the incident ray at the point of diffraction multiplied by a coefficient called the diffraction coefficient. That is, the diffraction coefficient is determined largely by the local properties of the field and the boundary in the immediate neighborhood of the point of diffraction. Since it is only the local conditions near the point of diffraction that are important, the diffracted ray amplitude may be determined from the solution of the appropriate boundary value problem having these local properties. Such a problem is called a canonical problem and wedge diffraction is one such canonical problem. Wedge diffraction is perhaps the most important canonical problem in the extension of GO as originally proposed by Keller. His theory is known as the geometrical theory of diffraction (GTD).

Through the use of GO and the solution to a number of canonical problems, such as those in Figure 7.9, we can construct solutions to more complex problems via the principle of superposition. To start, we will consider scalar diffraction by an infinitely conducting and infinitesimally thin half-plane sheet as shown in Figure 7.10. The half-plane is a wedge of zero included angles. To calculate the field in the region $z > 0$, we will use the Huygens principle in two-dimensions. Thus, each point on the primary wavefront along $z = 0$ is con-

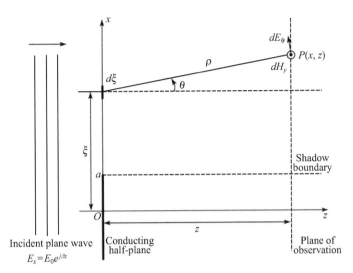

Figure 7.10   Field from a Huygens element in the aperture plane.

sidered to be a new source for a secondary cylindrical wave, the envelope of these secondary cylindrical waves being the secondary wavefront. We assume that $\rho \gg \lambda$ and $\rho \gg |x - \xi|$, the inclination of the elemental electric-field vector to the vertical may be ignored, and the x-component of the electric field at (x, z) point may be written as

$$E_x(x, z) = \int_{x=0}^{x=\infty} dE_x \qquad (7.4.1)$$

where dE_x is the electric field at (x, z) due to a magnetic line source parallel to the y-axis in the $z = 0$ plane, which actually is the $E_x \hat{a}_x \times \hat{n}$ on the aperture with the unit normal of $\hat{n}$. According to the distance and phase dependence, we have

$$dE_x(x, z) = \frac{C_1}{\sqrt{\rho}} e^{-j\beta\rho} d\xi \qquad (7.4.2)$$

where C_1 is a constant, $\rho = \sqrt{z^2 + (x - \xi)^2}$. With the assumptions of $z \gg \lambda$ and $z \gg |x - \xi|$, we may write for the contribution to $E_x(x, z)$ from those two-dimensional Huygens sources between $\xi = a$ and $\xi = \infty$

$$E_x(x, z) = \frac{C_1}{\sqrt{\rho}} e^{-j\beta z} \int_a^\infty e^{-j\beta(\xi - x)^2/(2z)} d\xi \qquad (7.4.3)$$

Letting $\gamma^2 = 2/\lambda z$, $u = \gamma(\xi - x)$, and assuming $\rho \approx z$ in the amplitude term, then we have

$$E_x(x, z) = C_1 \sqrt{\frac{\lambda}{2}} e^{-j\beta z} \int_{\gamma(a-x)}^\infty e^{-j\pi u^2/2} du \qquad (7.4.4)$$

The integral in (7.4.4) is called Fresnel integral.
   The Fresnel integral has different definitions such as[5, 6]

$$F_1(t) = \int_t^\infty e^{-ju^2} du, \quad F_2(t) = \int_0^t e^{-j\pi u^2/2} du, \quad F_3(t) = \int_t^\infty e^{-j\pi u^2/2} du \qquad (7.4.5)$$

The relationship between them is as follows

$$F_1(t) = \frac{\sqrt{\pi}}{2}e^{-j\pi/4} - \sqrt{\frac{\pi}{2}}F_2\left(\sqrt{\frac{2}{\pi}}\,t\right)$$

$$F_3(t) = \sqrt{\frac{2}{\pi}}F_1(\sqrt{\frac{\pi}{2}}\,t)$$

$$F_3(t) = \frac{\sqrt{2}}{2}e^{-j\frac{\pi}{4}} - F_2(t) \tag{7.4.6}$$

If the lower limit of the integral a in (7.4.4) goes to minus infinity, $E_x(x, z)$ will be equal to the field strength without the half-plane. Because of the following identity:

$$F_2(\infty) = \frac{\sqrt{2}}{2}e^{-j\pi/4} = -F_2(-\infty) \tag{7.4.7}$$

formula (7.4.4) yields

$$\lim_{x\to\infty} E_x(x, z) = C_1\sqrt{\frac{\lambda}{2}}(1-j)e^{-j\beta z} = E_0 e^{-j\beta z}$$

Solving for C_1 and substituting it into (7.4.4) gives the value of $E_x(x, z)$ in terms of E_0

$$E_x(x, z) \approx \frac{E_0 e^{j\pi/4}}{\sqrt{2}}e^{-j\beta z}\int_{\gamma(a-x)}^{\infty} e^{-j\frac{\pi}{2}u^2}\,du$$

If the conducting half-plane is located from $x = 0$ to $x = -\infty$, that is $a = 0$. The above expression results in

$$E_x(x, z) \approx \frac{E_0 e^{j\pi/4}}{\sqrt{2}}e^{-j\beta z}\int_{-\gamma x}^{\infty} e^{-j\frac{\pi}{2}u^2}\,du \tag{7.4.8}$$

The above formula may be written into the following forms based on (7.4.5) and (7.4.6)

$$E_x(x, z) = \frac{E_0 e^{j\pi/4}}{\sqrt{2}}e^{-j\beta z}F_3(-\gamma x) \tag{7.4.9}$$

$$= \frac{E_0}{2}e^{-j\beta z}\left(1 - \sqrt{2}e^{j\pi/4}\int_0^{-\gamma x} e^{-j\frac{\pi}{2}u^2}\,du\right) \tag{7.4.10}$$

$$= \frac{E_0}{2}e^{-j\beta z}\left(1 - \sqrt{2}e^{j\pi/4}F_2(-\gamma x)\right) \tag{7.4.11}$$

$$= \frac{E_0}{2}e^{-j\beta z}\left(1 + \sqrt{2}e^{j\pi/4}F_2(\gamma x)\right) \tag{7.4.12}$$

$$= E_0 e^{-j\beta z}W(\gamma x) \tag{7.4.13}$$

The Fresnel integral may be expressed in complex form, that is, $F_2(v) = \xi(v) - j\eta(v)$. A plot of $\xi(v)$ against $\eta(v)$ with $v = \gamma x$ as a parameter is shown in Figure 7.11. It is known as Cornu's spiral. The radius vector OR marked on the diagram represents the complex conjugate of $F_2(v)$ for a v-value of about 1.3. $W(\gamma x)$ is the (complex) ratio of the field at point P to its value in the absence of the conducting half-plane. Figure 7.12 uses the Cornu's spiral to represent $W(\gamma x)$ as the radius vector OQ. From (7.4.7), (7.4.12) and (7.4.13), we have

$$\lim_{\gamma x\to\infty} W(\gamma x) = 1, \qquad \lim_{\gamma x\to-\infty} W(\gamma x) = 0 \tag{7.4.14}$$

as is seen in Figure 7.12.

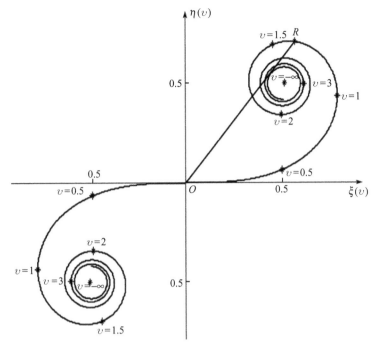

Figure 7.11   Cornu's spiral : a plot of $\eta(v)$ against $\xi(v)$.

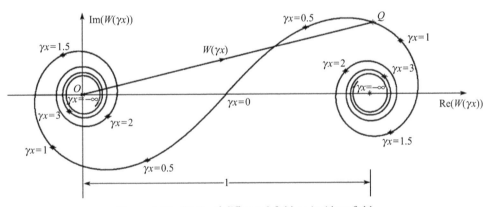

Figure 7.12   Ratio of diffracted field to incident field.

Notice the asymptotical formula

$$\lim_{v \to \infty} F_2(v) = \frac{\sqrt{2}}{2} e^{-j\pi/4} - \frac{1}{j\pi v} e^{-j\pi v^2/2} \qquad (7.4.15)$$

and

$$\lim_{\gamma x \to \infty} E_x(x, z) = \frac{E_0}{2} e^{-j\beta z} \left[2 - \frac{\sqrt{2}\exp(-j\pi/4)}{\pi \gamma x} e^{-j\pi(\gamma x)^2/2} \right] = E_0 e^{-j\beta z}, x > 0 \qquad (7.4.16)$$

From (7.4.11)

$$\lim_{\gamma x \to -\infty} E_x(x, z) = \frac{E_0}{\sqrt{2}} e^{-j\beta z} \left[\frac{\exp(j3\pi/4)}{\pi \gamma x} e^{-j\pi(\gamma x)^2/2} \right], \qquad x < 0 \qquad (7.4.17)$$

The detailed dependence of the magnitude of the diffracted field on x is plotted in Figure 7.13. We note that on the shadow boundary the value of the normalized field is $1/2$ and that in the lit region, the value of the field oscillates about the value of unity. This oscillation can be interpreted as being caused by interference between the diffracted field and the direct field. Since there is no direct field in the shadow region, we observe that no such oscillation occurs.

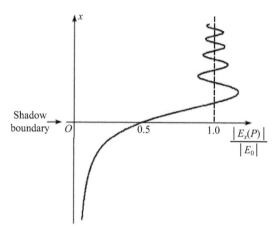

Figure 7.13   Graph of the magnitude of the diffracted field against distance from the shadow boundary.

The solution we have obtained for the field in the shadow region has an interesting interpretation in terms of Keller's geometrical theory of diffraction (GTD)[5]. In this theory, ordinary geometrical rays represent the plane wave incident on the conducting half plane. The theory applies in the limit of vanishingly small wavelength. Thus, the field in the illuminated region is due to the unobstructed incident rays, as indicated in (7.4.16).

In addition to this classical geometrical optics field, the geometrical theory of diffraction supposes that diffracted rays exist. The scalar diffracted field at the observation point P, in the zero-wavelength limit, is assumed to be of the form

$$E_P = DE_0 \frac{e^{-j\beta\rho}}{\sqrt{\rho}} \tag{7.4.18}$$

where D is termed the diffraction coefficient. To find the analytical form of D, starting from (7.4.17) and after some simple derivation we have

$$E_x(x,z) = \frac{jE_0}{2\pi \sin\theta} \sqrt{\frac{j\lambda}{z}} e^{-j\beta[z+x^2/(2z)]} \cos\theta, \quad x < 0 \tag{7.4.19}$$

This is well approximated for small θ by

$$E_x(\rho,\theta) = \frac{jE_0}{2\pi \sin\theta} \sqrt{\frac{j\lambda}{\rho}} e^{-j\beta\rho} \cos\theta, \quad x < 0 \tag{7.4.20}$$

To find E_θ, we need to know H_y. The Maxwell's curl equation gives

$$H_y = \frac{1}{j\omega\mu} \left[\frac{\partial E_z}{\partial x} - \frac{\partial E_x}{\partial z} \right] \tag{7.4.21}$$

Now we have E_x in hand, through Maxwell's divergence equation

$$\frac{\partial E_x}{\partial x} + \frac{\partial E_z}{\partial z} = 0 \tag{7.4.22}$$

combined with the approximation that the amplitude of E_z is a slowly varying function of z thus may be considered to be invariant with respect to z (it is the parabolic equation approximation, so named because the approximation has yielded a partial differential equation of the parabolic type), we have the relationship between E_z and E_x

$$E_z = \frac{1}{j\beta} \frac{\partial E_x}{\partial x} \tag{7.4.23}$$

and from (7.4.19) with $\tan\theta = x/z$, we have

$$E_z = -E_x \tan\theta \tag{7.4.24}$$

With (7.4.19), (7.4.21) and (7.4.24) we have

$$H_y = \frac{E_x}{Z}\left[1 + \frac{x^2}{2z^2} - \frac{1}{j\beta z}\right] \tag{7.4.25}$$

In the geometrical optics limit, the third term in (7.4.25) may be neglected, thus using the same approximation in deriving (7.4.20), we obtain

$$H_y = \frac{E_x}{Z}\sec\theta \tag{7.4.26}$$

Finally, from (7.4.20) and (7.4.26), the field at the point P in the shadow region is given by[5]

$$E_\theta(\rho,\theta) = ZH_y(\rho,\theta) = \sqrt{\frac{j\lambda}{\rho}}\,\frac{jE_0}{2\pi\sin\theta}e^{-j\beta\rho} \tag{7.4.27}$$

which is a cylindrical wave emanating from the edge O of the diffracting half plane. Comparing (7.4.18) with (7.4.27) results in

$$D = \sqrt{\frac{\lambda}{j}}\,\frac{1}{2\pi\sin\theta} \tag{7.4.28}$$

Although the derivation of the diffraction coefficient is for the specific case defined in Figure 7.10. it is not accurate enough, but through it, we can have some idea about GTD, which will be helpful in understanding the extended cases discussed in the next section. In using GTD, the field well into the illuminated region is identical to the incident field, and is given straightforwardly by classical geometrical optics. The field well into the dark region has the form of a cylindrical wave originating at the edge and may be obtained by the simple closed form through the diffraction coefficient D. However, in the vicinity of the shadow boundary, the diffraction coefficient D is singular. To obtain the complete solution in terms of Fresnel's integrals, any of the formulas (7.4.9)–(7.4.13) is available. An alternative, the uniform diffraction theory (UTD), that will be introduced in the next section, could be used.

7.5   Diffraction by a Conducting Half Plane With Arbitrary Incidence

Before considering the case of wedge, we will consider the case of a half-plane with arbitrary incidence shown in Figure 7.14 first. We can identify two shadow boundaries, the incident or direct field shadow boundary and the reflected field shadow boundary. Also, the arbitrary value of ϕ' will be considered. These two shadow boundaries serve to divide space into three regions where region (1) contains direct (D_i), diffracted (D_f) and reflected (R_e) rays; region (2) contains D_i, D_f but no R_e and region (3) contains only D_f. In Figure 7.14, $|\boldsymbol{\rho}_i| = |\boldsymbol{\rho}|\cos\phi_i = \rho\cos(\phi - \phi')$, $|\boldsymbol{\rho}_r| = |\boldsymbol{\rho}|\cos\phi_r = \rho\cos(\pi - (\phi + \phi')) = -\rho\cos(\phi + \phi')$.

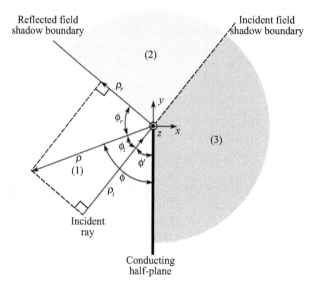

Figure 7.14   Diffraction by a conducting half-plane with arbitrary incidence.

For the field in any one of the three regions, let us write $E(\rho, \phi)$ as consisting of a reflected field $v_r(\rho, \phi + \phi')$ and an incident field $v_i(\rho, \phi - \phi')$. Thus

$$E(\rho, \phi) = \pm v_r(\rho, \phi + \phi') + v_i(\rho, \phi - \phi') \tag{7.5.1}$$

The choice of sign depends on the polarization of the incident field. If the electric field is perpendicular to the diffraction edge, it is called the perpendicular or TM polarization with respect to the xoy plane and it is also related to the Neumann or hard boundary condition, $E_z = 0$. If parallel, it is called parallel or TE polarization with respect to the xoy plane and it is also related to Dirichlet or soft boundary condition, $H_z = 0$. For the former, the plus sign is used and for the latter, the minus sign is used.

Both incident field and reflected field can be viewed as consisting of two parts, a GO incident field and D_i field. Thus, for the R_e field

$$\pm v_r(\rho, \phi + \phi') = \pm [v_{o,r}(\rho, \phi + \phi') + v_{d,r}(\rho, \phi + \phi')] \tag{7.5.2}$$

and for the incident field

$$v_i(\rho, \phi - \phi') = v_{o,i}(\rho, \phi - \phi') + v_{d,i}(\rho, \phi - \phi') \tag{7.5.3}$$

where, subscripts o and d denote optics field and diffraction field respectively. (7.5.1) may be thought of as being composed of four parts. Each of the terms on the right-hand side of (7.5.2) and (7.5.3) satisfies the wave equation individually. However, the sum of $v_{o,r}$ and $v_{d,r}$ makes v_r continuous across the reflected field shadow boundary and thus v_r satisfies the wave equation there. Similar comments apply to v_i. But, neither v_r nor v_i alone satisfies the boundary conditions at the edge. However, the sum of them in (7.5.1) does satisfy the boundary conditions as well as the wave equation.

From simple geometrical considerations:

$$v_{o,r}(\rho, \phi + \phi') = e^{j\beta\rho\cos(\phi+\phi')}, \quad 0 < \phi < \pi - \phi' \quad \text{in region (1)} \tag{7.5.4}$$

$$v_{o,i}(\rho, \phi - \phi') = e^{j\beta\rho\cos(\phi-\phi')}, \quad 0 < \phi < \pi + \phi' \quad \text{in region (1), (2)} \tag{7.5.5}$$

for other values of ϕ, $v_{o,i} = v_{o,r} = 0$. The edge is taken as the phase reference.

The term $v_d(\rho, \phi^\pm)$ can be derived using spectrum domain techniques[6, 7]

$$v_d(\rho, \phi^\pm) = -e^{-j\pi/4}\sqrt{\frac{2}{\pi\alpha}}e^{j\beta\rho\cos\phi^\pm}\cos\frac{\phi^\pm}{2}\int_{\sqrt{\alpha\beta\rho}}^{\infty}e^{-j\tau^2}\,d\tau \qquad (7.5.6)$$

where $\phi^\pm = \phi \pm \phi'$, $\alpha = 1 + \cos\phi^\pm$.

After some mathematical manipulations,

$$v_d(\rho, \phi^\pm) = -e^{j\pi/4}\sqrt{\frac{2}{\pi\alpha}}e^{j\beta\rho\cos\phi^\pm}\cos\frac{\phi^\pm}{2}F_1(\sqrt{\alpha\beta\rho}) \qquad (7.5.7)$$

Applying (7.4.6) and (7.4.15) to (7.5.7) results in

$$v_d(\rho, \phi^\pm) = -\frac{e^{-j(\beta\rho+\pi/4)}}{2\sqrt{2\pi\beta\rho}}\cdot\frac{1}{\cos(\phi^\pm/2)} = D(\phi^\pm)\frac{e^{-j\beta\rho}}{\sqrt{\rho}} \qquad (7.5.8)$$

where

$$D(\phi^\pm) = -\frac{e^{-j\pi/4}}{2\sqrt{2\pi\beta}\cos(\phi^\pm/2)} = -\frac{e^{-j\pi/4}}{2\sqrt{2\pi\beta}}\sec\left(\frac{\phi^\pm}{2}\right) \qquad (7.5.9)$$

In region (3) of Figure 7.14, $v_d(\rho, \phi^-)$ dominates. From Figures 7.10 and Figure 7.14, we see that $(3\pi/4) + \theta = \phi$. When we use the approximation in (7.4.20) we notice that $\phi' = \pi/2$ and $\sin(\theta/2) \approx (\sin\theta)/2$. If we also let $E_0 = 1$, then we find that the diffraction coefficient D in (7.4.28) and $D(\phi^-)$ in (7.5.9) are exactly the same.

For wedges as shown in Figure 7.15, the asymptotic results will be[1]

$$v_d(\rho, \phi^\pm) = \frac{e^{-j(\beta\rho+\pi/4)}}{\sqrt{2\pi\beta\rho}}\cdot\frac{(1/n)\sin(\pi/n)}{\cos(\pi/n) - \cos(\phi^\pm/n)} = D(\phi^\pm)\frac{e^{-j\beta\rho}}{\sqrt{\rho}} \qquad (7.5.10)$$

$$v_{d,i}(\rho, \phi^-) \mp v_{d,r}(\rho, \phi^+) = [D(\phi^-) \mp D(\phi^+)]\frac{e^{-j\beta\rho}}{\sqrt{\rho}} \qquad (7.5.11)$$

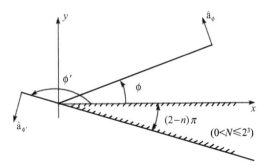

Figure 7.15   Diffraction from a wedge.

where

$$D(\phi^-) = \frac{e^{-j\pi/4}(1/n)\sin(\pi/n)}{\sqrt{2\pi\beta}\,[\cos(\pi/n) - \cos((\phi - \phi')/n)]} \qquad (7.5.12)$$

$$D(\phi^+) = \frac{e^{-j\pi/4}(1/n)\sin(\pi/n)}{\sqrt{2\pi\beta}\,[\cos(\pi/n) - \cos((\phi + \phi')/n)]} \qquad (7.5.13)$$

Up to now, we only considered the scalar diffracted field due to a plane wave normally incident upon a perfectly conducting infinite wedge whose edge is along the z-axis. Such a coordinate system is said to be an edge-fixed coordinate system. On the other hand, the obliquely incident and obliquely diffracted rays associated with the point Q in Figure 7.16 are more conveniently described in terms of spherical coordinates centered at Q. Such a coordinate system is said to is ray-fixed. Using spherical coordinates, the position of the source of the incident ray is defined by (S', γ_0', ϕ'), and the observation point by (S, γ_0, ϕ) as shown in Figure 7.17.

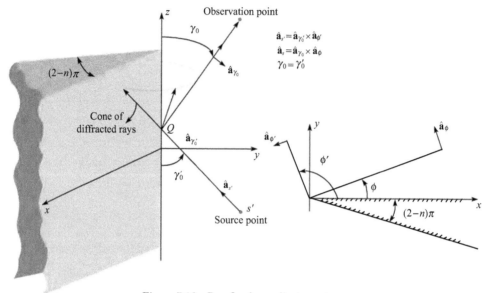

Figure 7.16   Ray fixed coordinate system.

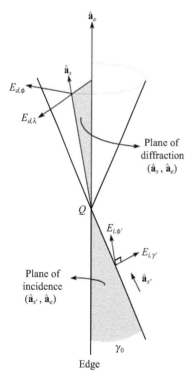

Figure 7.17   Plane of incidence
and diffraction.

Let us write a symbolic expression for the diffracted field in matrix form as

$$[E_d] = [D]\,[E_i]\,A(\rho)e^{-j\beta\rho} \qquad (7.5.14)$$

when $[E_d]$, $[E_i]$ are column matrices consisting of the scalar components of the diffracted and incident fields respectively. $[D]$ is a square matrix of the appropriate scalar diffraction coefficients, ρ is the distance from the wedge's edge to the observation point and $A(\rho)$ is a spreading factor. Now if the edge-fixed coordinates system is used, it is clear that $[E_d]$ will have in general three scalar components $E_{d,\rho}$, $E_{d,\phi}$, $E_{d,z}$ and that $[D]$ will be a 3×3 matrix. It can be shown that in such a situation, seven of the nine terms in $[D]$ are nonvanishing. However, when the ray-fixed coordinate system is used, there is no component of the diffracted field in the direction of the diffracted ray tube, since the incident field is not allowed to have a component in the direction of the incident ray tube. It follows that there are then only two possible components of the diffracted field, $E_{d,\gamma}$ and $E_{d,\phi}$, and only two components of the incident field, $E_{i,\gamma}$ and $E_{i,\phi}$. Clearly $[D]$ is a 2×2 matrix. In this case, $[D]$ has nonvanishing terms on the main diagonal. Thus, for plane wave incidence in the ray-fixed system (7.5.13) can be written as[1]

$$\begin{bmatrix} E_{d,\gamma}(S) \\ E_{d,\phi}(S) \end{bmatrix} = \begin{bmatrix} -D_{\parallel} & 0 \\ 0 & -D_{\perp} \end{bmatrix} \begin{bmatrix} E_{i,\gamma'}(Q) \\ E_{i,\phi'}(Q) \end{bmatrix} A(S)e^{-j\beta S} \qquad (7.5.15)$$

We use the notation $D_{\parallel}$ (or use D_s indicating the diffraction coefficient for soft boundary) in association with $E_{i,\gamma'}$ and $D_{\perp}$ (or use D_h indicating the diffraction coefficient for hard boundary) in association with $E_{i,\phi'}$, not because $E_{i,\gamma'}$ and $E_{i,\phi}$ are parallel and perpendicular, respectively, to the diffracting edge (which they are at normal incidence when $\gamma_0 = \pi/2$), but because $E_{i,\gamma'}$ and $E_{i,\phi}$ are parallel and perpendicular, respectively, to the plane of incidence as shown in Figure 7.17.

Since $E_{i,\gamma'}$, $E_{i,\phi'}$ are parallel and perpendicular, respectively, to the plane of incidence, we will let $E_{i,\gamma'}$ be written as $E_{i,\parallel'}$ and let $E_{i,\phi'}$ be written as $E_{i,\perp'}$. Similarly, $E_{d,\gamma}$ and $E_{d,\phi}$ are parallel and perpendicular respectively to the plane of diffraction as shown in Figure 7.17. Thus, we will let $E_{d,\gamma}$ be written as $E_{d,\parallel}$ and let $E_{d,\phi}$ be written as $E_{d,\perp}$. With these notation changes, (7.5.14) may be written as[1]

$$\begin{bmatrix} E_{d,\parallel}(S) \\ E_{d,\perp}(S) \end{bmatrix} = \begin{bmatrix} -D_{\parallel} & 0 \\ 0 & -D_{\perp} \end{bmatrix} \begin{bmatrix} E_{i,\parallel}(Q) \\ E_{i,\perp}(Q) \end{bmatrix} A(S)e^{-j\beta S} \qquad (7.5.16)$$

where the spatial attenuation factor $A(S)$ is defined as

$$A(S) = \begin{cases} \dfrac{1}{\sqrt{S}}, & \text{for plane, cylindrical, and conical wave incidence} \\[2ex] \sqrt{\dfrac{S'}{S(S'+S)}}, & \text{for spherical wave incidence} \end{cases} \qquad (7.5.17)$$

From uniform diffraction theory[8]

$$\begin{aligned} \frac{D_{\parallel}}{D_{\perp}} = \frac{-e^{-j\pi/4}}{2n\sqrt{2\pi\beta}\sin\gamma_0'} \Bigg\{ &\cot\left(\frac{\pi+(\phi-\phi')}{2n}\right) F(\beta La^+(\phi-\phi')) \\ &+ \cot\left(\frac{\pi-(\phi-\phi')}{2n}\right) F(\beta La^-(\phi-\phi')) \mp \left[\cot\left(\frac{\pi+(\phi+\phi')}{2n}\right) \times \right. \\ &\left. F(\beta La^+(\phi+\phi')) + \cot\left(\frac{\pi-(\phi+\phi')}{2n}\right) F(\beta La^-(\phi+\phi')) \right] \Bigg\} \end{aligned} \qquad (7.5.18)$$

where, if the argument of F is represented by X

$$F(X) = 2j\,|X|\,e^{jX} \int_{|\sqrt{X}|}^{\infty} e^{-j\tau^2}\, d\tau \qquad (7.5.19)$$

Again we see that a Fresnel integral appears in the expression for the diffraction coefficient. The factor F may be regarded as a correction factor to be used in the transition regions of the shadow and reflection boundaries. Outside the transition regions where the argument of F exceeds about 3, the magnitude of F is approximately equal to one. The argument of the transition function is $X = \beta La^{\pm}(\phi \pm \phi')$ and may be calculated for a known value of βL if $a^{\pm}(\phi + \phi')$ as a function of $(\phi \pm \phi')$ is known. $a^{\pm}(\phi \pm \phi')$ are determined by

$$a^{\pm}(\phi \pm \phi') = 2\cos^2\left[\frac{2n\pi N^{\pm} - (\phi \pm \phi')}{2}\right] \qquad (7.5.20)$$

in which $N^\pm$ are the integers which most nearly satisfy the following equation

$$2\pi n N^+ - (\phi \pm \phi') = \pi \tag{7.5.21}$$

where

$$N^+ = \begin{cases} 1, & (n-1)\pi < (\phi \pm \phi') \leqslant 4\pi \\ 0, & -2\pi \leqslant (\phi \pm \phi') \leqslant (n-1)\pi \end{cases} \tag{7.5.22}$$

$$2\pi n N^- - (\phi \pm \phi') = -\pi \tag{7.5.23}$$

where

$$N^- = \begin{cases} 1, & 2\pi + (n-1)\pi < (\phi \pm \phi') \leqslant 4\pi \\ 0, & -(n-1)\pi < (\phi \pm \phi') \leqslant 2\pi + (n-1)\pi \\ 1, & -2\pi \leqslant (\phi + \phi') \leqslant -(n-1)\pi \end{cases} \tag{7.5.24}$$

We note that N^+ and N^- may each have two separate values in a given problem. For exterior wedge diffraction where $1 < n < 2$, $N^+ = 0$ or 1, but $N^- = -1$, 0 or 1. The factor $a^\pm(\phi \pm \phi')$ may be interpreted physically as a measure of the angle separation between the field point and a shadow or reflection boundary. L is given by

$$L = \begin{cases} S\sin^2\gamma_0, & \text{Plane waves} \\ \rho\rho'/(\rho + \rho'), & \text{Cylindrical waves} \\ SS'\sin^2\gamma_0/(S + S'), & \text{Conical and spherical waves} \end{cases} \tag{7.5.25}$$

7.6   Applications of Geometrical Theory of Diffraction in Antennas

7.6.1   Radiation from a Slit Aperture

The simplest approach to this two-dimensional problem is to imagine a plane wave incident on a slit of width a cut in a thin, perfectly conducting sheet as shown in Figure 7.18[6]. The amplitude distribution of the aperture field can be taken as constant over the slit and zero outside; the phase distribution of the aperture field is constant for the normal incidence and is linear for the oblique incidence.

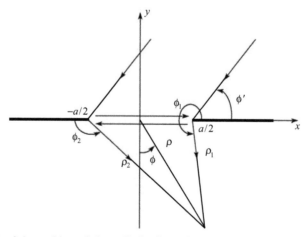

Figure 7.18   Analysis of the problem of the radiation by a slit aperture as that of the diffraction by a slit.

Case 1.   A TE-polarized (parallel polarized) plane wave

$$E_{i,z} = e^{jk(x\cos\phi' + y\sin\phi')} \tag{7.6.1}$$

is incident from $y > 0$ on the slit $|x| < a$ in a conducting plane in $y = 0$. Diffracted rays are excited at the two edges $x = \pm a$ and their amplitude and phase are given by

$$A^i D_{\parallel}(\phi', \phi) \frac{e^{-j\beta\rho}}{\sqrt{\rho}} \tag{7.6.2}$$

where A^i is the complex amplitude of the field on the incident ray at the point of diffraction. The singly diffracted field is the sum of the fields of these diffracted rays.

$$E_z^{sing.diff.} = e^{j\beta a \cos\phi'} D_{\parallel}(\phi', \phi_1) \frac{e^{-j\beta\rho_1}}{\sqrt{\rho_1}} + e^{-j\beta a \cos\phi'} D_{\parallel}(\phi'', \phi_2) \frac{e^{-j\beta\rho_2}}{\sqrt{\rho_2}} \tag{7.6.3}$$

where $\rho_1 = \rho - a\sin\phi$, $\rho_2 = \rho + a\sin\phi$, $\phi_1 = (3\pi/2) + \phi$, $\phi_2 = (\pi/2) + \phi$, $\phi'' = \pi + \phi'$.
Due to (7.5.9)

$$
\begin{aligned}
D_{\parallel}(\phi', \phi_1) &= D(\phi_1^-) - D(\phi_1^+) \\
&= -\frac{e^{-j\pi/4}}{2\sqrt{2\pi\beta}} \left[\sec\left(\frac{3\pi/2 + \phi - \phi'}{2}\right) - \sec\left(\frac{3\pi/2 + \phi + \phi'}{2}\right) \right]
\end{aligned}
$$

$$
\begin{aligned}
D_{\parallel}(\phi'', \phi_2) &= D(\phi_2^-) - D(\phi_2^+) \\
&= -\frac{e^{-j\pi/4}}{2\sqrt{2\pi\beta}} \left[\sec\left(\frac{-\pi/2 + \phi - \phi'}{2}\right) - \sec\left(\frac{3\pi/2 + \phi + \phi'}{2}\right) \right] \\
&\quad - \frac{e^{-j\pi/4}}{2\sqrt{2\pi\beta}} \left[-\sec\left(\frac{3\pi/2 + \phi - \phi'}{2}\right) - \sec\left(\frac{3\pi/2 + \phi + \phi'}{2}\right) \right]
\end{aligned} \tag{7.6.4}
$$

where $\phi_1^{\pm} = \phi_1 \pm \phi'$, $\phi_2^{\pm} = \phi_2 \pm \phi''$. Substituting (7.6.4), (7.6.4) into (7.6.3) and making the approximation of $\sqrt{\rho_1} \approx \sqrt{\rho_2} \approx \sqrt{\rho}$, we have

$$
\begin{aligned}
E_z^{sing.\ diff.} &= -\frac{e^{-j(\beta\rho + \pi/4)}}{2\beta\sqrt{\lambda\rho}} \left\{ e^{j\beta a(\cos\phi' + \sin\phi)} \left[\sec\left(\frac{3\pi/2 + \phi - \phi'}{2}\right) \right. \right. \\
&\quad \left. - \sec\left(\frac{3\pi/2 + \phi + \phi'}{2}\right) \right] + e^{-j\beta a(\cos\phi' + \sin\phi)} \\
&\quad \left. \times \left[-\sec\left(\frac{3\pi/2 + \phi - \phi'}{2}\right) - \sec\left(\frac{3\pi/2 + \phi + \phi'}{2}\right) \right] \right\} \\
&= \frac{e^{-j(\beta\rho + \pi/4)}}{\beta\sqrt{\lambda\rho}} f_s(\phi) \tag{7.6.5}
\end{aligned}
$$

where

$$f_s(\phi) = \frac{j\sin[\beta a(\sin\phi + \cos\phi')]}{\sin[(\phi - \phi' + \pi/2)/2]} - \frac{\cos[\beta a(\sin\phi + \cos\phi')]}{\cos[(\phi + \phi' - \pi/2)/2]} \tag{7.6.6}$$

On the shadow boundaries of the two half-planes $\phi - \phi' = -\pi/2$, the diffraction coefficients in (7.6.4) and (7.6.4) are singular. However, the singularities on the shadow boundaries cancel and produce the effect on the incident field. This fact is easily understood from the derivation of (7.6.5) and (7.6.6).

The singly diffracted ray from the right-hand edge of the slit in the direction $\phi_1 = \pi$ strikes the left-hand edge in Figure 7.18 when $\rho_1 = 2a$, and is diffracted again. So also is the singly diffracted field from the left-hand edge in the direction $\phi_2 = \pi$, and the doubly diffracted field is

$$
\begin{aligned}
E_z^{double\ diff.} &= e^{j\beta a \cos\phi'} D(\phi', 2a, \pi) D(\pi, \rho_2, \phi_2) \\
&\quad + e^{-j\beta a \cos\phi'} D(\pi + \phi', 2a, \pi) D(\pi, \rho_1, \phi_1) \tag{7.6.7}
\end{aligned}
$$

where

$$D(\phi', \rho, \phi) = D_\parallel(\phi', \phi)\frac{e^{-j\beta\rho}}{\sqrt{\rho}} \tag{7.6.8}$$

The multiple diffraction formula may be obtained by the same concept.

Case 2. A TM-polarized (perpendicular polarized) plane wave

For a TM-polarized plane wave incident on the slit the singly diffracted field is given by (7.6.3) with H_z replacing E_z and a diffraction coefficient

$$D_\perp(\phi', \phi) = -\frac{e^{-j\pi/4}}{2\sqrt{2\pi\beta}}\left[\sec\frac{(\phi - \phi')}{2} + \sec\frac{(\phi + \phi')}{2}\right] \tag{7.6.9}$$

This field vanishes in the aperture ($\phi = \pi$). If only the tangential magnetic field is considered, it might be concluded incorrectly that there is no interaction between the aperture edges. However, the tangential electric field $E_x = (j\omega\epsilon_0)^{-1}\partial H_z/\partial y$ does not vanish in the aperture. For interaction across the aperture, therefore, the diffracted field is proportional to the normal derivative $\partial A^i/\partial n$ of the field incident upon the diffracting edge. Rather than (7.6.2), the field on a diffracted ray is

$$\frac{\partial A^i}{\partial n}D'(\phi)\frac{e^{-j\beta\rho}}{\sqrt{\rho}} \tag{7.6.10}$$

where $D'(\phi)$ is a new diffraction coefficient[6]

$$D'(\phi) = \frac{1}{j\beta}\frac{\partial}{\partial\phi'}D_\perp(\phi', \phi)|_{\phi'=\pi} = \frac{-e^{j\pi/4}}{2\sqrt{2\pi}\beta^{3/2}} \cdot \frac{\cos(\phi/2)}{\sin^2(\phi/2)} \tag{7.6.11}$$

To calculate the field resulting from TM double diffraction of the field singly diffracted at $(a, 0)$, the normal derivative of the field on this ray at $(-a, 0)$ is required. It can be obtained from A_i by applying $(2a)^{-1}\partial/\partial\phi_1$ and then setting $\rho_1 = 2a$ and $\phi_1 = \pi$. This is

$$\frac{\partial A^i}{\partial n} = -e^{j\beta a\cos\phi'}\frac{1}{2a}\frac{\partial}{\partial\phi_1}\left[D_\perp(\phi', \phi_1)\frac{e^{-j\beta\rho}}{\sqrt{\rho_1}}\right]\Bigg|_{\substack{\rho_1 = 2a \\ \phi_1 = \pi}} \cdot$$

$$= \frac{\exp(-j\beta a(2 - \cos\phi') - j\pi/4)}{8(\pi\beta a)^{1/2}}\frac{\cos(\phi'/2)}{\sin^2(\phi'/2)} \tag{7.6.12}$$

and the doubly diffracted field from the left edge of Figure 7.18 is

$$H_z^{double\ diff.} = -\frac{\exp(-j\beta a(2 - \cos\phi') - j\beta\rho_2)}{16\pi(\beta a)^{3/2}(2\beta\rho_2)^{1/2}}$$

$$\times \frac{\cos(\phi'/2)\cos(\phi_2/2)}{\sin^2(\phi'/2)\sin^2(\phi_2/2)} \tag{7.6.13}$$

The other multiply diffracted field contributions are calculated similarly[6].

7.6.2 Edge Diffracted Fields from the Finite Ground Plane of a Microstrip Antenna

The GTD may be employed for calculation of the edge diffracted fields from the finite ground plane of a microstrip antenna. Because the surface wave effect of the dielectric substrate and the dielectric wedge diffraction have not been taken into consideration in

GTD, this method is valid only when the product of the substrate thickness (in wavelength) and dielectric constant is much greater than $0.1^{[9]}$.

As is seen in Chapter 3, the radiation from a rectangular microstrip patch is equivalent to that from two parallel slots adjacent to the metallic patch as shown in Figure 7.19(a). The width t of each slot is approximated by the thickness of the substrate, and the length l is equal to the length of the patch a plus the substrate thickness due to fringing effect.

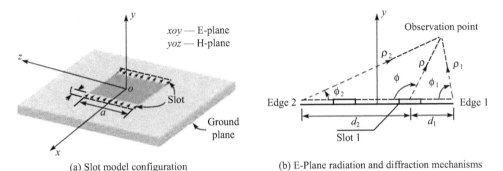

(a) Slot model configuration   (b) E-Plane radiation and diffraction mechanisms

Figure 7.19   Geometry and diffraction mechanisms.

The results in Section 7.6.1 may be used; however, the differences between problems in Figure 7.18 and Figure 7.19(a) should be noticed. First, in Figure 7.19, it is always grazing incidence but in Figure 7.18 the incident angle in upper space is arbitrary. Second, for the grazing incidence, in Figure 7.19(a), the incidence of the source is always behind the edges, but in Figure 7.18, the source is behind only one edge and is facing for the other edge. The results caused by the differences will be discussed shortly.

The E-plane pattern can be calculated by summing three rays from each of the two slots as illustrated in Figure 7.19(a). The distance and the phase dependence of the direct geometrical optics (GO) field are $\sqrt{\rho}$ and $e^{-j\beta\rho}$ respectively. The angle dependence is shown in (3.2.21). Therefore, the GO field is given by

$$\mathbf{E}_{GO} = \hat{\phi}\frac{\sin(\beta t \cos\phi/2)}{\beta t \cos\phi/2} \cdot \frac{e^{-j\beta\rho}}{\sqrt{\rho}} \tag{7.6.14}$$

where t is the slot width in terms of wavelength, and ρ is the distance from slot center to the observation point. The singly diffracted GTD field from each edge generated from the same slot is given by

$$\mathbf{E}_d = \hat{\phi}_i E_{inc} D_\perp \frac{e^{-j\beta\rho_i}}{\sqrt{\rho_i}} = \hat{\phi}_i\frac{\sin(\beta t/2)}{\beta t/2}\frac{e^{-j\beta d_i}}{\sqrt{d_i}} D_\perp \frac{e^{-j\beta\rho_i}}{\sqrt{\rho_i}} \tag{7.6.15}$$

where subscript i denotes the number of the edge. Due to the difference between Figures 7.18 and Figure 7.19(a), unlike the case in Figure 7.18, here $D_\perp \neq 0$ ($D_\perp$ is also represented by D_h in some literature). If a continuous pattern is desired at the two shadow boundaries ($\phi = 0°$ and $\phi = 180°$), the doubly diffracted fields need to be included. The finite ground plane effect on the H-plane radiation pattern may be also analyzed and the details are given in [9].

7.7   Fresnel Diffraction in Three Dimensions

Applying Fresnel transform in (1.3.31) and formula (1.3.51) to a circular aperture of radius a, assuming that the aperture field $E_x(\rho,\phi) = E_0$ is constant over a circular aperture and zero outside, the field at the z axis ($\theta = \phi = 0°$) may be written as

$$E_\theta(0,0,z) = E_x(0,0,z) = \frac{e^{-j\beta z}}{\lambda z} E_0 F_2 \qquad (7.7.1)$$

where

$$
\begin{aligned}
F_2 &= \int_0^{2\pi} \int_0^a e^{-j\beta \rho'^2/(2z)} \rho' d\phi' d\rho' \\
&= -j\lambda \left(1 - e^{-j\beta a^2/(2z)}\right) z \\
&= 2\lambda z e^{-j\beta a^2/(4z)} \sin\left(\frac{\beta a^2}{4z}\right) \qquad (7.7.2)
\end{aligned}
$$

Introducing the parameter $\xi = \beta a^2/4z$, (7.7.1) can be rewritten as

$$E_x(0,0,z) = 2j E_0 e^{-j\beta z} e^{-j\xi} \sin\xi \qquad (7.7.3)$$

With formula (7.7.3), we may observe the transition from near-field to far-field. The power flow $S_z(0,0,z)$ in the axial direction from (7.7.3) is

$$S_z(0,0,z) = \frac{|E_x(0,0,z)|^2}{2Z} = \frac{2|E_0|^2}{Z} \sin^2\xi \qquad (7.7.4)$$

where Z is the wave impedance in free space. The total power flowing through the aperture is

$$P_0 = \pi a^2 \frac{|E_0|^2}{2Z} \qquad (7.7.5)$$

and the (on-axis) gain is therefore

$$G = \frac{4\pi z^2 S_z}{P_0} \qquad (7.7.6)$$

which is, from (7.7.4) and (7.7.5)

$$G = \frac{4\pi(\pi a^2)}{\lambda^2} \cdot \frac{\sin^2\xi}{\xi^2} \qquad (7.7.7)$$

And in terms of the far-field gain G_0 of a uniformly illuminated aperture of area πa^2,

$$\frac{G}{G_0} = \left(\frac{\sin\xi}{\xi}\right)^2 \qquad (7.7.8)$$

for example, has the value of 0.987 at the Rayleigh distance $z = 8a^2/\lambda$[5].

The following construction, due to Fresnel, is useful in showing how the field arising from an extensive aperture can be thought of as being due just to a small part, known as the first Fresnel zone[5]. We take $x - y$ plane as a plane aperture which is illuminated by a normally incident plane wave from $z < 0$. The Fresnel zones are separated by circles which are formed by the intersection with the aperture plane of spheres, centered on the point of observation $F(0,0,f)$ whose radii are successively $z + \lambda/2$, $z + 2(\lambda/2)$, $z + 3(\lambda/2)$ and so on, as is shown in Figure 7.20. The first Fresnel zone is therefore a circle, and is denoted as zone 0, while the second and higher zones are annuli and are denoted as zone 1, zone 2, $\cdots$. Using the right triangle relationship, the radius of each zone, r_n, is given by

$$r_n = \sqrt{(f + n\lambda/2)^2 - f^2} = \sqrt{nf\lambda + (n\lambda/2)^2} \qquad (7.7.9)$$

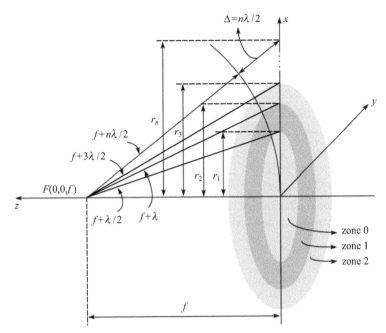

Figure 7.20   Geometry of Fresnel zones on the $x - y$ plane for the focal point.

If f is much larger than λ, according to (7.7.9), the radius of the first Fresnel zone is $(\lambda f)^{1/2}$. Inserting this into (7.7.3) and assuming $a = r_1$, $z = f$ give

$$E_x(0, 0, f) = 2E_0 e^{-j\beta f} \tag{7.7.10}$$

which is precisely twice the value of $E_x(0, 0, f)$ had the aperture been infinite, and the incident plane wave unobstructed. This phenomenon comes from the removing of more Fresnel zones being negatively contributed, which results in the focusing. This is also the basic principle of the Fresnel zone plate antenna (FZPA)[10].

Now we consider the contribution of the other Fresnel zones. For certain circular apertures, if odd-numbered zones are assumed to be covered with absorbing material, we just remove only the negatively contributed zones and keep the positively contributed ones. This forms the FZPA with the type of absorbing transparent (a/t) zones. Alternatively, we may use a dielectric plate and the altering of phase may be accomplished by cutting annular grooves into this plate. This forms the FZPA with the type of phase-correction (p/c) zones[10]. Point F is the focal point of either of the antennas.

Bibliography

[1] W. L. Stutzman and G. A. Thiele, *Antenna Theory and Design*, John Wiley & Sons, 1981.

[2] R. S. Elliott, *Antenna Theory and Design*, Prentice-Hall, 1981.

[3] J. Huang, "Microstrip reflectarray," *IEEE AP-S/URSI Symposium*, 1991, pp. 612–615.

[4] S. D. Targonski and D. M. Pozar, "Analysis and design of a microstrip reflectarray using patches of variable sizes," *IEEE AP-S/URSI Symposium*, 1994, pp. 1820–1823.

[5] R. H. Clarke and J. Brown, *Diffraction Theory and Antennas*, John Wiley & Sons, 1980.

[6] E. V. Jull, *Aperture Antennas and Diffraction Theory*, Peter Peregrinus, 1981.

[7] D. G. Fang, *Spectral Domain Approach in Electromagnetics*, Anhui Education Press, 1995.

[8] R. G. Kouyoumjian, "The geometrical theory of diffraction and its application," in *Numerical and Asymptotic Techniques in Electromagnetics*, Springer-Verlag, 1975.

[9] J. Huang, "The finite ground plane effect on the microstrip antenna radiation patterns," *IEEE Trans. Antennas Propagat.*, vol. 31, no. 4, pp. 649–653, Jul., 1983.

[10] L. C. J. Baggen and M. H. A. J. Herben, "Design procedure for a Fresnel-zone plate antenna," *International Jounal of Infrared and Millimeter Waves*, vol. 14, no. 6, pp. 1341–1352, 1993.

Problems

7.1 Based on the Snell's law

$$\hat{\mathbf{n}} \times \hat{\mathbf{S}}_i = \hat{\mathbf{n}} \times \hat{\mathbf{S}}_r$$
$$\hat{\mathbf{n}} \cdot \hat{\mathbf{S}}_i = -\hat{\mathbf{n}} \cdot \hat{\mathbf{S}}_r$$

where $\hat{\mathbf{n}}$ is the unit normal of plane boundary, $\hat{\mathbf{S}}_i$ and $\hat{\mathbf{S}}_r$ are the unit vectors of the directions of incident and reflected rays respectively, prove

$$\hat{\mathbf{S}}_r = (\underline{\mathbf{I}} - 2\hat{\mathbf{n}}\hat{\mathbf{n}}) \cdot \hat{\mathbf{S}}_i = \underline{\Omega} \cdot \hat{\mathbf{S}}_i$$

where $\underline{\Omega}$ is the dyadic reflection coefficient that relates to the image mapping orthogonal matrix.

7.2 Prove for the reflection from conducting plane is

$$\mathbf{e}_r = -\underline{\Omega} \cdot \mathbf{e}_i$$

where $\mathbf{e}_i$ and $\mathbf{e}_r$ are the incident and reflected vectors of electric fields respectively.

7.3 For a two mirror system with unit normal $\hat{\mathbf{n}}_1$ and $\hat{\mathbf{n}}_2$, $\underline{\Omega} = \underline{\Omega}_2 \cdot \underline{\Omega}_1$ prove

$$\hat{\mathbf{S}}_r = \hat{\mathbf{S}}_i - 2\hat{\mathbf{S}}_i \cdot \hat{\mathbf{n}}_2\hat{\mathbf{n}}_2 - 2\hat{\mathbf{S}}_i \cdot \hat{\mathbf{n}}_1\hat{\mathbf{n}}_1 + 4(\hat{\mathbf{n}}_1 \cdot \hat{\mathbf{n}}_2)(\hat{\mathbf{S}}_i \cdot \hat{\mathbf{n}}_1)\hat{\mathbf{n}}_2$$

7.4 For a lens antenna with equal phase and uniform amplitude on the aperture, if we need a null along the direction of axis, we may make a cylindrical cave on the lens as shown in the figure. Find the diameter d and the height t of the cylinder and give the sketch diagram of the field pattern. The dielectric constant of the lens is ϵ_r.

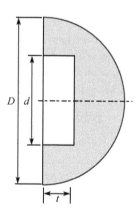

7.5 For a typical field distribution on the aperture of an antenna reflector shown in the figure, where the width of the reflector is a and the width of the feed blockage is $\delta(\delta \ll a)$, give the expression of the radiation pattern with blocked aperture distribution. If p is the normalized level of the first side lobe of the pattern without blockage, give the side lobe level for the pattern with blockage.

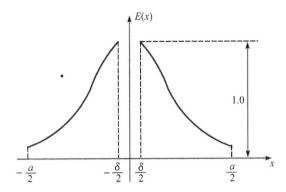

7.6 Eqn. (7.3.8) can be converted to a different and useful form by noting that $(\hat{\mathbf{a}}_z \cdot \hat{\mathbf{n}})dS$ is the projection of the element of surface area dS onto the x-y plane. Thus,

$$(\hat{\mathbf{a}}_z \cdot \hat{\mathbf{n}})dS = dS_z = \left(\frac{dS_z}{dl}\right)dl$$

where dS_z is the projection of dS onto the x-y-plane, then (7.3.8) becomes

$$\sigma = \frac{4\pi}{\lambda^2}\left(\int_0^a e^{-j2\beta z}\frac{dS_z}{dl}dl\right)$$

where l is the distance from the reference plane ($z = 0$) to the surface. Use the above expression for the radar cross section to derive the physical optics expression for the radar cross section of the sphere.

7.7 Evaluate the following Fresnel integrals:

$$(1)\ \int_0^\infty e^{-j\tau^2}\,d\tau, \quad (2)\ \int_0^5 e^{-j\tau^2}\,d\tau, \quad (3)\ \int_5^\infty e^{-j\tau^2}\,d\tau.$$

7.8 Examine the equivalence between the diffraction coefficient D in (7.4.28) and $D(\phi^-)$ in (7.5.9).

7.9 Show that the diffraction coefficient matrix $[D]$ in (7.5.16) will generally have seven nonvanishing coefficients if an edge-fixed coordinate system is used rather than a ray-fixed system.

7.10 Based on formulas (7.7.3) and (7.7.9), show that the odd-numbered zones and the even-numbered zones in Figure 7.20 are negatively contributed zones and positively contributed zones respectively.

CHAPTER 8

Planar Near-Field Measurement and Array Diagnostics

8.1 Introduction

The technique of near-field measurement has had a dramatic impact on antenna measurements. This technique requires a limited space yet is capable of yielding measurements of great accuracy and has allowed antenna measurements to move from an outdoor environment into an indoor laboratory[1].

The recent interest in near-field measurement has been generated primarily by the development of modern, specially designed antennas that are not easily measured on conventional far-field ranges. These antennas include, for example, electrically large antennas which require too large far-field ranges; physically large antennas which are difficult to rotate on conventional antenna pedestals; nonreciprocal antennas that must be measured in the transmitting mode and thus may be inconvenient for measurement on conventional far-field ranges; classified antennas that must be measured in a secure environment; and antennas with side lobes too low to be accurately measured on conventional far-field ranges[2].

The near-field data is useful not only to obtain the far-field patterns of the array under test, but also to reconstruct the aperture field for diagnostic purposes. This "backward" transform enables the near-field probe to identify accurately aperture faults at a distance free of interactions and couplings with the array elements[3].

A reasonable understanding of the theory of near-field measurements is a prerequisite to a successful near-field antenna measurements program. In this chapter, we are going to focus on the basic principles of the spectral domain approach introduced in Section 4.10.5. Some recent progress related to this topic will be briefly addressed as well.

8.2 Fundamental Transformations

The Fourier transform pairs related to the aperture-field, near-field and far-field are given in Section 4.10.5. Here we rewrite them by using the new formula numbers for convenience. Assume $E_x \neq 0$, $E_y = 0$, then we have

$$E_x(x, y, z) = \frac{1}{4\pi^2} \int_{-\infty}^{\infty} \int_{-\infty}^{\infty} \widetilde{E}_x(k_x, k_y) e^{-j(k_x x + k_y y + k_z z)} dk_x dk_y \quad (8.2.1)$$

$$E_x(x, y, 0) = \frac{1}{4\pi^2} \int_{-\infty}^{\infty} \int_{-\infty}^{\infty} \widetilde{E}_x(k_x, k_y) e^{-j(k_x x + k_y y)} dk_x dk_y \quad (8.2.2)$$

and

$$\widetilde{E}_x(k_x, k_y) = \int_{-\infty}^{\infty} \int_{-\infty}^{\infty} E_x(x, y, 0) e^{j(k_x x + k_y y)} dx dy \quad (8.2.3)$$

For a plane wave, electric-field vector is perpendicular to the direction of propagation. That is

$$\mathbf{k} \cdot \mathbf{E} = E_x k_x + E_y k_y + E_z k_z = 0 \quad (8.2.4)$$

(8.2.4) yields the following expression

$$E_z = -\frac{k_x}{k_z} E_x \tag{8.2.5}$$

The vector electric field is obtained as

$$\mathbf{E}(x,y,z) = \frac{1}{4\pi^2} \int_{-\infty}^{\infty} \int_{-\infty}^{\infty} (\hat{\mathbf{a}}_x k_z - \hat{\mathbf{a}}_z k_x) \tilde{E}_x(k_x,k_y) e^{-j(k_x x + k_y y + k_z z)} \frac{dk_x dk_y}{k_z} \tag{8.2.6}$$

Using the stationary phase evaluation of integrals presented in Chapter 4, we have the asymptotic form of (8.2.6) which is related to the far field

$$\mathbf{E}(r,\theta,\varphi) \overset{r \to \infty}{=} \frac{jk_z e^{-jkr}}{k\lambda r} \left[(\hat{\mathbf{a}}_x - \hat{\mathbf{a}}_z \frac{k_x}{k_z}) \tilde{E}_x(k_x,k_y) \right] \tag{8.2.7}$$

where $k_x = k\sin\theta\cos\varphi$, $k_y = k\sin\theta\sin\varphi$, $k_z = k\cos\theta$. All the angles are the observation ones, therefore (8.2.7) represents the radiation pattern.

Assume $E_x = 0$, $E_y \neq 0$, similarly we have

$$\mathbf{E}(x,y,z) = \frac{1}{4\pi^2} \int_{-\infty}^{\infty} \int_{-\infty}^{\infty} (\hat{\mathbf{a}}_y k_z - \hat{\mathbf{a}}_z k_y) \tilde{E}_y(k_x,k_y) e^{-j(k_x x + k_y y + k_z z)} \frac{dk_x dk_y}{k_z} \tag{8.2.8}$$

and

$$\mathbf{E}(r,\theta,\varphi) \overset{r \to \infty}{=} \frac{jk_z e^{-jkr}}{k\lambda r} \left[(\hat{\mathbf{a}}_y - \hat{\mathbf{a}}_z \frac{k_y}{k_z}) \tilde{E}_y(k_x,k_y) \right] \tag{8.2.9}$$

For arbitrary polarization, that is, $E_x \neq 0$, $E_y \neq 0$, from (8.2.6) and (8.2.8), we have

$$\mathbf{E}(x,y,z) = \frac{1}{4\pi^2} \int_{-\infty}^{\infty} \int_{-\infty}^{\infty} \left[(\hat{\mathbf{a}}_x k_z - \hat{\mathbf{a}}_z k_x) \tilde{E}_x(k_x,k_y) + (\hat{\mathbf{a}}_y k_z - \hat{\mathbf{a}}_z k_y) \right.$$
$$\left. \cdot \tilde{E}_y(k_x,k_y) \right] \frac{e^{-j(k_x x + k_y y + k_z z)}}{k_z} dk_x dk_y \tag{8.2.10}$$

According to (8.2.4), (8.2.10) may be rewritten as

$$\mathbf{E}(x,y,z) = \frac{1}{4\pi^2} \int_{-\infty}^{\infty} \int_{-\infty}^{\infty} \tilde{\mathbf{E}}(k_x,k_y) e^{-j(k_x x + k_y y + k_z z)} dk_x dk_y \tag{8.2.11}$$

where

$$\tilde{\mathbf{E}}(k_x,k_y) = \hat{\mathbf{a}}_x \tilde{E}_x(k_x,k_y) + \hat{\mathbf{a}}_y \tilde{E}_y(k_x,k_y) + \hat{\mathbf{a}}_z \tilde{E}_z(k_x,k_y) \tag{8.2.12}$$

is called the plane wave spectrum (PWS) of an antenna.

For arbitrary polarization, from (8.2.7) and (8.2.9), we have the expression for far field

$$\mathbf{E}(r,\theta,\varphi) \overset{r \to \infty}{=} \frac{jk_z e^{-jkr}}{k\lambda r} \left[(\hat{\mathbf{a}}_x - \hat{\mathbf{a}}_z \frac{k_x}{k_z}) \tilde{E}_x(k_x,k_y) + (\hat{\mathbf{a}}_y - \hat{\mathbf{a}}_z \frac{k_y}{k_z}) \tilde{E}_y(k_x,k_y) \right] \tag{8.2.13}$$

Using (8.2.4), (8.2.13) may be rewritten as

$$\mathbf{E}(r,\theta,\varphi) \overset{r \to \infty}{=} \frac{jk_z e^{-jkr}}{2\pi r} \tilde{\mathbf{E}}(k_x,k_y) \tag{8.2.14}$$

Similar to the formulas (8.2.2) and (8.2.3), the Fourier transform pair between the vector electric field on the aperture and the PWS may be written as

$$\mathbf{E}(x, y, 0) = \frac{1}{4\pi^2} \int_{-\infty}^{\infty} \int_{-\infty}^{\infty} \widetilde{\mathbf{E}}(k_x, k_y) e^{-j(k_x x + k_y y)} dk_x dk_y \qquad (8.2.15)$$

$$\widetilde{\mathbf{E}}(k_x, k_y) = \int_{-\infty}^{\infty} \int_{-\infty}^{\infty} \mathbf{E}(x, y, 0) e^{j(k_x x + k_y y)} dx dy \qquad (8.2.16)$$

In (8.2.16), $\mathbf{E}(x, y, 0)$ is the field at the antenna aperture. If in (8.2.15), we let $z = d$, then we have the field on the near-field measurement plane as shown in Figure 8.1.

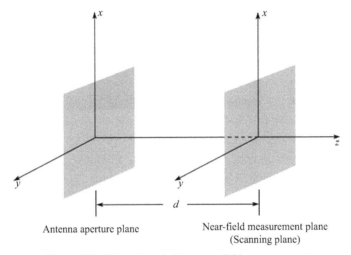

Antenna aperture plane

Near-field measurement plane
(Scanning plane)

Figure 8.1   Geometry of plane near-field measurement.

$$\mathbf{E}(x, y, z = d) = \frac{1}{4\pi^2} \int_{-\infty}^{\infty} \int_{-\infty}^{\infty} \left[\widetilde{\mathbf{E}}(k_x, k_y) e^{-j k_z d} \right] e^{-j(k_x x + k_y y)} dk_x dk_y \qquad (8.2.17)$$

The inverse Fourier transform of (8.2.17) is

$$\widetilde{\mathbf{E}}(k_x, k_y) = e^{j k_z d} \int_{-\infty}^{\infty} \int_{-\infty}^{\infty} \mathbf{E}(x, y, z = d) e^{j(k_x x + k_y y)} dx dy$$
$$= e^{j k_z d} \widetilde{\mathbf{E}}(k_x, k_y, d) \qquad (8.2.18)$$

(8.2.18) relates the measured near-field data at $z = d$ to the PWS. (8.2.17)–(8.2.18) form the basic formulas in near-field measurement based on PWS approach[4, 5]. In the measurement, the first step is to sample the fields at the scanning plane. The next step is the computation of the PWS using (8.2.18). The far fields are then calculated using (8.2.14). A planar near-field antenna-measurement and diagnostics setup is shown in Figure 8.2. Because actually only a finite amount of data can be measured and processed, the infinite continuous Fourier transforms (CFT) must be approximated by finite Fourier transforms, which are called discrete Fourier transforms (DFT). The selection of Δx and Δy, denoting sample spacings will lead to PWS equally spaced in the k space with $k_x \in [-k_{xm}, k_{xm}]$ and $k_y \in [-k_{ym}, k_{ym}]$. The sample spacing and spectral extents are related as follows:

$$\Delta x = \frac{1}{2(k_{xm}/2\pi)} = \frac{\pi}{k_{xm}}$$

$$\Delta y = \frac{1}{2(k_{ym}/2\pi)} = \frac{\pi}{k_{ym}} \tag{8.2.19}$$

Figure 8.2   Planar near-field antenna measurement and diagnostics setup
in Nanjing University of Science and Technology.

One must next select spectral spacings Δk_x and Δk_y in the k space. This selection will lead to inverse DFT equally spaced in the (x, y) space with $x \in [-x_m, x_m]$ and $y \in [-y_m, y_m]$. The spectral spacings and the spatial extents are related as follows:

$$\Delta k_x = \frac{1}{2(x_m/2\pi)} = \frac{\pi}{x_m}$$

$$\Delta k_y = \frac{1}{2(y_m/2\pi)} = \frac{\pi}{y_m} \tag{8.2.20}$$

Let the number of sample points on the x and y (or k_x and k_y) coordinates be N_x and N_y, respectively. There will then be a total of $N_x N_y$ sample points in this scheme. Geometry dictates that

$$2x_m = N_x \Delta x$$
$$2y_m = N_y \Delta y$$
$$2k_{xm} = N_x \Delta k_x$$
$$2k_{ym} = N_y \Delta k_y \tag{8.2.21}$$

Equations (8.2.19)–(8.2.21) are the fundamental relationship in DFT.

The bracketed term in (8.2.17) is the function which must be band-limited, in order to apply the sampling theorem. This band limitation arises from the rapid attenuation provided by the exponential term, whenever $k_x^2 + k_y^2 > k^2$, forcing k_z to become negative imaginary (in compliance with the radiation condition). This inequality defines the region outside of "real space," real space being the region of the far-field spectrum corresponding to the forward hemisphere into which the antenna radiates. The band limits in x and y can therefore be selected as $k_{xm} = k = 2\pi/\lambda$, and $k_{ym} = k = 2\pi/\lambda$, such that the sampling requirement in Eqn. (8.2.19) becomes

$$\Delta x \leqslant \lambda/2, \quad \Delta y \leqslant \lambda/2 \tag{8.2.22}$$

This is the basis for the half-wavelength-sample spacing used for planar-near-field measurement.

The physical implementation of the planar-near-field technique necessitates the use of a non-infinite or truncated-planar surface for near-field measurement. This truncation will cause an error in the calculated far-field pattern, the magnitude of which depends on the relative power of the fields neglected outside the measurement area. The concept of the valid angle in planar-near-field measurements has arisen as a convenient rule-of-thumb for predicting the region of validity of the calculated far-field pattern. The majority of the rays, which emanate from a directive antenna, will pass through the area defined by the measurement plane, and continue to travel to the far field. Geometrical considerations tell us that we should not expect to be able to predict the far-field pattern at an angle θ which is greater than θ_{valid} shown in Figure 8.3, since we have not included all the rays entering this region of the far-field in the near-field measurement. This is a highly simplified view of the actual mathematical formulation, which is used to compute the far-field from measurements made in the near field. However, it leads the way to a surprisingly simple and useful measurement criterion[4]. In Figure 8.3, L denotes the side length of measurement plane, D denotes the diameter of the aperture of the antenna under test (AUT), and d denotes the distance between the probe and the aperture of AUT. The valid angle, determined purely from geometrical considerations, is given by

$$\theta_{valid} = \tan^{-1}\left[\frac{L - D}{2d}\right] \qquad (8.2.23)$$

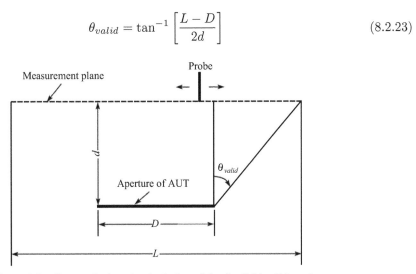

Figure 8.3   Geometrical-optics depiction of the far-field valid angle.

This simple result was developed empirically from extensive near-field measurement of the near field on a large scan plane, which is close to the AUT. A careful examination reveals that the evanescent PWS for practical antennas is very small, and it can practically be ignored at a distance of one wavelength or more from the aperture plane. The larger the antenna aperture is, the smaller the evanescent content is in the PWS representation. Thus in antenna near fields the evanescent modes can be ignored primarily because they are small, not because they are attenuated as it is commonly believed. The extensive simulated data clearly demonstrate that the effects of evanescent PWS are limited to a region less than 1λ from the aperture[5]. There is a lower limit to the separation distance d, as determined by creation of multiple reflections between the probe and the AUT. Usually, this distance is chosen to be $(3 \sim 5)\lambda$[6]. In reflector-antenna measurements, one must also allow ample separation to clear the feed and associated support, which may extend many wavelengths above the reflector surface. There exists a tradeoff among the maximum valid angle in the far field, scan-plane-truncation errors, and the errors induced by multiple reflections.

8.3   Probe Compensation

In measuring the near field, the probe has to be used as a detector. The probe is not a point, therefore its effect on the measurement should be taken into account. The purpose of this section is to derive an expression for probe compensation.

Assume the output of the probe is C, it can be proved that[7, 8]

$$C = \frac{\lambda^2}{j4\pi Z P_0}\mathbf{e}_i \cdot \mathbf{e}(\hat{\mathbf{a}}_i) \tag{8.3.1}$$

where $\mathbf{e}_i$ is vector amplitude of electric field due to the plane wave incident on an antenna from the direction $\hat{\mathbf{a}}_i$ as seen from the antenna. $\mathbf{e}(\hat{\mathbf{a}}_i)$ is the vector pattern function of the field in the direction $\hat{\mathbf{a}}_i$ when the antenna is used as a transmitter and is defined as

$$\mathbf{E}(x, y, z) = \frac{e^{-jkr}}{kr}\mathbf{e}(\hat{\mathbf{a}}_i) \tag{8.3.2}$$

Z is the plane-wave impedance, P_0 is the power delivered to the antenna to give rise to the radiated field $\mathbf{e}(\hat{\mathbf{a}}_i)$. In spectral domain, the plane wave related to $\mathbf{e}_i$ actually may be looked upon as the plane wave spectra of the transmitting antenna and the angle may be extended to the complex one to involve the evanescent modes. Therefore C in (8.3.1) contains the information of polarization, pattern functions and near field coupling of both transmitting and receiving antennas.

Formula (8.3.1) can be used to derive the general antenna coupling formula useful in near-field antenna measurement. In order to keep the analysis fairly simple, the geometry of transmitting antenna T and receiving antenna R is supposed to be as is shown in Figure 8.4. Let the (unprimed) coordinate system of the transmitter be based on o in its aperture, and the (primed) coordinate system of the receiver be based on o' in its aperture, and let the vector distance between o and o' be $\mathbf{r}_0$ directed from o to o'.

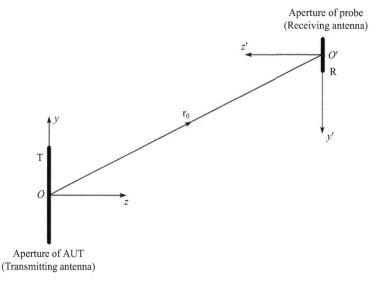

Figure 8.4   Geometry of transmitting antenna T and receiving antenna R in near-field measurement.

Assume that the transmitted angular spectrum of the transmitting antenna, with unit power delivered and radiated, is $\widetilde{\mathbf{E}}_T(k_x, k_y)$. From (8.2.11), a representative elemental plane wave radiated by the transmitter will therefore have a vector electric field

$$\mathbf{e}_i = \frac{1}{4\pi^2}\widetilde{\mathbf{E}}_T(k_x, k_y)dk_x dk_y \tag{8.3.3}$$

referred to the point o as the phase reference. Changing the phase-reference point from o to o'

$$\mathbf{e}_i = \frac{1}{4\pi^2}\widetilde{\mathbf{E}}_T(k_x, k_y)e^{-j\mathbf{k}\cdot\mathbf{r}_0}dk_x dk_y \tag{8.3.4}$$

From (8.2.14) and (8.3.2), $\mathbf{e}_i(\hat{\mathbf{a}}_i)$ in a primed coordinate system is given as

$$\mathbf{e}(\hat{\mathbf{a}}_i) = \frac{jk_{z'}}{\lambda}\widetilde{\mathbf{E}}_R(k_{x'}, k_{y'}) \tag{8.3.5}$$

where

$$\widetilde{\mathbf{E}}_R(k_{x'}, k_{y'}) = \widetilde{E}_{Rx}(k_{x'}, k_{y'})\hat{\mathbf{a}}_{x'} + \widetilde{E}_{Ry}(k_{x'}, k_{y'})\hat{\mathbf{a}}_{y'} + \widetilde{E}_{Rz}(k_{x'}, k_{y'})\hat{\mathbf{a}}_{z'} \tag{8.3.6}$$

Then, in the unprimed coordinate system of transmitting antenna, using the obvious transformations

$$\begin{array}{ccc}
\hat{\mathbf{a}}_x & \rightarrow & \hat{\mathbf{a}}_{x'} \\
\hat{\mathbf{a}}_y & \rightarrow & -\hat{\mathbf{a}}_{y'} \quad \text{and} \\
\hat{\mathbf{a}}_z & \rightarrow & -\hat{\mathbf{a}}_{z'}
\end{array}
\qquad
\begin{array}{ccc}
k_x & \rightarrow & -k_{x'} \\
k_y & \rightarrow & k_{y'} \\
k_z & \rightarrow & k_{z'}
\end{array}
\tag{8.3.7}$$

formula (8.3.5) becomes

$$\mathbf{e}(\hat{\mathbf{a}}_i) = \frac{jk_z}{\lambda}\widetilde{\mathbf{E}}_R(-k_x, k_y) \tag{8.3.8}$$

$$\widetilde{\mathbf{E}}_R(-k_x, k_y) = \widetilde{E}_{Rx}(-k_x, k_y)\hat{\mathbf{a}}_x - \widetilde{E}_{Ry}(-k_x, k_y)\hat{\mathbf{a}}_y - \widetilde{E}_{Rz}(k_x, k_y)\hat{\mathbf{a}}_z \tag{8.3.9}$$

The scalar product of (8.3.4) and (8.3.8) is

$$\mathbf{e}_i \cdot \mathbf{e}(\hat{\mathbf{a}}_i) = \frac{jk_z}{4\pi^2\lambda}e^{-j\mathbf{k}\cdot\mathbf{r}_0}\widetilde{\mathbf{E}}_T(k_x, k_y) \cdot \widetilde{\mathbf{E}}_R(-k_x, k_y)dk_x dk_y \tag{8.3.10}$$

Then, using (8.3.1) and integrating over all directions, the complex received signal with unit P_0 is given by

$$C(\mathbf{r}_0) = \frac{1}{8\pi^2\omega\mu}\int_{-\infty}^{\infty}\int_{-\infty}^{\infty}k_z\widetilde{\mathbf{E}}_T(k_x, k_y) \cdot \widetilde{\mathbf{E}}_R(-k_x, k_y)e^{-j\mathbf{k}\cdot\mathbf{r}_0}dk_x dk_y \tag{8.3.11}$$

In formula (8.3.11), $k_z\widetilde{\mathbf{E}}_R(-k_x, k_y)$ is proportional to the radiation pattern of the probe. When the plane wave $\widetilde{\mathbf{E}}_T(k_x, k_y)e^{-j\mathbf{k}\cdot\mathbf{r}_0}$ is incident on the probe, the output of the probe is proportional to $k_z\widetilde{\mathbf{E}}_T(k_x, k_y) \cdot \widetilde{\mathbf{E}}_R(-k_x, k_y)e^{-j\mathbf{k}\cdot\mathbf{r}_0}$. The integral with respect to $k_x k_y$ represents the superposition of the plane-wave response of the probe to all directions of incidence. If the radiation pattern of the probe is a constant vector, (8.3.11) degenerates to the form of (8.2.17) which is the solution without probe compensation.

Formula (8.3.11) is a two-dimensional Fourier transform which can be inverted immediately to give

$$k_z\widetilde{\mathbf{E}}_T(k_x,k_y)\cdot\widetilde{\mathbf{E}}_R(-k_x,k_y) = 2\omega\mu\int_{-\infty}^{\infty}\int_{-\infty}^{\infty}C(\mathbf{r}_0)e^{j\mathbf{k}\cdot\mathbf{r}_0}dx_0 dy_0 \qquad (8.3.12)$$

Formula (8.3.12) is actually the same as formula (36) in [9], except there is coefficient difference $(1/(4\pi^2))^2\cdot(1/4)$. The former factor is due to the different definition in Fourier transform pair and the latter is due to the different definition in $C = b/a$ (see (4.37) in [7] or (3.7-6) in [8]).

In (8.3.12), for far field approximation, θ and φ are observation angles. They are real numbers. From (8.2.14), it is seen that $k_z\widetilde{\mathbf{E}}_T(k_x,k_y)$ corresponds to the radiation pattern of the AUT and may be expressed as

$$E_\theta(\theta,\varphi)\hat{\mathbf{a}}_\theta + E_\varphi(\theta,\varphi)\hat{\mathbf{a}}_\varphi \qquad (8.3.13)$$

and $k_z\widetilde{\mathbf{E}}_R(-k_x,k_y)$ corresponds to the radiation pattern of the probe

$$E_{\theta'}(\theta',\varphi')\hat{\mathbf{a}}_{\theta'} + E_{\varphi'}(\theta',\varphi')\hat{\mathbf{a}}_{\varphi'} \qquad (8.3.14)$$

In the unprimed system of the transmitting antenna, using the obvious transformation

$$\begin{aligned}\hat{\mathbf{a}}_{\theta'} &\rightarrow \hat{\mathbf{a}}_\theta \\ \hat{\mathbf{a}}_\varphi &\rightarrow -\hat{\mathbf{a}}_{\varphi'} \\ \theta &\rightarrow -\theta' \\ \varphi &\rightarrow -\varphi'\end{aligned} \qquad (8.3.15)$$

expression (8.3.14) becomes

$$E_{\theta'}(-\theta,-\varphi)\hat{\mathbf{a}}_\theta - E_{\varphi'}(-\theta,-\varphi)\hat{\mathbf{a}}_\varphi \qquad (8.3.16)$$

From (8.3.13) and (8.3.16), formula (8.3.12) may be rewritten as

$$E_\theta(\theta,\varphi)E_{\theta'}(-\theta,-\varphi) - E_\varphi(\theta,\varphi)E_{\varphi'}(-\theta,-\varphi) = c_1\cos\theta e^{jkz_0\cos\theta}$$
$$\cdot\int_{-\infty}^{\infty}\int_{-\infty}^{\infty}C(\mathbf{r}_0)e^{jk(x_0\sin\theta\cos\varphi+y_0\sin\theta\sin\varphi)}dx_0 dy_0 \qquad (8.3.17)$$

where c_1 is a constant.

In (8.3.17), $C(\mathbf{r}_0)$, $E_{\theta'}$ and $E_{\varphi'}$ are obtained through measurement; E_θ and E_φ are to be determined. It is impossible to find them through only one equation (8.3.17). To solve this problem, it is required to do the orthogonal test twice, for example, to use both horizontal and vertical polarization of the probe in the test.

We use H and V to denote horizontal and vertical respectively, and let

$$I_H(\theta,\varphi) = e^{jkz_0\cos\theta}\int_{-\infty}^{\infty}\int_{-\infty}^{\infty}C_H(\mathbf{r}_0)e^{jk(x_0\sin\theta\cos\varphi+y_0\sin\theta\sin\varphi)}dx_0 dy_0 \qquad (8.3.18)$$

$$I_V(\theta,\varphi) = e^{jkz_0\cos\theta}\int_{-\infty}^{\infty}\int_{-\infty}^{\infty}C_V(\mathbf{r}_0)e^{jk(x_0\sin\theta\cos\varphi+y_0\sin\theta\sin\varphi)}dx_0 dy_0 \qquad (8.3.19)$$

Then from (8.3.17), we have

$$E_\theta(\theta,\varphi)E_\theta^V(-\theta,-\varphi) - E_\varphi(\theta,\varphi)E_\varphi^V(-\theta,-\varphi) = c_1\cos\theta I_V(\theta,\varphi) \tag{8.3.20}$$

$$E_\theta(\theta,\varphi)E_\theta^H(-\theta,-\varphi) - E_\varphi(\theta,\varphi)E_\varphi^H(-\theta,-\varphi) = c_1\cos\theta I_H(\theta,\varphi) \tag{8.3.21}$$

The solutions of the set of linear equations are[6]

$$E_\theta(\theta,\varphi) = \frac{c_1\cos\theta}{\Delta(\theta,\varphi)}\left[I_H(\theta,\varphi)E_\varphi^V(-\theta,-\varphi) - I_V(\theta,\varphi)E_\varphi^H(-\theta,-\varphi)\right] \tag{8.3.22}$$

$$E_\varphi(\theta,\varphi) = \frac{c_1\cos\theta}{\Delta(\theta,\varphi)}\left[I_H(\theta,\varphi)E_\theta^V(-\theta,-\varphi) - I_V(\theta,\varphi)E_\theta^H(-\theta,-\varphi)\right] \tag{8.3.23}$$

where

$$\Delta(\theta,\varphi) = E_\theta^H(-\theta,-\varphi)E_\varphi^V(-\theta,-\varphi) - E_\theta^V(-\theta,-\varphi)E_\varphi^H(-\theta,-\varphi) \tag{8.3.24}$$

Formulas (8.3.18)−(8.3.24) form the basic ones in radiation pattern measurement from planar near-field scanning with the probe compensation.

It is seen from (8.3.24) that if the circular polarization is used as the probe, it is impossible to obtain different results through the rotation of the probe. Therefore (8.3.24) will be always zero. Consequently, it is impossible to find the solution by using the probe with circular polarization.

In (8.3.22) and (8.3.23), we only need to know the radiation pattern of the probe facing the direction of (θ,φ) in the measurement-range of AUT. To insure the non-zero of (8.3.24), the radiation pattern of the probe should be no zero-point in this range.

Good polarization-purity of the probe is important to the precision of the measurement. For simplicity, the maximum radiation of the probe should be directed parallel to the z axis and the orientation of the polarization should be either parallel to x axis (for H-polarization) or parallel to y axis (for V-polarization).

For a linearly polarized antenna, performance is often described in terms of its principal $E-$ and $H-$ plane patterns. We consider two simple cases

1. $\varphi = 0°$ plane, H-polarization: in this case, $E_\varphi^H = 0$, (8.3.22) yields

$$E_\theta(\theta,\varphi) = \frac{c_1\cos\theta I_H(\theta,\varphi)}{E_\theta^H(-\theta,-\varphi)} \tag{8.3.25}$$

2. $\varphi = 90°$ plane, V-polarization: in this case, $E_\theta^V = 0$, (8.3.23) yields

$$E_\varphi(\theta,\varphi) = -\frac{c_1\cos\theta I_V(\theta,\varphi)}{E_\varphi^V(-\theta,-\varphi)} \tag{8.3.26}$$

$I_H(\theta,\varphi)$ and $I_V(\theta,\varphi)$ in (8.3.18) and (8.3.19) may be calculated by using fast Fourier transform (FFT). Dividing the measurement plane into grids with the coordinates $(z_0, m\Delta x, n\Delta y)$, where $0 \leqslant m \leqslant M - 1$, $0 \leqslant n \leqslant N - 1$, M and N are the numbers of the measurement-points along x and y directions respectively, they depend on the size of the measurement-plane (X, Y) and samplings

$$M = \frac{X}{\Delta x} + 1$$

$$N = \frac{Y}{\Delta y} + 1 \tag{8.3.27}$$

Then, (8.3.18) and (8.3.19) may be written in the form of FFT

$$I\left(\frac{2\pi L_1}{M\Delta x}, \frac{2\pi L_2}{N\Delta y}\right) = e^{j\left\{k^2 - [2\pi L_1/(M\Delta x)]^2 - [2\pi L_2/(N\Delta y)]^2\right\}^{1/2}}$$

$$\cdot \sum_{m=0}^{M-1} \sum_{n=0}^{N-1} \left[C(z_0, m\Delta x, n\Delta y) e^{j\frac{2\pi L_1}{M}m} \cdot e^{j\frac{2\pi L_2}{N}n}\right] \qquad (8.3.28)$$

where

$$\frac{2\pi L_1}{M\Delta x} = k_x = k\sin\theta\cos\varphi, \quad -\frac{M}{2} \leqslant L_1 \leqslant \frac{M}{2} - 1$$

$$\frac{2\pi L_2}{N\Delta y} = k_y = k\sin\theta\sin\varphi, \quad -\frac{N}{2} \leqslant L_2 \leqslant \frac{N}{2} - 1 \qquad (8.3.29)$$

the corresponding angles are

$$\theta = \cos^{-1}\left[\frac{\lambda L_2}{N\Delta y}\right]$$

$$\varphi = \sin^{-1}\frac{L_1\lambda/(M\Delta x)}{\left\{1 - [L_2\lambda/(N\Delta y)]^2\right\}^{1/2}} \qquad (8.3.30)$$

8.4   Integral Equation Approach

The basic formula of near-field measurement based on PWS approach as is given in (8.2.18) involves the integral with limits of infinity. Since in practice the measurement plane is finite, when making the Fourier transform, it is clear that the result will be only approximate. It has been shown in Figure 8.3 that the smaller the measurement plane, the smaller the valid angle. To solve this problem, the near-field measurement based on the integral equation approach was proposed[10, 11].

For simplicity of illustrating the principle, the probe compensation is not included here. The fundamental relationships used in this section are based on Section 8.2. For the integral equation approach, a fictitious source plane is considered of the same dimension as the measurement plane but translated a distance d towards the source with the source plane located at $z = 0$. On this source plane, we put fictitious magnetic currents. If one knows the complex values of the magnetic currents on the source plane, one can evaluate the fields at the measurement plane. Conversely, if the measurement fields are known, then one can find the complex amplitudes of the magnetic currents put on the source plane mathematically. The philosophy of this approach is to recover the aperture fields, and magnetic current density, from the measurement fields through the integral equation which relates the aperture and the measurement fields.

Consider an arbitrary shaped antenna radiating into free space with the aperture of the antenna being a plane surface, which separates the space into left-half and right-half spaces. Consider the general equivalent problem as shown in Figure 8.5. Because it is postulated that the electromagnetic fields in the left-half space are zero, a perfect electric conductor can be placed on the xoy plane. If it is further assumed that the tangential component of the electric field on the total plane is zero except on S_0 then $\mathbf{m}$ exists only on S_0. Using image theory, the equivalent magnetic current $\mathbf{m}$ can be expressed as

$$\mathbf{m} = 2\mathbf{E} \times \hat{\mathbf{n}} \quad \text{on} \quad S_0 \qquad (8.4.1)$$

The measurement electric field $\mathbf{E}_{meas}(\mathbf{r})$ produced by the radiation of $\mathbf{m}$ in free space can be obtained through the integral equation derived in the following.

Now the magnetic current is the only source, from the equivalence principle we have

$$\mathbf{E}_{meas} = -\nabla \times \mathbf{A}_m \tag{8.4.2}$$

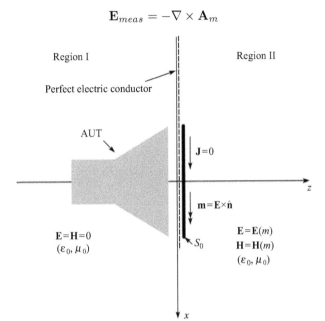

Figure 8.5   Equivalent magnetic current approach.

where $\mathbf{A}_m$ is the vector electric potential and is expressed as

$$\mathbf{A}_m = \int_{-\infty}^{\infty} \int_{-\infty}^{\infty} \mathbf{m}(x', y', z' = 0) G(\mathbf{r}, \mathbf{r}') dx' dy' \tag{8.4.3}$$

$G(\mathbf{r}, \mathbf{r}')$ is the Green's function in free space

$$G(r, r') = \frac{e^{-jk|\mathbf{r}-\mathbf{r}'|}}{4\pi|\mathbf{r}-\mathbf{r}'|} \tag{8.4.4}$$

where $|\mathbf{r} - \mathbf{r}'|$ is the distance between the source point and the field point. Substituting (8.4.3) into (8.4.2) yields

$$\mathbf{E}_{meas}(\mathbf{r}) = -\iint_{S_0} \mathbf{m}(\mathbf{r}') \times \nabla' G(\mathbf{r}, \mathbf{r}') dx' dy' \tag{8.4.5}$$

where

$$\nabla \times G(\mathbf{r}, \mathbf{r}')\mathbf{m}(\mathbf{r}') = \nabla G(\mathbf{r}, \mathbf{r}') \times \mathbf{m}(\mathbf{r}') + G(\mathbf{r}, \mathbf{r}')\nabla \times \mathbf{m}(\mathbf{r}') = \nabla G(\mathbf{r}, \mathbf{r}') \times \mathbf{m}(\mathbf{r}')$$

and $\nabla = -\nabla'$ are used.

For the planar scanning, the x and y components of the electric near fields are usually measured and are taken into account in (8.4.5). In this case, the following integral equation can be obtained for the equivalent magnetic currents

$$\begin{bmatrix} E_{meas,x}(\mathbf{r}) \\ E_{meas,y}(\mathbf{r}) \end{bmatrix} = -\iint_{S_0} \begin{bmatrix} 0 & \frac{\partial G(\mathbf{r},\mathbf{r}')}{\partial z'} \\ \frac{\partial G(\mathbf{r},\mathbf{r}')}{\partial z'} & 0 \end{bmatrix} \begin{bmatrix} m_x(\mathbf{r}') \\ m_y(\mathbf{r}') \end{bmatrix} dx' dy' \tag{8.4.6}$$

It is seen from (8.4.6) that the obtained integral equation is a decoupled one with respect to the two components of the magnetic currents. So instead of solving (8.4.6), the following two simple integral equations can be solved separately:

$$E_{meas,x}(\mathbf{r}) = - \iint\limits_{S_0} \frac{\partial G(\mathbf{r}, \mathbf{r}')}{\partial z'} m_y(\mathbf{r}') dx' dy' \tag{8.4.7-a}$$

$$E_{meas,y}(\mathbf{r}) = - \iint\limits_{S_0} \frac{\partial G(\mathbf{r}, \mathbf{r}')}{\partial z'} m_x(\mathbf{r}') dx' dy' \tag{8.4.7-b}$$

The explicit expression is:

$$\frac{\partial G(\mathbf{r}, \mathbf{r}')}{\partial z'} = \frac{e^{-jk_0|\mathbf{r}-\mathbf{r}'|}}{4\pi|\mathbf{r}-\mathbf{r}'|^2}(z - z') \left[jk_0 + \frac{1}{|\mathbf{r}-\mathbf{r}'|} \right] \tag{8.4.8}$$

In the present case, $z' = 0$ and $z = d$.

After formulating the E-field integral equations, the moment method is used to transform them into matrix equations.

If the number of measured near-field points is the same as that of current elements then the solution of the matrix equation is unique. If the number of measured near-field points is larger than that of current elements a least-squares solution is obtained. The reduction of size S_0 in Figure 8.5 results in the higher condition number of the matrix. The proper choice of it is important.

Instead of using equivalent magnetic currents as sources in the general magnetic current approach, an equivalent magnetic dipole array can also be used to replace the aperture of the test antenna. This is a good approximation as long as the source and the measurement planes are separated by a wavelength (i.e., $d = \lambda$). The use of dipoles eliminates the need for integration of the matrix elements in the evaluation of the impedance matrix.

In solving the matrix equation, the matrices involved could be very large. The equations can be solved very efficiently utilizing the FFT and the conjugate gradient method. Since per iteration, two two-dimensional FFT is computed, the integral equation approach is about 20 times slower than the PWS approach for 6400 unknowns. However, once these magnetic currents are known, the far field can easily be computed utilizing the FFT without suffering from the truncation error caused by the finite measurement plane.

To make the comparison between these two approaches, we take the two-dimensional Fourier transform of both sides of (8.4.7-a) and extend the limits of the integral from $-\infty$ to $+\infty$. The application of convolution theorem results in

$$\int_{-\infty}^{\infty} \int_{-\infty}^{\infty} E_{meas,x}(x, y, z = d) e^{j(k_x x + k_y y)} dx dy = \tilde{m}_y(k_x, k_y) \frac{\partial \tilde{G}(k_x, k_y)}{\partial z'} \tag{8.4.9}$$

where $\tilde{m}_y$ is the two-dimensional Fourier transform of the magnetic currents located at the source plane, $\tilde{G}(k_x, k_y)$ is the two-dimensional Fourier transform of $G(\mathbf{r}, \mathbf{r}')$ and has been given in Chapter 4 formula (4.2.3) as:

$$\tilde{G}(k_x, k_y, z, z') = \frac{-j}{2k_z} e^{-j|z-z'|(k_0^2 - k_x^2 - k_y^2)^{1/2}} \tag{8.4.10}$$

with

$$\text{Re}\left\{(k_0^2 - k_x^2 - k_y^2)^{1/2}\right\} \geqslant 0 \tag{8.4.11}$$

and

$$\text{Im}\left\{(k_0^2 - k_x^2 - k_y^2)^{1/2}\right\} < 0 \tag{8.4.12}$$

In (8.4.9), the identity $\widetilde{\partial G/\partial z'} = \partial \widetilde{G}/\partial z'$ is used.
   From (8.4.10), we have

$$\frac{\partial}{\partial z'}\widetilde{G}(k_x, k_y, z, z') = \frac{1}{2}\,\text{sgn}(z - z')e^{-j|z-z'|(k_0^2 - k_x^2 - k_y^2)^{1/2}} \tag{8.4.13}$$

where $sgn(z)$ is the signum function. Since $z' = 0$ and $z = d$.

$$\frac{\partial}{\partial z'}\widetilde{G}(k_x, k_y, z, z') = \frac{1}{2}e^{-jd(k_0^2 - k_x^2 - k_y^2)^{1/2}} \tag{8.4.14}$$

   Substituting (8.4.14) into (8.4.9) results in

$$\int_{-\infty}^{\infty}\int_{-\infty}^{\infty} E_{meas,x}(x, y, z = d)e^{j(k_x x + k_y y)}dxdy = \frac{1}{2}\widetilde{m}_y(k_x, k_y)e^{-jk_z d} \tag{8.4.15}$$

The left side of (8.4.15) is defined as $\widetilde{E}_{meas,x}(k_x, k_y, d)$. Using (8.4.1), (8.4.15) may be written as

$$\widetilde{E}_x(k_x, k_y) = e^{jk_z d}\widetilde{E}_{meas,x}(k_x, k_y, d) \tag{8.4.16}$$

or, in the vector form, as

$$\widetilde{\mathbf{E}}(k_x, k_y) = e^{jk_z d}\widetilde{\mathbf{E}}_{meas,x}(k_x, k_y, d) \tag{8.4.17}$$

Hence (8.4.17) is equivalent to (8.2.18). So for the information on the measurement plane, the integral equation approach is basically to transfer the measurement data to the source plane and then take the Fourier transform to the far field. However, if the measurement plane is finite in size, then the transfer of the data from $z = d$ plane to $z = 0$ plane utilizing the Fourier transform is not accurate because of the truncation error. Therefore in order to go to the source plane, an alternate transformation is used in the integral equation approach through the Green's function. Theoretically, this reduces the truncation error problem introduced by the two-dimensional Fourier transform.
   Although the integral equation approach takes more CPU time to produce the far fields for the same number of data points, this approach requires fewer measured data points than the conventional PWS approach to provide comparable numerical accuracy in the far fields when applied to the same near-field data. So the total measurement time in the integral equation approach is less to achieve equivalent numerical accuracy in the far-field result compared to the PWS approach[11].

8.5   Array Diagnostics

8.5.1   Theory

   Any microwave reconstruction (microwave holographic) method can be considered as a special case of a more general topic in the area of the inverse scattering. Generally speaking, from the knowledge of the complex measured scattering (or radiation) data, one wants to identify the object[12]. This is a very difficult problem. Many attempts have been made to develop a variety of techniques. The basic idea is to reconstruct the induced surface or volume currents from the integro-differential equations.

For the array diagnostics, the assumption is made that we know shape of the array and element locations and geometry. The object of the diagnostics is to identify the array element excitation coefficient variations, both in the amplitude and in the phase. It is clear that this special inverse problem is better defined than the general case where not much is known about the object.

Actually, the aperture distribution can be recovered from the near-field measurement by using formulas (8.2.15) and (8.2.18). The aperture distribution may be used for the purpose of diagnostics. The common technique used to perform the inverse Fourier transform in (8.2.15) is to use the fast Fourier transform (FFT). The FFT requires that the far-field be sampled on a regularized grid of points. If the location points fall on the element locations, they will accurately reconstruct the array current distribution. If the points do not fall on the element locations, the reconstruction will be poorer in quality because a discrete element grid is being sampled by a discrete reconstruction grid. The FFT offers some control over the separation and location of the reconstruction points, but it is difficult to match reconstruction points with array element points, and these reconstructions tend to be less accurate. A better technique is to use the discrete Fourier transform (DFT), which performs the integration of (8.2.15) by using the double summation[13]

$$\mathbf{E}(x, y, 0) = \frac{1}{4\pi^2} \sum_{i,j} \widetilde{E}(k_{xi}, k_{yj}) e^{-j(k_{xi}x, k_{yj}y)} \Delta k_{xi} \Delta k_{yj} \tag{8.5.1}$$

Note that (8.5.1) allows the user to choose the reconstruction location point $(x, y, 0)$. The FFT requires less computer time to perform the reconstruction than does the DFT, but the increased accuracy provided by the DFT warrants its use in cases where the element anomalies are small, or a more accurate determination of the amplitude and phase needs to be made. A possible algorithm for reconstruction would be to quickly locate the excitation anomalies with the FFT, and then to use the DFT to recover more precise values of the anomalies at those specific locations[13]. As a physical analogy of this procedure, consider star gazing with a telescope. The lens of the telescope acts as the FFT and the sky as the input data. When the telescope is out of focus so as to view the entire sky, the telescope does not need to be moved at all to characterize all of the sky at this coarse "focus." This smeared view cannot resolve closely spaced stars or planets. As the telescope is focused to its fine resolution, so as to examine a smaller region of the sky, it must be moved. In this case the telescope acts as the DFT and the removal is implemented through the choice of the reconstruction location point.

In transforming the field from the plane $z = d$ to the plane $z = 0$ by using Eqn. (8.2.18), which is called the backward transform, the visible part of $\widetilde{E}(k_x, k_y)$ is multiplied by a factor whose modulus is unity, whereas its invisible part is multiplied by a factor that is exponentially varying with z and is an amplification for the backward transform. As a consequence, when the Plane Wave Spectrum (PWS) on the aperture plane is calculated, large errors can be produced in the invisible part. The invisible part corresponds to the evanescent waves which vary rapidly on the plane. Therefore, to ignore the invisible part of the spectrum leads to a "smoothed" version of the actual aperture current distribution. This solution can be adequate for many purposes; however, if a more accurate reconstruction of the current distribution is required in the near-field diagnostics, the invisible PWS should be involved[14]. In fact, the near-field measurement approach based on the integral equation as described in Section 8.4 may be used to take into account the effect of the evanescent waves as is done in [15]. The effect of the evanescent waves is examined by defining the ratio q of the evanescent wave spectrum energy to the whole energy[15]

$$q = \frac{\iint\limits_{k_x^2+k_y^2>k_0^2} |\widetilde{E}(k_x, k_y)|^2 dk_x dk_y}{\int\limits_{-\infty}^{\infty}\int\limits_{-\infty}^{\infty} |\widetilde{E}(k_x, k_y)|^2 dk_x dk_y} \tag{8.5.2}$$

The numerical simulation of two examples shows that when ratio q is -33dB, the difference of the reconstructed amplitude distribution between those with and without evanescent waves is -30dB; when ratio q is -95dB, the difference is less than -100dB.

Several other techniques are available to include the evanescent waves[16–18]. For example, by steering an array, the pattern characteristics outside of real space can be brought into view. Thus, with several (e.g., four) tests, we can observe a complete picture of the antenna "spectrum" well beyond real space limits and fine details of the distribution computed with the merged spectrum. In merging the spectral regions, the average of the regions is taken where the regions overlap. A related method artificially turns on a fraction of the elements. If, for example, every other column were turned on, this sparsely populated array would have a spacing much larger than needed for the azimuth spectral data. The entire spectrum is visible in that case, and the distribution on the odd numbered columns can be calculated. Switching on the even numbered columns would provide the remaining information.

In the conventional near-field measurement, both the amplitude and the phase should be measured. However, in some cases the phase information is either unavailable or erroneous. The prohibitive cost of vector measurement equipment and high frequency measurements are two examples of applications in which phaseless near-field antenna measurements through phase retrieval methods may be attractive. The phase retrieval algorithm is based on an iterative Fourier method. The procedural steps required for executing this algorithm are depicted in Figure 8.6[19] where AUT denotes antenna under test. The algorithm requires near-field amplitude measurements on two planes (steps 1 and 2) which are separated by just a few wavelengths. A geometric description of the AUT's aperture plane, also commonly

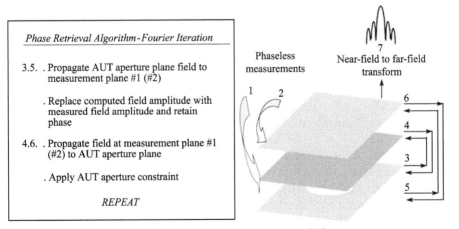

Figure 8.6   Phase retrieval algorithm[19]

referred to as the object or aperture constraint, is also required. The amplitude data on each measurement plane and the AUT constraint comprise the inputs to the Fourier iteration (steps 3−6) used for the phase retrieval. The product of the Fourier iteration is the complex near-field distribution on the AUT aperture plane and each of the two measurement planes. The Fourier iteration ensures, assuming successful retrieval of the phase, that the complex field distribution on these three planes are related by the Fourier transform. This relationship allows the far-field pattern of the AUT to be computed from the complex field distribution on any one of these three planes (step 7), using standard planar near-field techniques.

An initial guess for the amplitude and phase in the aperture plane of the AUT is made and truncated to the known physical extent of the AUT aperture constraint. This estimate is then propagated, using PWS techniques implemented by FFT, out to the first measurement plane. An error metric at this measurement plane is then computed by summing the squared difference of the calculated modulus and measured modulus at each point on the measurement plane. The calculated amplitude error metric is stored, the measured modulus replaces the calculated modulus, and the result is propagated back to the AUT aperture plane.

The calculated amplitude and phase at the AUT aperture plane is again truncated to the known physical extent of the AUT and the result is propagated out to the second measurement plane. An error metric, identical to that computed at the first measurement plane, is calculated and stored, the measured modulus replaces the calculated modulus, and the result is propagated back to the AUT aperture plane where the calculated amplitude and phase is again truncated to the known physical extent of the AUT.

The computed error metrics on the two measurement planes, at this point, are examined to determine whether iterations should continue. Appropriate stopping criteria include both an absolute error limit and an error convergence limit. If a stopping criterion is met, then the retrieved amplitude and phase on the AUT aperture plane and the two measured planes are stored and the iterations terminate. If a stopping criterion is not met, then the process is repeated until a stopping criterion is met.

The comprehensive study on the comparison between the conventional and phaseless planar near-field antenna measurements done in [19] has demonstrated that the phase retrieval method can produce the true pattern with greater accuracy than that of the conventional method when the measurement is contaminated with probe position errors. Examples of the successful diagnostics by using the phase retrieval method are also given in [20]. Distinguishing advantages of the phase retrieval processing are the absence of a requirement to know the actual measurement locations (other than the nominal location of the two required measurement planes) and the ability to obtain the true antenna pattern even when subject to relatively large probe position errors. These advantages offer the possibility of performing high frequency measurements with ordinary measurement apparatus and performing measurements "in the field" where measurement conditions are less controlled than in a laboratory environment.

8.5.2　Diagnostics Example of Microstrip Antenna Array

We used the compact microstrip antenna array shown in Figure 6.21 as an example to do the diagnosis. To simulate the failure elements, all the elements in the fourth column from the left are blocked by thin absorbing patches as shown in Figure 8.7. Figure 8.8 shows the diagnosis result by using FFT. The bright part is the aperture. The blocked part can be clearly seen in the Figure.

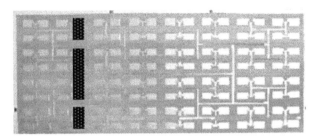

Figure 8.7   Blocked array of Figure 6.21.

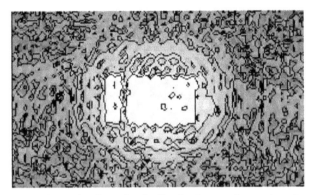

Figure 8.8   Two dimensional amplitude distribution on the aperture in gray scale.

Bibliography

[1] E. S. Gillespie, "Preface," *IEEE Trans. Antennas Propagat.*, vol. 36, no. 6, pp. 725–726, Jun., 1988.

[2] A. D. Yaghjian, "An overview of near-field antenna measurements," *IEEE Trans. Antennas Propagat.*, vol. 34, no. 1, pp. 30–45, Jan., 1986.

[3] J. J. Lee, E. M. Ferren, D. P. Woollen and K. M. Lee, "Near-field probe used as a diagnostic tool to locate defective elements in an array antenna," *IEEE Trans. Antennas Propagat.*, vol. 36, no. 6, pp. 884–889, Jun., 1988.

[4] Y. R. Samii, A. I. Williams and R. G. Yaccarino, "The UCLA bi-polar planar-near-field antenna-measurement and diagnostics range," *IEEE Antennas & Propagation Magazine*, vol. 37, no. 6, pp. 16–35, Dec., 1995.

[5] J. H. Wang, "An examination of the theory and practices of planar near-field measurement," *IEEE Trans. Antennas Propagat.*, vol. 36, no. 6, pp. 746–753, Jun., 1988.

[6] N. H. Mao and X. D. Ju, *Handbook of Antenna Measurement*, Defence Industry Press, 1987.

[7] R. H. Clarke and J. Brown, *Diffraction Theory and Antennas*, Ellis Horwood Limited, 1980.

[8] D. G. Fang, *Spectral Domain Approach in Electromagnetics*, Anhui Education Press, 1995.

[9] D. T. Paris, W. M. Leach, Jr. and E. B. Joy, "Basic theory of probe-compensated near-field measurements," *IEEE Trans. Antennas Propagat.*, vol. 26, no. 3, pp. 373–379, May, 1978.

[10] P. Petre and T. K. Sarkar, "Planar near-field to far-field transformation using an equivalent magnetic current approach" *IEEE Trans. Antennas Propagat.*, vol. 40, no. 11, pp. 1348–1356, Nov., 1992.

[11] P. Petre and T. K. Sarkar, "Difference between modal expansion and integral equation methods for planar near-field to far-field transformation," in a book *Progress In Electromagnetics Research*, PIER 12, 37–56, Elsevier, 1996.

[12] Y. R. Samii, "Microwave Holographic Metrology for Antenna Diagnosis," *SPIE*, vol. 1351, pp. 237–251, 1990.

[13] B. Toland and Y. R. Samii, "Application of FFT and DFT for image reconstruction of planar arrays in microwave holographic diagnostics," *IEEE Antennas Propagat. Symp.Dig.*, pp. 292–295, 1990.

[14] J. D. Hanflling, G. V. Borgiotti and L. Kaplan, "The backward transform of the near field for reconstruction of aperture fields," *IEEE Antennas Propagat. Symp. Dig.*, pp. 764–767, 1979.

[15] H. F. Hu and D. M. Fu, "A near-field diagnostic technique based on equivalent magnetic currents," *Proceedings of the Fifth Inter. Symp. on Antennas Propagat and EM Theory*, pp. 508–511, 2000.

[16] G. E. Evans, *Antenna Measurement Techniques*, Artech House, 1990.

[17] W. T. Patton and L. H. Yorinks, "Near-field alignment of phased-array antennas," *IEEE Trans. Antennas Propagat.*, vol. 47, no. 3, pp. 584–590, Mar., 1999.

[18] R. C. Wittmann, A. C. Newell, C. F. Stubenrauch et al., "Simulation of the merged spectrum technique for aligning planar phased-array antennas, part 1," *NISTIR 3981*, Oct., 1992.

[19] R. G. Yaccarino and Y. R. Samii, "A comparison of conventional and phaseless planar near-field antenna measurements: the effect of probe position errors," *Proceedings of Phased Array Systems and Technology Inter. Conference*, pp. 525–528, 2000.

[20] R. G. Yaccarino and Y. R. Samii, "Phaseless bi-polar planar near-field measurements and diagnostics of array antennas," *IEEE Trans. Antennas Propagat.*, vol. 47, no. 3, pp. 574–583, Mar., 1999.

[21] D. G. Fang, C. Liu, W. X. Sheng, J. Z. Xu, and Z. Y. Liu, "Aperture extrapolation in planar near-field measurement," *Inter. Conference on Microwave and Millimeter Wave Technology Proceedings*, pp. 536–530, 2002.

Problems

8.1 Rewrite (8.3.10) in terms of the receiving antenna's coordinate system being referred to.

8.2 For a ten-element electric dipole linear array in free space with the spacing of half wavelength, find the current distribution to produce the array factor being Tschebyscheff type of −30dB. Based on this array, do the simulation of near-field diagnostics by using the approach given in Section 8.4.

8.3 With the same scenario as Problem 8.2, do the simulation by using the aperture extrapolation approach proposed in [21], and compare the results with those obtained in Problem 8.2.

8.4 With the same scenario as Problem 8.2, do the diagnostics by using the combined FFT/DFT method.

8.5 With the same scenario as Problem 8.2, examine the phase retrieval method.

8.6 Discuss the scheme of phased array diagnosis by using the phase weighting technique in Section 2.6.2.

Index

9 780367 384678